AF522146

ENCYCLOPEDIA OF BIOPROCESS TECHNOLOGY

ENCYCLOPEDIA OF BIOPROCESS TECHNOLOGY

Vol. 5

LABORATORY BIOPROCESS TECHNOLOGY

By

Dr. Arvind N. Shukla

School of Studies of Zoology & Biotechnology

Vikram University

Ujjain

(India)

DISCOVERY PUBLISHING HOUSE PVT. LTD.

NEW DELHI-110 002

Published by:
Tilak Wasan
DISCOVERY PUBLISHING HOUSE PVT. LTD.
4383/4A, Ansari Road, Darya Ganj
New Delhi-110 002 (India)
Phone: +91-11-23279245, 43596064-65
Fax : +91-11-23253475
E-mail: parul.wasan@gmail.com
discoverypublishinghouse@gmail.com
web : www.discoverypublishinggroup.com

***First Edition:* 2012**

ISBN: 978-93-5056-026-6 (Set)
978-93-5056-027-3 (Vol. 1)
978-93-5056-028-0 (Vol. 2)
978-93-5056-029-7 (Vol. 3)
978-93-5056-030-3 (Vol. 4)
978-93-5056-031-0 (Vol. 5)

Encyclopedia of Bioprocess Technology

Printed at:
Shree Balaji Art Press
Delhi

Preface

The present title **"Encyclopedia of Bioprocess Technology"** has been written for undergraduate, post-graduate students and those engaged in pharmaceutical, environmental biotechnology, industrial and clinical research. Actually the explosion of new technologies with their vast potential has brought with it the need for emerging scientists to equip themselves and their laboratories with a whole array of new expertise. With the high discriminating power of the DNA systems has come high potential in evidentiary terms, high profile status for many investigations, and not least, a high degree of professional scruting of evidence produced by such technology.

This book is addressed to all who are curious and who want to understand fundamental biological processes that ensure our survival. The person who might be interested is probably someone just starting research who wants to learn about aspects of regulation, outside his chosen field. It could become a supplementary text for graduate students and post-graduate in biochemistry and molecular biology.

To make the work more comprehensive and informative, the author has consulted many authoritative books, research journals, abstracts, monographs etc., so there can be no claim to originality except in the manner of treatment.

The author express his thanks to his friends and colleagues whose continue inspirations have initiated him to bring out this book.

The author expresses her gratitude to Mr. Wasan and staff of M/s Discovery Publishing House Pvt. Ltd. for their whole hearted co-operation in the publication of this book.

The author will remain sincerely responsible for any shortcomings of the book and be grateful to the readers for their suggestions and constructive criticism for the continuous betterment of the book. He takes this opportunity to appeal to the readers to send their suggestions straightaway to this Publisher.

Arvind N. Shukla

CONTENTS

1

COMMERCIAL PRODUCTION OF CELLS

The process of choosing and optimizing a cell culture medium for a large-scale production process shares many biological issues and experimental approaches with the process of optimizing medium for any in vitro model system at the laboratory scale. However, there are some significant differences. The cell culture aspect of any large-scale process should fit in seamlessly with the upstream process of the transfection and selection of the protein of interest in the desired parental cell line, the equipment and scale available and desirable to run the process, and the downstream purification process. Issues surrounding the speed of development and the cost, safety, and reproducibility of the process are of major importance in developing a commercial process. The importance of seeing the entire process—from selection of the host cell line through transfection and selection of the production line, to the growth of the seed train cells and the actual production culture, harvest, and purification of the protein—as an integrated whole cannot be overemphasized. Failure to do so will result in an unworkable process, at worst, and missed opportunities for maximizing the reliability and minimizing the cost, at best.

If the end goal is kept in mind from the first, many opportunities for trade-offs in various aspects of the process will arise that allow decisions to be made in a timely manner. For example, if two candidate production lines exist, one with high specific productivity but slow doubling time and low saturation density and the other with lower specific productivity but better growth characteristics, the choice between the lines should be made taking into consideration the titer from each line and the total tank and equipment usage (seed train, production, and purification) required to obtain a given final yield of protein. It might be surprising to some how taking this broader view will change decisions made early in the process that are much harder to change later.

In another case, one medium formulation was found to increase the stability of a cell line even though the titer was reduced compared to the standard medium formulation. The advantages of the increased stability were seen to far outweigh the minor loss of product in the short term. This chapter will concentrate on specific aspects of choosing and optimizing the cell culture medium for large-scale processes. However, as will be emphasized again in the section at the end, this cannot be achieved optimally without taking into consideration the product (as well as the cell line) to be secreted and the purification process and equipment to be used.

METHODOLOGY

The aim of optimizing the growth and survival of cells in production culture is to obtain the maximum average viable cell number per volume of medium over the duration of the culture (viable cell days [VCD]). When selecting the production line to be used for large-scale culture, these parameters and specific productivity should be taken into consideration. A transfected cell line with high specific productivity is useless as a production line if it cannot be grown in the conditions necessary for large-scale culture. For example, a tetraploid line may have a high specific productivity but be too fragile to do well in a fermenter. Or, a low-cell inoculum may be acceptable with a rapidly growing cell line, but not with a slow growing line producing a labile product.

Continuous culture systems strive to produce a steady-state situation where the cell number is constant and the growth rate and death rate are balanced. Ideally, one would wish to achieve something like a contact-inhibited G_0-arrested state in which the cells are neither growing or dying. This is based on the as-yet-unproved assumption that suspension cells producing recombinant proteins will have a specific productivity in G_0 equal to or higher than that in log phase growth. In our experience, however, it is difficult to achieve a true G_0 growth arrest state in suspension-cultured cells. Although a great deal is known about the optimal growth conditions for specific types of cells, including Chinese hamster ovary (CHO), baby hamster kidney (BHK), and transformed human embryonic kidney cells, it is still impossible to exactly predict the optimal medium formulation for every cell type in every culture configuration. In addition, the process of obtaining a stably transfected cell line expressing a recombinant protein can change the growth phenotype of the cells. Thus, the empirical approach is still the only method of ensuring that the medium being used is optimal for a particular production line.

The optimal medium components or concentrations may vary from one CHO production line to the next. This is not surprising given that these lines are transfected with foreign DNA, selected using rather stringent conditions, and frequently amplified in the presence of a mutagen such as methotrexate. Additionally, the lines are engineered to secrete biologically active molecules that may act on the cell or the medium components, altering the environment of the culture. Generally speaking, however, two media optimal for the growth of two CHO-derived cell lines will be similar to each other and differ from the medium optimized for the growth of a dissimilar cell line such as a 293-derived line. For this reason, it is best to concentrate on one or a few cell lines for large-scale production efforts. Currently, most recombinant proteins produced commercially in mammalian cells use CHO as the host cell line.

Detecting the Viability

Accurate measurement of cell viability or characterization of cell death is key to the optimization of any production cell line. Viability can be readily measured by a variety of well-known techniques; however, it has more recently been established that many cell lines grown under batch culture conditions undergo cell death via apoptosis. Apoptosis is an active programmed cell death that exhibits specific morphological traits, such as nuclear condensation and fragmentation, maintenance of organelle and membrane integrity, and gross cell shrinkage. This contrasts with necrotic cell death, which is passive in nature and recognized by the degradation of cellular membranes and organelles and cell swelling. These different forms of cell death are distinct in morphology, mechanisms, and the way in which they can be identified.

An understanding of such is required for accurate reciprocal viability quantification, because the observation of simple loss of membrane integrity (e.g., using trypan blue exclusion) may be insufficient to account for all cell death caused by apoptosis. This may also be important in the accurate calculation of specific productivity. Finally, if one desires to improve the yield in an industrial process by decreasing cell death in the culture, it is crucial that one understand the mechanism of cell death in the cultures and the extent to which death decreases total VCD. Viability can be assessed most simply by the exclusion of charged dyes such as trypan blue or propidium iodide (PI) that move freely into dead cells and exhibit minimal uptake into live cells.

The extent of cell death can then be quantified by microscopy, or in the case of PI, by flow cytometry. Loss of plasma membrane integrity also results in the release of intracellular contents into the media that may then be quantified, such as the enzyme lactate dehydrogenase (LDH). LDH content can be assayed in both the cellular and supernatant fractions, allowing relative quantification of cell death. Apoptotic morphology specifically can be readily identified by staining fixed cells with a fluorescent dye such as acridine orange, which fluoresces green when bound to double-stranded nucleic acids. The classic apoptotic morphology of condensed or fragmented nuclear material can then be observed. From day 7 there is an abrupt increase in the percentage of cells that show the characteristics of apoptotic cell death even though the total cell number in the culture does not change appreciably over this time period. However, the percentage of apoptotic cells increased from day 7 to a value of 50% apoptotic cells on day 11. Clearly, this large proportion of apoptotic cells will adversely affect the titers late in the run. An additional characteristic of apoptotic cell death that can be used for quantification is that of DNA strand breaks, which are introduced at the internucleosomal linker sections of DNA in apoptotic cells.

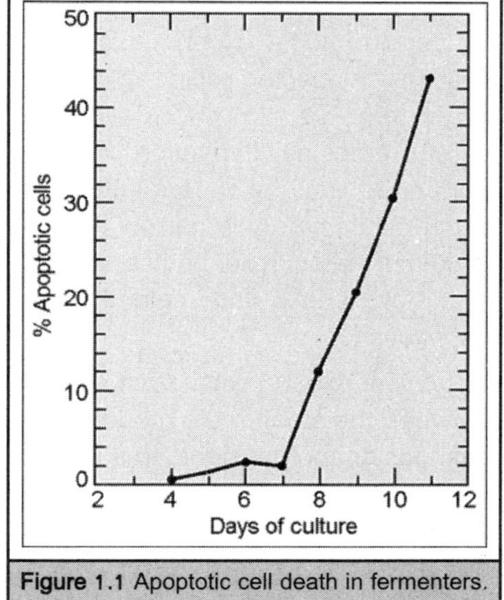

Figure 1.1 Apoptotic cell death in fermenters.

The ultimate loss of short DNA fragments, coupled with reduced accessibility to intercalating fluorochromes such as PI because of the condensation of chromatin, often results in the congregation of apoptotic cells within a fixed cell population to a hypodiploid peak, when analyzed by flow cytometry. The percentage of cells localized to this peak can be quantified if not obscured by excessive amounts of cellular debris. The free 3′-hydroxyl groups at the ends of the apoptotic DNA strand breaks may also be labeled with a fluorescent nucleotide using terminal deoxynucleotidyl transferase or annexin. These techniques allow the fluorescelnation of apoptotic cells with only a low level (or no) labeling of nonapoptotic cells. Cells labeled in such a way may be counterstained with PI, allowing concurrent cell cycle analysis, so that exit from the cell cycle can be tracked by flow cytometry. These methods can be used to trace the effects of manipulating the culture parameters on the amount of cell death in the cultures. Such experiments are an invaluable adjunct to direct empirical determination of conditions that optimize titer because they can help in deciding which experimental manipulations, among the many possibilities, might yield benefits.

Optimization

Virtually all industrial productions of recombinant proteins or antibodies in mammalian cell culture are done on a large scale (100 to 12,000 L). The few exceptions may be production of antibodies or proteins for research use only. Although even these small- and intermediate-scale processes will benefit from using an optimized medium, any medium optimization will obviously be aimed at use in large-scale processes. We have found that suspension-adapted cells and suspension processes scale well from plates to fermenters. Spinners are actually less predictive, in some cases, than static plate cultures using suspension cells. (Suspension-adapted cells grown in tissue culture plates in medium containing additives such as F-68 frequently will not attach but remain in suspension.) The experiments described in this article can be performed, preferably using the suspension-adapted cells, in plates, including 96-well plates for high throughput screening. However, larger (e.g., 60 mm) plates allow multiple sampling from the same plates over time. The type of careful empirical derivation of optimal process parameters described herein really cannot be performed entirely in fermenters because too many conditions are required. However, these experiments can help narrow the number of parameters to be tested on a larger scale. The cells in 60-mm plates, roller bottles, and spinners all performed similarly to those in the 1,000-L fermenter. The static shake flask cells that performed very poorly may have been suffering from oxygen deprivation.

A few experiments, such as adjustments in medium volume, might allow closer congruence between the small-scale model selected and the performance in large-scale fermenters. All major changes or improvements to a process based on results from small-scale work should be confirmed using fermenters. The use of 2-L minifermenters allows replication of conditions and good statistical analyses of results. If these titers are then confirmed in larger-scale fermenters, one can be confident that a rational approach using small-scale cultures, which allows complete analysis of the interacting variables in the medium, has led to a robust and optimal process. Although attached cell culture systems can be scaled up via roller bottle microcarrier culture, hollow fiber systems, and so forth, these devices have generally been used for medium-scale rather than very-large-scale production. Each introduces a different set of variables that should be understood to design a small-scale equivalent system for optimizing medium, as described later. Again, any improvements should be compared directly to the standard conditions before adoption. Some parameters change, of necessity, when going from static cultures to fermenters.

The method of pH control differs because base and acid addition are used to regulate pH in fermenters in addition to the bicarbonate buffer. This results in a more stable pH over the course of a run in fermenters, compared to plates, but at the cost of a greater increase in osmolarity in the fermenters because of the added salts. Shear and foaming are also factors in fermenters, but not plates. Components of the medium that are added to fermenters to prevent shear and foaming should also be added to plate experiments when optimizing medium to allow for better concurrence between large- and small-scale results.

Finally, the control of P_{O_2} is quite different in plates (diffusion) and fermenters (air or oxygen sparging). Adjusting the total surface area to volume in plates and spinners can have a significant effect on protein production and cell number and help in creating closer agreement between small- and large-scale data. Some parameters such as pH, temperature, and P_{O_2} can be adjusted or controlled more easily in a fermenter. These parameters are best optimized directly in experiments performed in fermenters after the nutrient and growth supplements have been optimized at small scale and tested at large scale. It is important to optimize these parameters both for improved cell growth and, on occasion, to improve product quality. With a few experiments to directly compare small- and large-scale culture, and an understanding of the principle underlying differences in plate and fermenter cultures, the former can be an invaluable tool in the process of obtaining optimized media for large-scale industrial processes.

Rich Culture Medium

For convenience, the culture medium can be divided into two components: the nutrient portion and the added growth factors and growth-promoting substances. The nutrient components are generally part of the commercially available powdered nutrient mixtures, such as MEM (minimal essential medium), DMEM, Ham's F12 nutrient mixture, RPMI 1640, or F12/DME. These media were all empirically derived for very specific cell types and growth conditions, none of which are very similar to a suspension culture of genetically modified cells in a fermenter. However, they still make a good starting point for optimizing the culture medium for the specific production line of choice. An initial screen should compare several commercially available or proprietary media to determine which is optimal for the production line. If the medium is being optimized before the cells are transfected, cell number can be used as the end point. If the line is already producing the protein of interest, protein titers should be measured. If the conditions for obtaining maximal titers and maximal VCD are the same, this is a good starting point for optimizing the nutrient portion of the medium. If different media give optimal VCD and titer, then more studies should be undertaken to understand the biology behind this phenomenon. This will prove useful in designing the optimal production system. Once a medium is found that works well, there is a temptation to move on the assumption that this adequate. However, large-scale commercial culture is one situation in which it is almost always worth the time and effort required to obtain a truly optimal medium.

Optimizing the nutrient portion of the medium is important not only for the increased yields it will afford directly but because optimizing nutrients will lead to a decreased requirement for parts of the growth supplements such as serum or growth factors. Additionally, optimizing VCD will lead to the ability to prolong culture time while minimizing damage to the product caused by enzymes released from dying cells. Even a 10% increase in VCD can be significant, and improvements of 2-fold to 10-fold in titer are frequently seen with optimized media. Because it is very difficult to change a process after clinical trials have begun, these early improvements will result in accrued benefits extending over many years that are often worth millions of dollars during the life of a production process. The available nutrient mixtures generally contain amino acids, a carbohydrate source (usually glucose), inorganic salts, vitamins, some type of buffering system (usually C_{O_2}, bicarbonate, and phosphate), fatty acids or more complex lipids, and other components that may vary from medium to medium (e.g., nucleosides, nucleotides, coenzymes, detergents). These components interact with each other so that the optimal concentration will depend to some extent on concentrations of other components in the mixture.

Many producers of cell culture media and components have developed their own proprietary cell culture media for specific cell types used in production of recombinant products. These media may offer a shortcut when proteins must be produced in short order; however, incomplete disclosure of the medium components can be a cause for concern, and costs are frequently higher than those of media formulated in-house. Optimization of the medium starts with choosing the best existing commercial or proprietary nutrient mixture. Then, media components are made up individually and in groups, and individual media are prepared missing only one component. That component is then titrated using the desired end point of viable cell number or product titer, if a good high throughput assay is available. It is obvious that the omission of KCl or isoleucine results in cell death over the production period and consequently very low protein yields. However, the omission of inositol had, if anything, a positive effect.

Conversely, increasing the inositol concentration 20-fold had little detrimental effect, whereas increasing isoleucine was beneficial and high levels of KCl were toxic. Using these data, one might then adjust the inositol concentration downward by 10-fold, the KCl downward by 2-fold, and the isoleucine upward. By using the data from all the titrations, one can then determine the optimal concentrations for each component and make up the best medium. This titration can then be performed again, using the best medium as a starting point. Each successive round of optimization would be expected to yield less benefit as the optimum is approached. This one round of optimization more than doubled the titers during the 8-day production period. This benefit was achieved in large part by increasing the VCD in the culture. Note that significant changes in the culture process, such as changing cell lines, changing inoculum density, changing the supplements, or going from attached to suspension culture, might well require reoptimization of the medium. With experience, it is possible to identify a few medium components that are most important and check only these after the initial optimization. Conversely, once an optimal medium is designed for a given cell line (e.g., CHO), the cells can be grown in that medium during transfection and amplification in order to select for cells that grow optimally in the medium designed.

Factors Affecting the Growth

The nutrient mixtures discussed in the previous section are generally supplemented with either undefined mixtures, such as serum, that promote the growth of cells or defined mixtures of growth factors, transport proteins, and hormones. The choice of whether one will attempt to reduce or eliminate serum and which supplements to use is based on the characteristics of the product and the cell line as well as cost and safety considerations. Most established cell lines have been carried in serum for years and are therefore well adapted to growth in serum. It may be difficult to achieve equivalent growth rates or titers in a serum-free medium. Even though higher specific productivities can be reached in defined medium, when compared to serum-containing medium, the total yield is lower because of the much-reduced cell growth. Note also that some of the additions primarily affect cell number, whereas others (e.g., nerve growth factor) increase specific productivity with little effect on cell yield. In this case, a subset of factors will increase cell number (via decreased cell death or increased cell division), and another set of factors might increase titer via increases in specific productivity of the cells or increase protein stability.

The former factors might be useful for all CHO-derived production lines, whereas the latter may be product specific. However, there are several reasons to eliminate serum, where possible, even if this results in a decrease in titer. Serum is expensive, contains high levels of many unknown proteins (about 40 mg/mL) that must then be purified away from the recombinant product, and contains enzymes and binding proteins that may affect product quality. For example, the production of tPA in the presence of serum will yield predominantly two-chain cleaved tPA rather than single-chain material. This is because the recombinant plasminogen activator produces plasmin from the plasminogen in the serum, which in turn cleaves the tPA. If single-chain material is desired, it is best to try and improve the cell yield by adding other supplements or adjusting the nutrient mixture. One straightforward method for reducing contaminating proteins from serum is to simply reduce the serum concentration. This makes the medium less expensive in many cases, simplifies purification by eliminating serum proteins, and often improves product quality or cell growth.

The general practice is to wean the cells by reducing the serum concentration over several passages (e.g., from 10% to 5%, 2%, 1%, 0.5%, 0.1%, etc.) allowing two or more passages at each concentration if a reduction in growth rate is observed. This method often works, but it is exceptionally time consuming. Eliminating serum altogether over one to two passages is often a reasonable solution for cells that do not have very fastidious growth requirements, although the resultant titers might not be optimal. A better solution is to derive a set of supplements that substitute for serum without any adaptation being required. In our experience, one can usually obtain titers in serum-free medium similar to, or even better than, those obtained in serum, although some thought must be taken as to the parental cell lines used for expression. With the advent of bovine spongiform encephalitis (BSE), it is likely that more processes will be driven to become serum free for safety considerations. Because testing for BSE on each lot of serum is time consuming and expensive, elimination of serum from the process is an easy way to avoid this problem. However, the entire process must be carried out serum free if one is to avoid the chance of infection. This is a more stringent growth condition than growing cells for only one passage serum free and requires more complete supplementation.

The question of which supplements to add is again one that can only be answered in the context of the desired end point. It should be emphasized in this context that optimizing the nutrient portion of the medium can reduce growth factor (and serum) requirements. Optimizing the nutrient portion of the medium is generally less expensive than adding supplements such as serum or growth factors. For production, minimizing expense and optimizing product quality may be more important than maximizing titer. It is the final yield of purified product as a function of fermentation plus purification cost that is the determining factor in deciding what the best medium will be. In an optimized serum-free culture process, the recombinant product may comprise 10 to 80% of the total protein in the harvest medium. This makes purification easier and less costly. In contrast, if one is growing up cells for the seed train, minimizing the doubling time to have cells ready for inoculation may be of more importance than the cost of the medium because cells are grown in smaller volumes. The quality of the protein produced is also not an issue in the early passages after thaw. However, it is generally best to not make too many major changes in media while bringing cells from the thawed vial to the production tank in order to avoid lag time as the cells adjust to new conditions. Much has been written concerning optimizing the supplements for serum-free media. Our preference is to use small-scale culture experiments to obtain defined supplements that will provide cell numbers and product titers equivalent to or higher than those obtained in serum- supplemented medium.

The elimination of serum necessitates the addition of hormones, growth factors, trace elements, and lipids. Because of the low concentrations of these components, the absence of binding and carrier proteins provided by serum, and the instability of the components in culture medium, these must be added to the media after filtration and just before use. The most commonly required additives are insulin (1 to 10 μg/mL), transferrin (1 to 100 μg/mL), and selenium (10 to 30 nM); however, some cell lines have an added requirement for lipids. One can then determine, based on experimental data, whether the increased product yield obtained by adding the most expensive components of the medium (for example, an epidermal growth factor requirement) are worth the expense. Using such data, one can then determine a minimum required medium that contains the most cost-effective supplements.

A more complete description of deriving the appropriate hormone supplement in order to eliminate serum is given by Barnes and Sato and Hewlett. For cell lines that are particularly fragile, Pluronic F-68 or Pluronic F-127 can be added to the medium to prevent shear-associated cell lysis. Finally, if the product is sensitive to proteolysis, one might wish to add protease inhibitors, as discussed more fully later. The empirical derivation of medium formulations optimizing both the nutrient portion of the medium and the supplements is time and labor consuming. It is seldom undertaken in a thorough fashion. One often-expressed wish is the desire to be able to predict what the optimal medium will be for a given cell line. Unfortunately, at this time, it is not possible to do this with any degree of assurance. We can make good guesses, largely based on experience gained by performing previous experiments on similar cells, but this will probably result in a good enough, not an optimal, medium. The only cases in which a full optimization of the medium is undoubtedly justified are (i) those research laboratories interested in understanding the nutritional and hormonal requirements of cells in vitro and (ii) in the development of a commercial process where small gains in titer, viability, and stability would, over many years of production, be expected to repay the initial effort expended in the optimization. Most large-scale culture processes fall into this category.

RECOMBINANT PRODUCTIVITY

Specific productivity is the amount of recombinant protein produced per cell per specified time period. Given that the cells are growing in an optimal medium, this will be a function of the expression system (promoter, enhancer, etc.) chosen and how well that particular system works in the host cell chosen, the integration site and orientation of the recombinant plasmid, the copy number of the gene to be expressed, and any specific or nonspecific inducers used. A discussion of the many expression systems available is beyond the scope of this article. However, the method used to select and clone transfected cells and amplify gene copy number frequently impacts the performance of the cells in the production culture medium and will be discussed briefly. The process of obtaining a stable cell line expressing a foreign protein involves a method of permeabilizing the cells to allow the plasmid containing the foreign DNA to enter the cell, a method of selecting for the minority of cells that have integrated the DNA into the nucleus, cloning to select a clonal cell line producing high levels of protein, and frequently several rounds of amplification of the DNA, each followed by cloning. This process allows ample opportunity for both genetic changes in the host line and the selection of a subpopulation from the initial population.

Usually, the primary, if not the sole, concern is to maintain the gene containing the recombinant protein in the cells. However, the phenotype of a good production cell line also includes the ability to grow rapidly and to high saturation density while maintaining a high viability. If one has made the effort to optimize the medium for the growth of the parental cell line, it is desirable to maintain the ability to grow well in this medium, including serum-free and suspension growth. To maintain the desired phenotype, selective pressure should be maintained for these growth characteristics during the process of transfection, amplification, and cloning. If the cells are grown in the process medium whenever possible, they will be much easier to handle when they reach the production setting. To the extent that this is not possible, the production cell line selected might benefit from a period of adaptation to the process medium and growth conditions before being banked.

PRODUCTION MEDIA

The production medium is designed to provide the maximum number of cells secreting protein over the maximum amount of time. The longer the productive run time, the longer the product can accumulate in the medium and the higher the resultant titer. However, this medium, along with the metabolites and proteins being secreted into it by the cells, is also the storage medium for the proteins being produced and secreted. An environment that is 37°C, neutral pH, fully oxygenated and stirred and where proteases and other enzymes are separated from the proteins by, at best, a layer or two of cell membrane would hardly be considered optimal for protein storage by any biochemist. This is, however, what recombinant proteins produced in the first days of the production run have to withstand for the duration of the run. Therefore, some thought should be given to protein stability in the production medium. This will be in large part dictated by the characteristics of the protein being produced. One is stable and accumulates linearly during the run; the other is easily proteolyzed and

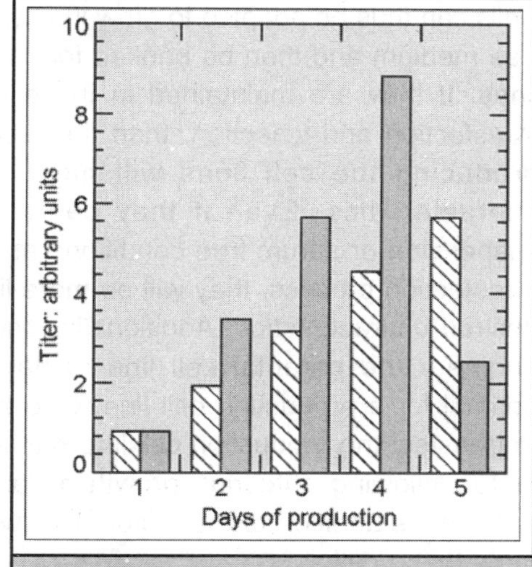

Figure 1.2 Titer as a function of run length.

begins to degrade faster than it is produced when the cells in the culture begin to die. Continuation of the culture at this point leads to loss of product.

The most important change in this process would be to optimize the medium components to prolong cell viability. One might then adjust the inoculum density so that the number of viable cell days is optimized. Too high an initial inoculum can result in early cell lysis and degraded product, whereas too low an inoculum results in too long a production period to reach the desired titer, resulting in overly long product residence time in the medium. One might also adjust the temperature to minimize cell death, shorten the run (including using processes that shorten the product residence time, such as perfusion), or add protease inhibitors to the medium. Protease inhibitors can be nonspecific (e.g., serum albumin) or specific (e.g., aproteinin or small molecular weight inhibitors). The small molecular weight inhibitors are frequently toxic to cells in culture, whereas inhibitors such as aproteinin are well tolerated but very expensive for production use.

INTERACTIVE PROCESS

As discussed in the introduction, medium formulation and optimization should not proceed in a vacuum but should be an integral part of the overall process design. From the initial choice of cell line through to the final purification, there should be consideration of the entire final process that will result.

Production Cell Line

The production cell line should be chosen not only for its protein secretion rate measured under laboratory conditions but also for its ability to perform well in the special conditions that occur in large-scale culture. The cells may be adapted to these conditions after transfection, as discussed earlier, or the parental cell line may be engineered to optimize its performance as a production cell line before transfection. Cells can thus be adapted to grow in suspension or serum-free medium and then be banked for use as parental cell lines. If they are maintained in these conditions during transfection and selection, then the recombinant protein producing the cell lines will have these important characteristics. Even if they cannot be handled in suspension or serum-free conditions at every step of the transfection process, they will be more likely to retain the desired characteristics. Additionally, one can genetically engineer the parental cell line to have traits that are optimal for a production cell line. One example of this is to transfect the production cell line with a required growth factor, allowing autocrine growth at large scale without exogenous addition of the factor. This will, of course, alter the composition of the medium used to grow these cells.

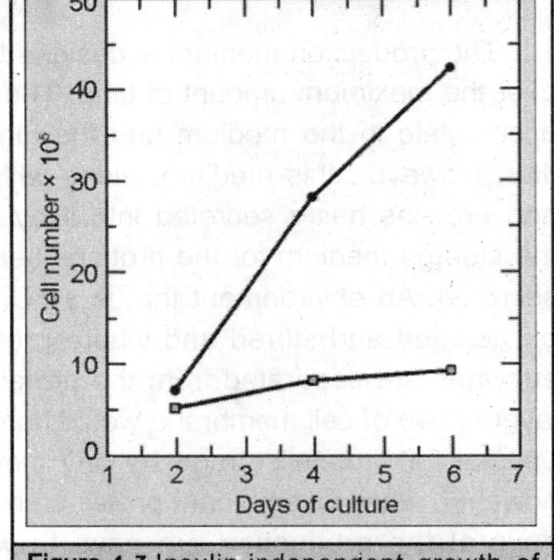

Figure 1.3 Insulin-independent growth of CHO cell transfected with the proinsulin gene.

Configuration of Process

It is obvious that the optimal medium for a process will depend, in part, on the equipment used and the configuration of the process. Clearly, suspension cultures and attached cells might require different media. The optimal nutrient mixtures for batch cultures and for perfusion cultures would be quite different. The inoculum density and amount of growth that will be required of the cells at each stage of the scale-up process will depend on the tank sizes used to inoculate each stage. The equipment to be used for the final process should be considered from the first stages of medium optimization so that one is sure that the medium designed does, in fact, support the cell growth and viability required in the process.

Method of Downstream Purification

Because the final goal of the process is to obtain a highly purified active protein, the downstream purification methods to be used should be considered from the beginning. Some protein characterization should be performed throughout the medium development process, in addition to a simple measurement of titer. Medium changes can alter protein characteristics as well as titers. It is therefore important to make certain that the desired product characteristics are maintained when medium changes are made to improve titer. Finally, everything that is put into the medium by the formulation or by the cells in the culture must be removed from the final protein product. Wherever possible, it is always best to use supplements that will not raise safety considerations.

Additionally, it does not make sense to add supplements to increase titer if the resulting product will be lost because of the extra purification steps that must be added. There is not as much room to alter pH, salt concentration, and so forth in a cell culture medium as there is in protein purification processes, but some understanding of the constraints and problems of both ends of the process will help make for a smooth transition and a reproducible process. In summary, the time and effort required to optimize all components oı the production medium will pay off many times over in a mammalian cell culture process. Optimal medium not only optimizes protein titers but can lead to improvements in protein quality and stability, more stable cell lines, more reproducibility from run to run, and improved use of facilities. To obtain maximal benefit from medium design it must be done in concert with the transfection and selection of the production cell line, the configuration of the manufacturing plant, and the design of the downstream purification process. All these factors working together have made large-scale mammalian cell culture an economically important part of the modern biotech industry.

2

PRESERVATION TECHNIQUES

Microorganisms have a significant effect on human life, whether as a result of the diseases or benefits they cause humans or to their food source (plants and animals) and through their effect on the environment (food spoilage, water pollution). In addition, microorganisms and cell lines are used increasingly within biotechnology and are therefore an integral part of the infrastructure of microbiology and fundamental in underpinning scientific research. They are used in crop improvement, food production, pharmaceutical production, and treatment of waste. Industry has the capacity to screen isolates for particular biochemical properties and metabolite production, to evaluate and develop commercial kits and rapid detection kits for microbially produced compounds, and to evaluate organisms' capacity for use in bioremediation and as biocontrol agents. With the decline in the use of animals in scientific experimentation over recent years, both for economic and ethical reasons, many modern standards and specifications require the use of microorganisms and cell lines in the testing of products and quality control procedures.

Culture collections, therefore, supply microorganisms and cell lines for the British, European, and U.S. testing standards, disinfection testing, the assessment of mutagenicity against a range of microorganisms, toxicity testing, NAMAS accreditation, quality accreditation, water and environmental reference strains, and biological and biomedical research. The supplier is responsible for adequate quality control procedures to ensure the authenticity of these cultures. Failure to do so may result in potential loss incurred as a result of contamination, work on the wrong organism or biochemical product, or invalid manufacturing quality control procedures.

MICROBIAL BIODIVERSITY

Over recent years, there has been increased concern regarding the conservation of biodiversity and the wealth of genetic information still to be harnessed. This matter has been found to be so important that it has been taken up governmentally at an international level. The Convention on Biological Diversity was signed by the governments of 155 countries and the European Community at the Earth Summit in Rio de Janeiro in 1992. The objectives of the convention were the conservation of biological diversity, the sustainable use of its components, and the fair and equitable sharing of benefits arising from the use of biological resources. The consequences of the introduction of this complex convention has been to generate interest in the scientific aspects of diversity and to begin to establish a number of major programs and projects aimed at increasing knowledge and improving conservation procedures, particularly in situ methodologies. At a legal level, the common inheritance principle regarding ownership of biological resources has changed to sovereign rights, and access to genetic resources will be controlled by the contracting parties and subject to national legislation.

Scientifically, countries are required to establish research and training programs, develop guidelines and procedures for conservation (ex situ and in situ), introduce procedures for assessing and monitoring environmental impact, establish emergency response procedures, and establish partnerships that support capacity building and technology transfer in developing countries. These regulations are far from being neatly juxtaposed to the Budapest Treaty legislating the patenting of microorganisms used in the industrial microbiology industry because the patent system limits the access to biological resources. The role of the Budapest Treaty is discussed later. Conservation of microorganisms is obviously a very important component of the biological diversity initiative because they are the most biochemically adaptable organisms on the planet on which they are ubiquitous, occupying every ecological niche. Thus, culture collections have a primary role in conserving microorganisms and cells.

TECHNIQUE OF CULTURE COLLECTIONS

Since Krael established the first culture collection in Prague in 1890, many others have been set up, not only to conserve living examples of microbes (e.g., bacteria, fungi, protozoa, algae, viruses), but also cell lines (human, animal, and plant cells) and DNA probes. Although the preservation of genetic diversity will fundamentally benefit humankind anyway, it is not known which materials may be useful or, indeed, for what purpose they may be required in the future. It is also impossible to predict changes in special habitats or in the general environment, nor what effect such changes may have on the bacterial genome or chromosomal elements. Therefore, some cultures that are of no importance today may be of significance in the future, and it is therefore of fundamental importance to preserve today's biodiversity. There are three main types of culture collections, the classification of which largely depends on the size of the collection, its documentation, and the service provided.

The first type is one where conservation of a small collection is a secondary activity of the laboratory. As such culture collections evolve and increase in size, a second type that is larger and more specialized will arise. These collections often result from specific research programs that are vulnerable when projects come to an end or individual senior scientists retire. Such collections need correct documentation to be a valuable resource and a greater effort to maintain the resource successfully. The third type of culture collection is the nationally and internationally recognized service-supply culture collections where conservation is the primary function. In order that such service-culture collections can become a recognized international resource, three principal elements must be satisfied. First, the culture collection has to be preserved in a viable state. Second, the history, original source, and properties of the culture must be documented. Third, the cultures must be available to the wider scientific community.

With the increasing demand on such culture collections for authenticated, reliable biological material, the World Federation for Culture Collections (WFCC) has published recommendations for good practice in culture collections so that new and existing collections have guidance and approved standards of operation. Service-supply culture collections exist primarily to maintain and supply on request cataloged, authenticated cultures and to give expert advice on cultivation and preservation to scientists, technologists, industrialists, and teachers.

The primary task of the staff therefore is curatorial, thus distinguishing it from secondary collections. Through continuity of work, these culture collections become expert in documentation and preservation techniques and in the knowledge of the behavioral properties of the holdings. For a service collection to be a truly valuable resource center, it must keep abreast of the developing science it serves. There must be an active accessioning policy, ongoing reassessment of preservation procedures, awareness of taxonomic changes, identification of new requirements, and an original research program. Recently, as a consequence of this, other services have become equally important, including cryopreservation, bulk supply of cultures for screening for products and quality control, patent and safe deposit facilities, identification of cultures, and the delivery of training courses.

OPERATIONAL CULTURE

It is important that all culture collections operate good practice in all aspects of their function, from microbiology to dispatch of samples. The various activities of culture collections and their successful management are discussed in this section.

Fundamental Importance

It is of fundamental importance that the establishment that houses a culture collection realizes its importance, accepts the responsibility inherent in maintaining a public service to appropriate standards, and is committed to its long-term maintenance. In the case of existing collections, this aspect should be continually clarified with the director of the parent institute, its scientific council, senior university officials, governing board, or other appropriate authorities.

Intrinsic to such a commitment is appropriate funding. The operation of such culture collections is a long-term commitment and requires core-funding from the overseeing parent organization. Support solely in the form of short-term contracts is totally inappropriate, because necessary levels of funding need to be guaranteed for the future activities of the culture collection, including the services being planned and the standards users would expect.

All culture collections should have a mission statement and long- and short-term objectives relating to the scope of the holdings and to the range of services that it will offer. If these statements are not agreed upon by the parent institute and other funding bodies, then the culture collection will not be able to operate effectively. A considerable degree of income generation is possible from culture collection activities, varying from as little as 25% to as much as 85%. This can depend on a mixture of sources of income, culture sales, and contract services.

Maintaining the Collection

The diversity of microorganisms and cells and their number should be continually addressed by a culture collection because appropriate funding will be needed to maintain the collection. Microorganisms that require particular containment facilities are those that are potentially pathogenic to humans, animals, or plants or those that are toxic. When a new culture collection is being established, it often relates to the interests of the parent organization. However, it is advisable to aim to complement existing collections rather than duplicating existing ones. It is necessary, therefore, to have a clearly defined accession policy on which new strains are to be added to the collection. Lack of such a policy will result in many unsolicited strains entering a collection uncritically without due regard to collection objectives, storage capacity, personnel, and financial resources. Conversely, the range of a collection should not be restricted as to limit the effectiveness of the service provided to the scientific community. New accessions may arise as a result of any of the following: (i) From the collection's own research; (ii) From samples sent to the collection for identification; (iii) From active solicitation by the collection; (iv) They may arise unsolicited; (v) As a formal patent deposit under the Budapest Treaty.

Routine Requirements

The effective curation, management, and staffing of a culture collection is a demanding task. Routine accessions, preservation, maintenance, viability checking, culture supply, and additional services are all extremely time consuming. The staff have to have considerable knowledge not only about the organisms themselves, but also their growth and preservation requirements, properties, and potential applications. Where a culture collection offers other services, such as identification and taxonomic expertise, additional specialized staff are required. Experience in culture collection management in addition to taxonomy is a fundamental prerequisite to expertise, and it is therefore important to attract and retain staff of a high caliber.

Preservation of Microorganisms

There are a number of techniques available for the preservation of microorganisms, each of which has particular advantages and disadvantages. In addition, different organisms require special preservation techniques in order to ensure optimal storage and maintain purity. The choice of technique should be determined by relating the features of each method to the needs of the user. A number of considerations must be taken into account when deciding which method is most appropriate. Purity of samples is of paramount importance, and the preservation method should minimize the chance of contamination. The process of preservation and storage should be fully validated. Because preservation could result in loss of viability and hence cell death, a process causing minimum damage should be chosen. Viability should be ascertained immediately after preservation and reassessed during storage. Preservation may introduce the possibility of a change in the characteristics of a culture. This can happen either as a result of a significant number of cells dying during the preservation process, thus selecting for a resistant population of surviving cells, or by genetic drift as a result of mutation or, for example, loss of plasmids. The expense of preserving and storing cultures may also influence the choice of method, and this may be further influenced by the number of cultures and factors, including staffing, equipment, materials, and storage space. The operation time for the preservation and maintenance of a small collection may be too labor intensive for a large collection.

The high capital cost for methods such as freeze-drying might be considered unsuitable for a small collection, although once preserved the samples require little maintenance. Storage in liquid nitrogen vapor may be cheaper, but for large collections, a sizeable storage space is required. Considerable attention must therefore be taken when the choice of preservation is being made, particularly with regard to the long-term objectives and size of a collection. The consequence of loss of a culture is another factor to be considered when choosing a preservation method. Valuable samples should be preserved by methods that minimize loss and should ideally be preserved by more than one technique. Finally, the choice of preservation technique will depend on the frequency of distribution of a sample. Samples that are regularly distributed should be preserved by a method amenable to bulk storage and that does not put them at risk to loss of viability or contamination as a result of frequent manipulation. Cultures that are to be supplied through the mail should be in a form suitable for packaging and must survive transportation time and temperature in accordance with IATA Dangerous Goods Regulations for Transport of Infectious Material.

Repeated culture

Preservation of samples by continued subculture is open to the risk of complete loss on the one hand and contamination on the other. Samples are incubated on a suitable medium, at an appropriate temperature, and for a suitable time. The procedure is repeated at intervals that ensure that a fresh culture is prepared before the old one dies. Samples are therefore continually in danger through loss of viability, particularly because the survival time of a particular microorganism varies considerably. *Staphylococcus* spp. and coliforms will survive for several years, but *Neisseria* spp. require subculture within a few weeks. The majority of media used for the storage of microorganisms are unenriched and have limited nutrients.

Excess carbohydrates should not be included in the medium because the acid produced might kill the culture. Storage periods can be extended by reducing the metabolic rate of the microorganism. This can be achieved by reducing the storage temperature to 5°C in many instances. However, exceptions occur; *Neisseria* spp. survive better at 37°C. Restricting the availability of air with parafilm can also limit metabolism. Thus, repeated subculturing, together with mislabeling or transposition of cultures, can result in significant contamination of deposits. Genetic drift and instability of characters also increases with each subculture. A large inoculum reduces the risk of selection but increases the risk of contamination. Therefore, this method is not to be recommended.

Thermal preservation

Freeze-drying has found popularity with many service culture collections primarily because it is suitable for batch production and distribution. Maintenance of viability requires very little attention, storage being relatively simple and effective and allowing preservation for 50 years for some microorganisms. The capital cost of freeze-drying equipment is initially high, and the preparation of samples is quite labor intensive, but the fact that large batches of samples can be prepared, significantly reducing labor hours per ampule, quickly recoups the initial financial outlay. The process of freeze-drying is the removal of water from frozen samples by sublimation. Cultures are suspended in a suitable medium (generally mannitol), frozen, and exposed to a vacuum. The water vapor is trapped in phosphorous pentoxide or a refrigerated condenser.

The microorganisms are stored in individual ampoules or vials in an inert gas or under vacuum. There are two types of freeze-dryers in general use; however, the centrifugal rather than shelf dryer is more popular and has more advantages. Although shelf dryers automatically stopper the ampules, the bungs can be naturally permeable to air or water vapor or theromas simply leak. The glass ampoules produced by centrifugal freezing are completely constricted, preventing leakage, and they can be plugged with cotton-wool, eliminating cross-contamination and acting as a filter to prevent scatter of microorganisms into the environment when the ampoules are opened. Even though freeze-drying has been widely used to preserve a variety of microorganisms, the process always needs refining to optimize preservation in a particular microorganism. Factors that can be adjusted and controlled are composition of medium, growth temperature, growth phase of the culture, rate of freeze-drying, final temperature, duration of drying, and the final moisture content of the culture. In general, freeze-drying does not cause instability of characteristics and genetic drift.

Dessication

Dessication has been used across a range of microorganisms, although the process is rather a specialized application and there is less information available on the method. Long-term viability appears to be good, when samples are stored between 5 and 10°C, and contamination is not a major problem, although strain stability is less well documented. The simplicity of the process makes it suitable for storing large numbers of cultures.

Drying from liquid state

The term *L-drying* originates from the term *drying from the liquid state.* Suspensions of microorganisms are dried in ampoules under pressure, but the vacuum is adjusted to allow rapid drying without freezing. The method is widely applied and allows a significant reduction in drying time and thus labor costs when a number of microorganisms are preserved. Samples should be stored at 5°C.

Preservation of freezing

Preservation by freezing has been carried out for a variety of different microorganisms over temperatures ranging from –20°C (refrigerator) to –140°C (vapor phase nitrogen) to –196°C (liquid nitrogen). Freezing and thawing can cause injury to cells by the formation of crystals that may damage the cells' integrity or by the concentration of electrolytes through the removal of water as ice. Eutectic mixtures form at –30°C and above, exposing cells to high salt concentrations. Temperatures of –70°C and below have been found to be increasingly successful, and storage in the vapor phase of liquid nitrogen is now ubiquitous as a preservation method. Storage in the liquid phase of nitrogen results in considerable safety problems resulting from nitrogen invading vials that subsequently become explosive. The freezing and thawing process itself may cause a significant loss in microorganisms. However, once stored, virtually no further loss occurs. The addition of cryoprotectants, the adjustment of growth conditions, the rate of cooling and warming, and freezing methodology can reduce losses.

The temperature of samples should be lowered at a controlled rate of 1 to 3°C/min to –30 °C followed by a more rapid rate of 15 to 30°C/min to –100 °C or lower. Ideally, this should be carried out using a commercial programmable freezer, which usually operate on a differential thermocouple principle. If such equipment is not available, many microorganisms, including bacteria and fungi, can be successfully frozen by placing the ampules in a dry chest, in a mechanical refrigerator set at below –65°C, or in the top of a liquid nitrogen storage tank for 30 minutes, and then placing the samples at the required temperature. Preprepared plastic ampoules containing beads and cryopreservant are now commercially available to enhance preservation. A vial of 25 beads is inoculated and agitated and the microorganisms attach to the beads. Excess cryoprotectant is then removed and the ampoule frozen. Beads can then be removed when a culture is required. The vials are placed in a cryoblock before they are removed from the storage vessel. This ensures that the remaining beads do not thaw out while one bead is removed. Thus, loss of viability does not result from repeated thawing. The major disadvantage of freezing is refrigerator failure or interruption to the supply of liquid nitrogen. Both will result in the loss of cultures; however, this can largely be overcome by 24-h alarm systems warning of refrigerator failure or loss of liquid nitrogen supply. The initial capital cost is high, but the process is not labor intensive.

Cell Lines Stocks

It is impractical for most laboratories to maintain cell lines in culture indefinitely; moreover, cell cultures will undergo genetic drift with continuous passage and risk losing their differentiated characteristics. Therefore, it is necessary to store adequate cell stocks for future use. It is recommended practice that a limit is set for the number of passages any cell line should undergo before replacing from cryopreserved stocks. This is of less importance for undifferentiated cell lines having infinite life span, although it still remains a consideration. Nearly all cell lines can be cryopreserved successfully in liquid nitrogen at –196 °C. A few simple criteria must be adhered to in order to provide sufficient cells for all future needs. The essentials for efficient cryopreservation are slow freezing (i.e., at a rate between 1 and 3°C per minute) and fast thawing, which is achieved by placing ampoules in a water bath at the temperature required for growth. The addition of a cryoprotectant, such as glycerol or dimethyl sulphoxide (DMSO), enhances survival. Nevertheless, DMSO does have certain toxic properties. Therefore, care must be taken in handling the cryoprotectant, and some consideration may be necessary as to whether cells should be washed immediately on thawing.

Cryopreservation and lyophilizaton

For security, and in order to minimize the possibility of strains being lost, each culture should be maintained by at least two different procedures. Cryopreservation and lyophilization are the best methods for minimizing the risk of genetic drift, therefore, at least one of these methods should be used. Where only freezing is available, duplicates should be stored with different electrical supply. Duplicates of important or irreplaceable strains should be securely held in different buildings or ideally on a different site. The documentation of samples is equally important because loss owing to natural disaster or fire would make the corresponding cultures meaningless and worthless.

Accessioning culture

When a culture is received in the laboratory, it is important that a strict accessioning scheme be followed. The method described prevents substantial genetic variation from the original deposit. Starter cultures or ampoules received should be propagated according to the conditions provided by the originator, and a token freeze should be produced (usually up to 5 ampoules). Cultures derived from this token freeze should then be subjected to detailed quality control and characterization. If these results are satisfactory, then a master (or seed) stock should be made.

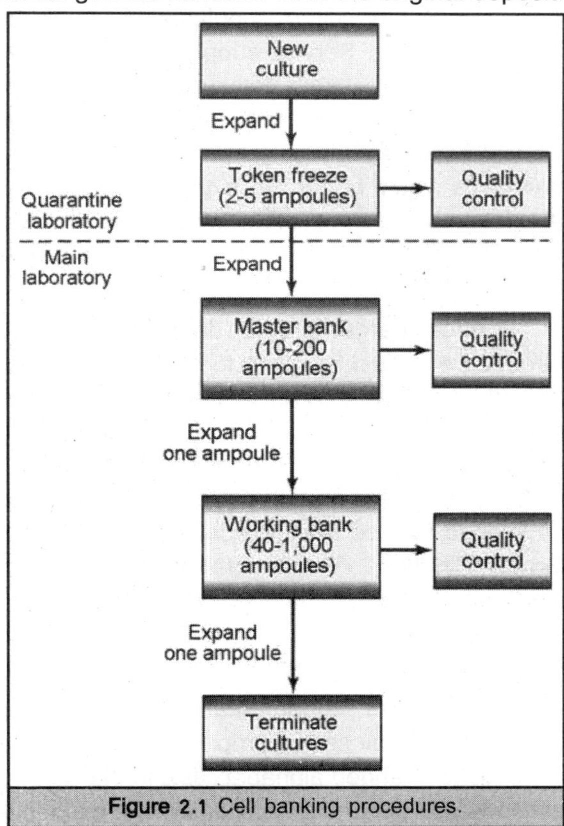

Figure 2.1 Cell banking procedures.

Major authentication of the culture should then be undertaken on the master cells because it is these ampoules that will be used to provide the working or distribution stock. One ampoule from the master stock is used to produce a working or distribution stock.The number of ampoules in each bank is dependent on how often it will be used. Because the production of these banks is both expensive and time consuming, the numbers required should not be underestimated. For some industrial processes, master banks may contain up to 200 ampoules, and working banks may contain 1,000 ampoules. Once the working bank is depleted, a fresh one is made from an ampoule from the master bank; similarly, a new master bank is made from the frozen freeze stock. If no more frozen freeze stock remains, then one of the current master bank ampoules should be used. The level of characterization and quality control that should be carried out at each stage will depend on what it is to be used for and what the necessary regulatory bodies require. Quality control implications are obviously more intense for cultures used in the manufacturing industry.

OBJECTIVE OF CULTURE COLLECTION

The aim of culture collections is to supply authenticated cultures to bona fide scientists, on request, promptly and without restriction on their ultimate application. The staff of culture collections have expertise and knowledge of the appropriate scientific literature and as such can advise on strain selection. In addition, they provide a range of scientific services, largely in response to the needs of the user. These include safe and patent deposit facilities and identification and preservation services. The expertise gained by culture collections and their staff enables them to run training courses and seminars for other scientists.

Economic Values

Culture collections should be able to readily distribute cultures listed in their catalog on request. Arrangements for culture supply vary according to the financial basis and policies of the legal owner of the collection. Service-supply collections usually levy fees for supplying cultures. This charge should strike a balance between definable measurable costs and those that yield an income that is significant in relation to costs, but that do not result in a declining effective use of the collection. Consideration should be taken as to the fact that many collections are partly government funded, unit costs would be far too high for what is a reusable commodity (i.e., they can be subcultured), and fees should not discourage the use of the collection. Therefore, culture collections operate on a nonprofit basis. It is often common practice to charge commercial companies higher fees than nonprofit organizations.

Further charges may be levied for postage and packaging and any special documentation. Great care should be given to distribution of pathogenic or toxic cultures. Originators of a request should be able to guarantee that their establishment has stipulated equipment and staff to handle pathogenic or toxic cultures. Dispatch of cultures should be in accordance with IATA (1988) Dangerous Goods Regulations for Transport of Infectious Material. Restrictions to which countries cultures can be distributed also exist. Because the threat exists that microorganisms could be developed for use in biological warfare, cultures can only be issued to countries in the Australia Group of nations. The nations belonging to this group and restricted groups are detailed in *The End-Use Control*.

Cataloging the Culture Collections

Catalogs are fundamental to the operation of culture collections. They describe the strains available and their salient properties, including literature citations, propagation properties, etc. Stocks are usually arranged alphabetically, in the case of bacteria by genus and species name. Increasingly catalogs are becoming available in computerized form, either on-line or on disk, because, they can easily be updated whereas paper catalogs cannot. Such catalogs adopt a field structure and definitions that enable the data to be integrated into the international and major regional schemes that are in operation, including the Internet, on which many resource centers are accessable through the World Wide Web, Microbial Strain Data Network (MSDN), World Data Centre on Micro-organisms (WDC), Microbial Information Network Europe (MINE), and more recently CABRI (Common Access to Biotechnological Resources and Information). CABRI is researching and developing the formation of an integrated resource center service linking databases of different organism type, genetic materials, and other biologicals in Europe so that users worldwide can access these relevant catalogs during one searching session through a common entry point and request or order products to be delivered to their place of work.

Taxonomic Status

Safe deposits are held for a variety of reasons, including as a back-up to existing collections and for the retaining of ownership and confidentiality of strains. These collections are not listed in catalogs and their taxonomic status may be unclear. In some cases, the service collection may hold stocks in its collection and market them on behalf of the owner.

Industrial Potential

The rapid advance in genetic manipulation techniques and the allied boom in biotechnology over recent years has led to increased awareness of the enormous industrial potential of microorganisms. Competition is fierce in the race to exploit this potential, and large sums of money are being risked in commercial development of new microbial processes. In such circumstances, patent protection naturally becomes a major concern of any organization involved in modern industrial microbiology. An important principle of the patent system is that an invention should be disclosed to the public. Although most kinds of invention can be disclosed simply by means of written description, those involving the use of new microorganisms cannot. No matter how detailed the written description may be, a new microbiological process cannot be tested without a culture of the organism itself.

Therefore, in addition to a written disclosure, the patent offices of most industrial countries require the strain used in a microbiological invention to be deposited in a recognized public culture collection, where it will eventually be available to the public. Patent deposits are now regulated internationally by the Budapest Treaty on the International Recognition of Deposit of Microorganisms for the purpose of patent procedures and patent depository facilities in Europe, under which certain culture collections of a generally acknowledged high scientific standard are officially recognized as international depository authorities (IDAs). The main feature of the treaty is that all countries party to it, and hence their patent offices, must recognize a deposit that is made in any one of these IDAs as being valid for the purposes of their patent procedure. This means that one deposit is all that is needed, regardless of the number of countries in which patent applications citing the strain are filed. The Budapest Treaty contains detailed provisions on the procedure to be followed by the applicant in making a deposit and by the IDA in accepting, storing, and subsequently making the material available to the public. Broadly speaking, these provisions are designed to ensure that the applicant makes a viable deposit and provides enough information on it to enable the IDA to handle and maintain it safely and correctly.

The IDA, for its part, must process the deposit promptly, issue the depositor with a receipt and certificate as proof of deposit and viability, keep the deposit for at least 30 years, and provide cultures to those parties entitled to receive them but most importantly, *to no one else.* Entitled parties in this respect are those who have the depositor's written authorization or who have an official certificate of entitlement from the appropriate patent office. The IDA is also bound by strict rules of secrecy and may not give information about a deposit to anyone not entitled to receive the patent-deposited cultures. Microorganisms must become available to the public at some stage of the patenting procedure. To date, there are three kinds of systems in operation. In the Netherlands and Japan, patent applications are published twice, first 18 months after the filing date of priority date and before the application has been examined, and second when the application has been accepted. Once the patent office has decided to grant a patent, after the second publication date, the microorganism must be made available.

The major advantage of this system to inventors is that their microorganisms do not have to be released until they has an enforceable right. In Japan, the patent owner is given a further measure of protection in that the recipients of cultures must not pass them on to a third party and must use them only for experimental purposes. In the United States, patent applications are not published until the patent is granted, and any microorganism deposited for patent purposes need not be available until then. Thus, under the U.S. system, inventors again have an enforceable right at the time they are required to make the organism available.

Additionally, under the U.S. mechanism an inventor is never put in the position of having to allow access to the organism when no legal protection is available because if a microorganism is not granted a patent, then it need never become available. The EPC (European Patent Convention) also operates a dual publication system, but stipulates that a microorganism for which a patent is being sought must be made available at the date of first publication, before any enforceable right exists. This mechanism reflects the principle that the microorganism is an integral part of the patent application and should therefore be available at the same time as the written description. Availability at first publication can be restricted to an independant expert acting on behalf of a third party. This expert has to be selected from a list held by the EPC and is not permitted to pass the strain on to anyone else. On second publication, the strain becomes generally available, whereupon the inventor has an enforceable right.

Authentication of Microorganisms

Culture collections have a critical responsibility when it comes to the authentication of microorganisms and cells from their initial receipt through preservation procedures and to dispatch to external scientists. Quality control of cultures should be as extensive as possible and have three primary targets; authenticity, purity, and viability. Research using the wrong organism wastes time, is expensive, and leads to publication of invalid results. The identification of a new accession should be ideally confirmed by a competent specialist (e.g., the depositor) or confirmed by the culture collection itself to ensure that it agrees with published descriptions. Culture collections with specialized expertise in identification can also offer an identification consultancy to the scientific community. The microorganisms identified would not necessarily enter the culture collection.

Quality Management

Increasingly, many culture collections operate using a quality management system such as BS EN ISO 9001. Essentially, these regulations ensure that all procedures have proper protocols that are fully documented, thus allowing all activities to be fully traced if the need arises. All equipment has to have records of calibration and servicing, and all laboratory records, worksheets, and equipment log books are regularly audited. All staff are trained in these procedures. In order for a culture collection to run completely successfully, it has to keep good records about the microorganisms and cells it holds, including the following categories of information, place, substrate or host, date of isolation, name or person isolating the strain, depositor, name of person identifying the strain, preservation procedure used, optimal growth media and temperature, any data on biochemical or other characteristics, and any regulatory conditions applying (e.g., contamination level). Duplicate copies of computer files or paper records of all information should be kept to prevent unexpected loss of information.

Research Program

It is important that culture collections have an ongoing research program. This ensures that staff are kept abreast of current developments and are aware of the needs of the wider scientific community. Lack of adequate research programs discourages high-caliber staff from joining a culture collection and may encourage the loss of existing quality staff. As well as the obvious research and development into preservation and taxonomy, culture collections have the necessary skills to undertake a diverse spectrum of research, whether by themselves or acting as consultants. As discussed earlier, culture collections have the capacity to screen isolates for particular biochemical properties and metabolite production, evaluate and develop commercial kits and rapid detection kits for microbially produced compounds, evaluate organisms' capacity for use in bioremediation and as biocontrol agents, and select test organisms used in the evaluation of materials, media, culture vessels, and diagnostic reagents.

Relevant Guidelines

It is the responsibility of all culture collections to handle cultures and carry out procedures as safely as possible and within the relevant guidelines. This is particularly important with regard to microorganisms that are pathogenic to man or are genetically manipulated. Within the United Kingdom, all laboratory procedures should be performed in line with COSHH (Classification of Substances Hazardous to Health), and all matters of safety are enforced by the HSE (Health and Safety Executive), which also offers an advisory service. All culture collections are responsible to their own staff, the delivery services, and receiving laboratories and must therefore address all matters of safety properly.

All service collections have regulations for both the deposit and dispatch of cultures that have to be fully abided by before samples can be transported following the IATA (1988) Dangerous Goods Regulations for Transport of Infectious Material. Within the United Kingdom, microorganisms are categorized in accordance with the Advisory Committee on Dangerous Pathogens Categorization of Pathogens According to Hazard and Categories of Containment. In the United States, a similar categorization has been devised. Both sets of guidelines have been accepted by the WHO, and other countries tend to adopt these systems.

Information is given on the degree of containment and protective clothing, which should be applied during handling of such organisms in the laboratory. The control of genetically manipulated organisms is somewhat more complicated in the United States than it is in the United Kingdom. U.S. researchers must adhere to the policy that research in this field must conform to the requirements of the coordinated framework, such as the National Institutes of Health Recombinant DNA Guidelines. In the United Kingdom, genetic manipulations are the responsibility of the Advisory

Committee in Genetic Manipulations. Other countries likewise tend to adopt these systems. All culture collections must additionally follow the regulations of their own countries and the institutes within which they work.

Management System

Culture collection staff require appropriate training for all techniques and procedures they carry out. If an organization has a quality management system, such as ISO 9000, then this will be performed as a mandatory requirement. Once the staff have become skilled, they are then in an ideal position to train others in techniques relating to culture preservation, growth, and identification given that they have adequate teaching facilities and supervision. Regular publication relating to culture collection activities and the research activities carried out within an organization is necessary to enhance its activities.

Federations and Associations

Microbiologists having interests in culture collections have banded and assembled in formal or informal national federations or associations of the culture collections within them, such as UKFCC. These federations tend to be further affiliated with international bodies, such as ECCO (European Culture Collections Organisation), which comprises 54 collections from 21 European countries. This organization itself is affiliated to WFCC (World Federation for Culture Collections). These federations provide excellent opportunities for discussion and exchange of information in all aspects of culture collection curation.

European Culture Collection Organization (ECCO)

ECCO was established in 1981. The aim of the organization is to promote collaboration and trade ideas and information about all aspects of culture collection activity. Membership is open to representatives of any resource centers that provide a professional service on demand and without restriction, that accept cultures for deposit, that provide catalogs, and that are housed in countries with microbiological societies affiliated to the Federation of the European Microbiological Society (FEMS). After 14 years of activity, ECCO comprises 54 members from 21 European countries. The total holdings of the collections are about 300,000 deposits representing a large pool of biodiversity from different sources (from humans, animals, plants, foodstuffs, environmental) as well as various sources of biotechnological interest, including GMO strains and organisms of interest for genetic engineering.

World Federation for Culture Collections (WFCC)

Most of the culture collections organized in national or regional federations are members of the WFCC. This international organization has been set up as a multidisciplinary commission of the International Union of Biological Science and is a federation within the International Union of Microbiological Societies (IUBS). Various committees exist within the WFCC for carrying out specific tasks in the different aspects of culture collection work. These include the publication of books, provision of training courses, collaboration with postal and patent organizations, setting up databases, providing help for endangered culture collections, and designating quality standards and guidelines for collections. One of the major activities of WFCC is to organize its quadrennial meetings. The World Data Centre (WDC), now the World Centre on Microorganisms (WDCM), covers a central role in WFCC. Initiated in 1972 by Prof. V.B.D. Skerman and sponsored by UNEP and UNESCO, the database pioneered the organization of microbial culture collections worldwide. Relocated to the RIKEN Institute in Japan in 1986, the WDC, under the directorship of Dr. H. Sugawara, developed into a valuable information resource at the microbial level.

Several databases are available as hard copies and are also accessible through the Internet. The core databases, CCINFO and STRAIN, are regularly updated and published as the WDC Directory. CCINFO covers information on nearly 500 microbial resource centers in the world, their organization, the kinds of cultures held, the expertise available, and the services offered. Strain lists all microbial species held by registered collections. The WDCM activity of continual data collection and dissemination plays an important role in the improvement of data quality of collections and for making the resource centers known to the scientific community. Any individual with an interest in culture collection activities is eligible for membership. Culture collections as institutions may become affiliate members if they meet the standards established by the federation. Individuals or organizations may be accepted as sustaining members on the basis of extraordinary support of the objectives and activities of the Federation. Further information can be obtained from the WFCC.

3

INDUSTRIAL APPLICATIONS AND PRODUCT DEVELOPMENT

The ubiquitous presence of microorganisms, as well as their metabolic diversity, has made them important factors in industrial applications and product development. Bacteria, fungi, and yeasts have been used for the production of food, medicinals, solvents, enzymes, and numerous other products. They have also been used as standards for bioassay and bioremediation applications. In addition, advances in bioengineering have enabled the incorporation of genes into microorganisms for the production of mammalian proteins and peptides. During the past 20 years, cell line technology grew dramatically to enable the isolation and maintenance of a wide variety of tissue lines from any species of interest.

The initial use of cell lines for vaccine production was expanded by developments in cell hybridization, transduction and transfection, and genetic manipulation to include the production of monoclonal antibodies, cytokines, and other pharmaceutical products. Lines that have been genetically engineered have been successfully expanded and grown in fermenters for large-scale production. The development of microbial and cell cultures represents an enormous investment in time and money and is an important asset to industry. These cultures must be protected from accidental loss by means of standardized preservation techniques. The primary aim of culture preservation is to maintain the culture such that it is as close as possible to the original strain or line exhibiting the desired phenotypic or genetic traits.

Often, industrial cultures are not analogous to taxonomic strains in that they have been modified to have special properties, either rapid growth, more active metabolic rates, or other features. Many methods have been used to preserve bacteria and fungi, but not all species respond in a similar way to a given method. Unfortunately there is no universal method that can be applied to all microorganisms. It should be emphasized, however, that the success or failure of any preservation technique also depends on the use of the proper growth medium and cultivation procedures and on the age of the culture at the time of preservation. This is particularly true when working with microorganisms that contain plasmids or recombinant DNA, or that exhibit growth phases such as morphogenesis or spore formation. The following provides practical information for the maintenance and preservation of bacteria, fungi, yeast, and cell lines. It should be noted, however, that not all microorganisms will be preserved by these methods, and some experimentation with cryoprotective agents and other factors may be necessary for success. Additional information on bacterial, fungal, and cell-line preservation may be found elsewhere.

BACTERIAL PRESERVATION

Most bacteria can be preserved by periodic serial transfer to fresh medium. The period between transfers varies with the organism, medium employed, and the external conditions. In general, minimal media are preferred for subculturing because they lower the metabolic rate of the organism and thus prolong the interval between transfers. In rich media, growth is often much faster, with the accumulation of metabolic products. These compounds may alter environmental conditions such as pH and gas phase and shorten the interval between transfers.

Some bacteria such as pathogens need complex media for growth, or their retention of specific physiological properties require the addition of complex compounds to the medium. The subculturing interval for these organisms must be determined by experience. When preserving bacteria by serial transfer, it is important to use containers that can be securely sealed. Glass test tubes or 6-mL bottles with rubber-lined screw caps are widely used. The bacteria can be grown with slightly loose caps until adequate growth is obtained and then tighten to prevent dehydration.

Solid media are preferred to broth because contaminants can be readily seen. Duplicate tubes should be maintained as a precaution against loss, at least until the subculturing interval is determined. The cultures should be examined for purity after each transfer, and abbreviated characterization tests must be run periodically to ensure retention of desired traits. Storage in a refrigerator is the preferred method for subcultures. Many bacteria can be kept for 3 to 5 months between transfers if proper precautions are taken to avoid dehydration of the medium. It should be noted, however, that the maintenance of cultures by serial transfer is precarious, time-consuming, and expensive.

The frequent transfer of cultures can result in contamination, mislabeling, and the selection of strain variants. Other methods involve storing cultures at refrigeration temperatures, drying in sterile soil, glass beads, filter paper, and other substrates. Short-term maintenance and preservation procedures are not recommended for industrial strains because a number of problems can occur using these methods. The main hazards are contamination, population changes through selection resulting in the loss of key genetic and physiological properties, and the expense of maintaining cultures under these conditions.

Preservation at Freezing Temperature

The use of the freezer compartment of a refrigerator or an ordinary freezer with a temperature range of 0 to –20°C to preserve bacteria is not recommended because the freezing process damages the cells. In addition, most refrigerators that have a freezer compartment in the range of 0 to –20°C are frost free. The frost-free state is achieved by an alternating warming and cooling cycle, which can result in further damage to the cells and a loss of viability.

Freeze-drying, or Lyophilization

Freeze-drying, or *lyophilization,* is one of the most economical and widely used methods for long-term preservation of bacterial cultures and other microorganisms. Many metabolically diverse bacterial species have been successfully preserved by this technique and have remained viable and unaltered for more than 60 years. Freeze-drying is a complex process that involves the removal of water and other solvents from a frozen product by sublimation. Sublimation occurs when a frozen liquid goes directly to a gaseous phase without passing through the liquid phase. There are three stages in the freeze-drying process: (i) prefreezing of the suspension to ensure a solid frozen starting material, (ii) primary drying during which most of the water is removed through sublimation, and (iii) secondary drying to remove the bound water. The rate of cooling during the prefreezing step and the final temperature of the frozen material can affect the freeze-drying process. The removal of water during the primary drying stage requires an environment that allows the water molecules to migrate from the frozen material. This is achieved by using a vacuum pump. A large-capacity pump must be used to prevent the water vapor load from compromising the efficiency of the system.

The procedure also requires a moisture trap or condenser to collect and remove the water molecules before they enter the vacuum system. The condenser temperature must be lower than the frozen material for sublimation to occur. The sublimation rate is dependent on the differential vapor pressure between the frozen material and the condenser. Because the vapor pressure varies with the temperature, heat can be carefully applied to the frozen cell suspension to increase the vapor pressure and promote a faster primary drying rate. Precautions must be taken during the application of heat to prevent the frozen cell suspension from thawing. Most commercial instruments have sophisticated controllers that allow the input of heat to the frozen material to achieve maximum drying rates without damaging the product. When the primary drying is complete, some bound water remains. This water cannot be eliminated by sublimation and requires the application of external heat to the material under low pressure and low condenser temperature. The moisture remaining in freezed-dried cells is termed residual moisture. A residual moisture content of 1-3% seems to be required for the long-term shelf life of bacteria.

Centrifugal freeze-drying and prefreezing are two of the most commonly used methods of lyophilization of bacteria. In centrifugal freeze-drying, the suspensions are initially frozen by evaporative freezing under a vacuum while centrifuging to prevent frothing due to removal of dissolved gases. After primary freeze-drying, the vials are constricted using a narrow flame and then placed on a manifold for secondary freeze-drying. A detailed discussion of centrifugal freeze-drying can be found in articles by Rudge and Lapage. The prefreezing method employs a controlled rate of cooling to freeze the cell suspension and then a primary drying phase, as discussed earlier. The simplest system consists of a vacuum pump capable of an ultimate pressure of less than 10 μmHg (a pump rated at 35–50 L/min is usually adequate), a small stainless steel condenser for cooling with dry ice, a thermocouple vacuum gauge to monitor the vacuum system, and heavy-walled vacuum/pressure tubing to connect the components. The connection of these modules into a single system is often termed a component freeze-dryer.

The main components are the vacuum pump, thermocouple vacuum gauge, and a condenser that can be attached to several manifolds, to which glass vials can be attached by means of vacuum-tubing nipples, or the manifolds can be connected to a stainless steel pan containing the vials. The manifold system is relatively simple, and it has been used successfully by many laboratories to preserve cultures. A small amount of cell suspension (0.2 mL) containing a cryoprotectant is dispensed into sterile ampules, which are plugged with sterile cotton. A 1-inch piece of nonpowdered amber latex IV tubing is attached to the rim of each ampule (by using a tube stretcher) and then to the manifold. The ampules are then immersed in an ethylene glycol (50%) dry-ice bath at –40°C to freeze the cell suspension.

The freeze-drying cycle is initiated by starting the vacuum pump and allowing the temperature of the bath to rise of its own accord to ambient temperature. In general, runs can be started early in the afternoon and allowed to proceed overnight. At the end of the drying cycle, the ampules are sealed using a double-flame air/gas torch and stored at 2–8°C. For specific details on the equipment and methods see articles by Gherna and Simione and Brown. The American Type Culture Collection (ATCC) has successfully lyophilized many of the bacteria described in the literature. It currently uses three methods:

1. *Component freeze-dryer.* Samples are freeze-dried in cotton-plugged inner vials, which are then sealed in outer vials under vacuum.
2. *Commercial freeze-dryer.* Cultures are freezed-dried in the double-vial system in a commercial freeze-dryer.
3. *Preceptrol cultures.* Cultures are lyophilized in a commercial freeze-dryer using glass serum vials that are sealed with a rubber stopper and metal cap.

It should be noted that after the drying cycle is complete in the commercial freeze-dryer and Preceptrol methods, the vacuum in the drying chamber is replaced with cooled nitrogen gas (2–8°C) that has been passed through a 0.2-μm Pall filter. The condenser temperature must not be allowed to rise above –50 °C. The nitrogen gas may have to be passed through a copper coil immersed in liquid nitrogen.

Healthy cell growths

Successful freeze-drying depends on using healthy cells grown under optimum conditions on the medium of choice for each strain, which ensures the retention of the desired features of the bacterium. Bacteria can be grown on agar or in shaken or static broth cultures. The cells are harvested at maximum stability and viability, in the late logarithmic or early stationary phase, and suspended in a medium. The cell suspension should contain at least 10^7–10^{10} cells/mL in order to achieve optimal results. Anaerobic cultures must be grown, harvested, and dispensed under anaerobic conditions. The viability of freeze-dried cultures is greatly influenced by the suspending medium. The choice of suspending medium for freeze-drying depends on the type of bacteria and on the method used.

The ATCC uses two suspending media for the commercial freeze-drying method:

1. Reagent 18

   0.75 g trypticase soy broth

   10.0 g sucrose

5.0 g bovine serum albumin fraction V

100.0 mL distilled water

Filter sterilized through a 0.2-μm filter

2. Reagent 20

10.0 g bovine serum albumin fraction V

20.0 g sucrose

100 mL distilled water

Filter sterilized through a 0.20-μm filter Reagent 18 is more commonly used for freeze-drying bacteria and fungi. Reagent 20 is employed for bacteria that are adversely affected by trypticase soy broth and is added to an equal volume of cell suspension in the broth. In the component freeze-drying method (double-vial), the additive is a 20% (wt/vol) sterile solution of skim milk, prepared by autoclaving a 20% skim milk solution at 116°C for 20 min in 10-mL tubes. For cultures grown on agar surfaces, the cells are washed off with the 20% skim milk solution.

Broth cultures are centrifuged, and the cell pellet is resuspended with the sterile skim milk to give a suspension of 10^8 cells/mL. Skim milk should not be used for bacteria that are inhibited by milk, instead use a 24% (wt/vol) sterile sucrose solution diluted equally with growth medium to yield a 12% (wt/vol) sucrose solution (final concentration) for the single-vial manifold or commercial freeze-dryer method. As soon as the cell suspension is prepared, it should be dispensed into the ampules. The interval between dispensing and freeze-drying should be kept to a minimum to avoid possible alteration of the culture. Other investigators have used 10% (wt/vol) dextran, horse serum, inositol, raffinose, trehalose, and other cryoprotective chemicals in the suspending medium. Most bacteria can be lyophilized using these methods.

The same methods have been employed successfully to freeze-dry most bacteriophages; exceptions are stored in liquid nitrogen. In general, bacteriophages can be grown in a soft agar layer or in broth. The phages are harvested by scraping off the soft agar with a sterile glass rod or rubber policeman, macerating and dispensing into a sterile centrifuge tube, and centrifuging at low speed to sediment the agar and most of the unlysed bacteria. Broth-gown cultures are centrifuged at low speed to remove the unlysed bacteria. The supernatant is then filtered through a 0.45-μm and then through a 0.2-μm Millipore filter. The filtrate should be titrated to determine the concentration of phages; a titer of 10^8 pfu/mL is desirable. The phage can be freeze-dried using 20% skim milk and the component freeze-dryer method.

Freeze-dried bacteria and bacteriophages can be stored at 2–8°C. A longer shelf life has been obtained when cultures are stored at –30 or –70°C in a mechanical freezer. Freezed-dried cultures should not be stored in a freezer compartment of a frost-free refrigerator because of the alternate heating and freezing cycles, nor should they be stored at room temperature. Although lyophilization has facilitated the long-term preservation of bacteria, viability checks must be done before and after lyophilization to determine the effectiveness of the process. Quality control procedures must be in place to check for purity, retention of essential characteristics and periodic viability to determine the shelf life of the culture.

Freeze-drying

The long-term preservation of bacterial species not amenable to freeze-drying has been achieved through storage in the frozen state at a temperature of –70 °C or, more advisable, at the temperature of liquid nitrogen (–196°C) or liquid vapor phase (–150°C). The freezing process results in several events that can be destructive to the cells. As the bacterial suspension begins to cool, the liquid water turns to ice at 0°C (the exact temperature depending on the nature of the bacterial suspension). Ice formation occurs first in the external aqueous environment, resulting in an increased concentration of solutes outside the cells. The differential osmotic pressure causes water to migrate from the cells. The rate and extent of the water loss are dependent on the rate of cooling and the permeability of the cells. Removal of too much water leads to a concentration of solutes within the cells that may be harmful. If too much water is left in the cells, internal ice crystals may form, which would result in intracellular damage.

It is important to maintain the critical balance between the two events by controlling the rate of cooling while freezing the cells, and warming as rapidly as possible when the cells are thawed. Chemical compounds such as glycerol and dimethyl sulfoxide (DMSO) can be added to the cells to minimize the damage caused by freezing. These compounds, labeled cryoprotective agents, exhibit certain properties that are essential for good cryoprotection. They must be nontoxic to the cells, show good permeability, and bind either the electrolytes that accumulate or the water molecule to delay freezing. The physiological condition of the cells plays an important role in survival of freezing. In general, actively growing cultures harvested at the mid to late logarithmic phase of growth will survive the freezing process better than those harvested at an earlier or later phase. The bacteria should be grown under optimum conditions in the appropriate medium that best retains the prominent characteristics of the strain. Bacteria can be grown on broth that is static or shaken, or on agar. Shaken cultures generally reach optimal cell density faster than agar-grown cultures. Some strict anaerobes require prereduced media and anaerobic procedures for harvesting and dispensing.

Cells grown in broth cultures are harvested by aseptic centrifugation, and the resultant pellet is resuspended in sterile broth containing 30–50 % (vol/vol) sterile glycerol. Bacteria grown on agar are harvested by aseptically washing the growth with sterile broth containing the cryoprotectant. Dispense 0.4 ml of the cell suspension, containing at least 10^8 cell/mL into each sterile, prelabeled vial. Plastic presterilized screw-capped vials are recommended for bacteria. Place the filled vials into a mechanical freezer set at –70°C. The cells are recovered by rapidly thawing the frozen cell suspensions in a 37°C water bath. A wide variety of bacterial have been successfully preserved at –70°C using glass beads and 15% glycerol as the cryoprotectant. The glass beads are placed in 2-mL glass screw-cap vials and sterilized for 15 min at 121°C. The vials must be prelabeled with ink that will not come off at ultralow temperatures.

Cell suspensions are prepared containing approximately 10^8 organisms/mL in sterile 15% glycerol, and 0.5 mL is dispensed into each vial. The vials are shaken gently to ensure that all the beads are wetted with the bacterial suspension, and then the excess liquid is aseptically removed with a sterile pasteur pipette. The vials are placed into a mechanical freezer at –70°C. The cells are recovered by aseptically removing a bead from the vial with a sterile spatula or forceps and placing the bead in sterile broth. It should be noted, however, that the contents of the vial should not be allowed to thaw; this can be prevented by placing the vial in crushed dry ice and removing the bead after ensuring the beads will remain frozen. This technique has been used to preserve bacteria obtained from extreme environments such as hypersaline pools. Precautions must be taken against electrical shutdowns and compressor malfunctions. Adequate back-up systems (such as alarms, back-up freezers, or an electrical generator) will help to prevent the loss of a valuable collection. However, additional alarm systems are recommended, such as the sound/off power-temperature monitor, which has a temperature range of –75 to + 200°C. The unit is battery powered and can be mounted on a wall.

Cryopreservation

Although the long-term maintenance of bacteria in liquid nitrogen has been considered expensive, the successful preservation of physiologically diverse bacteria by this method, along with minimal handling and lower labor cost, makes this method feasible to use. The ATCC has successfully preserved many bacteria and bacteriophages, including fastidious ones, in liquid nitrogen for over 36 years without the loss of phenotypic properties. A large variety of liquid-nitrogen refrigerators are now commercially available, with a wide assortment of features and storage capacities. The sizes range from 10 to 1,000 L, allowing storage of 300–40,000 ampules. The physiological condition of the cultures plays an important role in the survival of bacteria under liquid-nitrogen freezing. Bacteria are grown in the appropriate medium and harvested at mid-late logarithmic phase of growth. Cultures grown in broth are harvested by aseptic centrifugation, and the resultant pellet is resuspended with sterile fresh medium containing either glycerol or DMSO. For agar-grown cells, the growth is washed off from the agar surface with sterile broth containing the appropriate cryoprotectant.

Cryoprotective agents must be added to the suspending medium to protect cells from freeze damage. Cryoprotectants fall into two main classes: compounds such as glycerol and DMSO, which are permeable and appear to provide both intracellular and extracellular protection against freezing; and agents such as dextran, glucose, lactose, mannitol, polyglycol, polyvinyl pyrrolidone, and sucrose, which seem to exert their protective effect external to the cell membrane. The first class of cryoprotectants has proven to be the most effective, and in general, glycerol and DMSO seem to be equally effective in protecting a wide range of bacteria. It is advisable to conduct a tolerance test on new species to ascertain whether the cryoprotectant is toxic or beneficial. DMSO and glycerol are routinely used at concentrations of 5% (vol/vol) and 10% (vol/vol), respectively, in the proper suspending medium.

Glycerol is normally prepared as a double-strength (20%, vol/vol) solution and then mixed with an equal amount of the cell suspension. Glycerol is sterilized by autoclaving at 121°C for 15 min and stored in 6-mL volumes at 2–8°C. DMSO is filter sterilized with a 0.22-μm-pore-size Teflon (polytetrafluoroethylene) membrane that has been prewashed with methanol and DMSO, collected in 10 to 15-mL quantities in sterile test tubes, and stored in the frozen state at 5°C (DMSO freezes at 18°C) and protected from light. An opened bottle of DMSO should not be used for more than 1 month because of the accumulation of oxidative breakdown products. Cell suspensions containing at least 10^8 cells/mL are dispensed into prelabeled sterile vials. There are a variety of vials available commercially, ranging from prescored glass to sterile plastic. With glass vials, care must be taken when sealing and storing because improperly sealed glass vials can explode when retrieved from the liquid nitrogen as a result of the rapid expansion of the liquid nitrogen that enters the vials through microscopic holes.

Plastic screw-cap vials that are not properly sealed can fill with liquid nitrogen if stored in liquid phase; retrieval of the vials to a warmer temperature can result in liquid nitrogen expansion, causing the contents of the plastic vials to spray into the laboratory environment. Plastic screw-cap vials should always be stored in the vapor phase. Protective gloves and face shields should be worn when handling frozen liquid-nitrogen vials. The rate of cooling is important in liquid-nitrogen storage. In general, the best results for preservation of bacteria have been achieved with slow cooling. This step can be controlled using programmable controllers such as the Linde BF 3-2, which has an adjustable cooling rate. The cells are frozen at a controlled rate of 1–3°C/min to –40 °C and then at a more rapid rate of 10°C/min to –90°C. After this temperature is reached, the vials are transferred to a liquid-nitrogen tank and stored in the liquid phase at –196°C or above in the vapor phase at –150°C.

If a programmable freezer is not available, slow cooling can be obtained by placing the filled glass or plastic vials in a stainless steel pan at the bottom of a mechanical freezer at –60°C for 1 h and then plunging the vials into a liquid-nitrogen bath for 5 min. The rate of cooling to –60°C is approximately 1.5°C/min. The frozen vials can then be stored in the liquid-nitrogen tank in the liquid or vapor phase. It has been reported that the best recovery of cultures from liquid-nitrogen storage is obtained by rapid thawing. This has also been the experience with cultures frozen at the ATCC. The events during thawing are complex, and slow warming rates can lead to ice crystal formation; a rapid warming rate of approximately 100°C/min will minimize the damage due to ice formation. A rapid rate of warming can be achieved by quickly immersing the frozen vials in a 37°C water bath with moderate agitation until all the ice melts. This usually takes about 50 s for glass vials and 90 s for polypropylene ones. The successful cryopreservation of plasmid-containing bacteria depends on the stability of their plasmids. These organisms are less genetically stable and more likely to lose their foreign genetic material. In general, bacteria containing unstable plasmids should be grown on antibiotic-containing liquid medium at 30°C, frozen with 10% (vol/vol) glycerol, and stored in the vapor over liquid nitrogen.

YEAST AND FUNGAL PRESERVATION

Many filamentous fungi can be successfully maintained by serial transfer on suitable media. It is important, however, to use optimal media and growth conditions to ensure healthy cells that retain their morphological and physiological characteristics. A list of suitable media and incubation temperatures for a variety of fungi is provided by Onions and Pitt. In general, a medium that promotes good sporulation is considered to be the most desirable. When possible, it is best to cultivate the fungi on agar slants in test tubes or culture bottles. Transfers should be made only from the growing edges of the culture, taking precautions to ensure that contamination is prevented. Some fungi degenerate when maintained on the same medium for extended periods, thus media should be alternated from time to time. A minimal medium such as potato/carrot agar can often be used to induce sporulation in a culture that has deteriorated. Also, the continual transfer of spores from some aging fungal cultures such as *Mucorales* can result in a deterioration of the culture; the transfer of mycelium and spores seems to minimize this problem. The interval between transfers varies from fungus to fungus; some require transfer every 2–4 weeks, the majority every 2–4 months, while others may survive for 12 months without transfer. Cultures grown on agar slants can be stored at room temperature. Storage in the refrigerator or cold room at 5–8°C can extend the transfer period to 4–6 months. However, there are some fungi, such as the thermophiles, that are sensitive to storage at these temperatures. The maintenance of fungi by serial transfer has many disadvantages.

The most notable are contamination, selection of variants, mislabeling of cultures, infestation with mites, and labor intensity. Yeasts have been successfully maintained by serial transfer on either solid or liquid media. In general, many yeasts will survive for longer periods when grown on solid media rather than in broth, especially nonfermentative yeast. Fresh medium is inoculated with actively growing cells and incubated at the appropriate temperature. After sufficient growth occurs, the cells are stored at refrigerated temperature (+4°C). The majority of yeast species will remain viable for a least 6 months. The interval between transfers must be determined by experimentation. Serial transfer is not recommended for extended maintenance because genetic drift has been observed. In addition, ascosporogenous strains may sporulate on agar slant media stored for prolonged periods, resulting in strain variability. Many fungi and yeast can be preserved for years by drying on a suitable menstruum such as soil, silica gel, and filter paper, and drying under vacuum from the liquid state (L-drying).

Sterile Anhydrous Silica Gel

Sterile anhydrous silica gel (6–22 mesh, non-indicator) is prepared by dispensing the gel into growth bottles so that they are one-quarter filled, and sterilizing in an oven at 180°C for 2–3 h. The bottles containing the sterile gel are cooled by placing in a pan containing ethylene gycol and dry ice to the depth of the gel layer or by placing the bottles in water in a mechanical freezer at a temperature of –17 to –24°C. Fungal spores are suspended in a sterile 5% skim milk solution that has been cooled to 4°C. The spore suspension is added to the cooled silica gel until three-quarters of the gel has been moistened. It is important to avoid saturating the gel. The growth bottles are left in the ice bath for approximately 20–30 min and then agitated to ensure thorough dispersion of the suspension. The bottles are held at 25°C for approximately 1–2 weeks until the crystals easily separate when shaken. Inoculated bottles are sealed with screw caps and stored over indicator silica gel in an airtight container at 4°C. To recover the fungal spores, a few crystals are sprinkled on an appropriate medium and incubated. This method has been used successfully for a variety of fungi. This method has produced variable results with yeast; some are not viable after 3 months of storage, whereas other survive after 2–5 years.

Preservation and Soil

A variety of fungi have been maintained in soil. The method involves the inoculation of approximately 1 mL of a spore suspension into soil that has been sterilized by autoclaving twice,

24 h apart, at 121°C for 15 min. The soil is incubated at room temperature for 5–10 days and then stored in a refrigerator at 4–8°C. Cultures are recovered by inoculating a suitable medium with a few grains of soil.

Preservation by Paper

Yeasts have been preserved by drying on paper discs or squares and storage over the desicant silica gel. Whatman No. 4 filter paper is cut into small squares or discs about 10 mm across and placed on a small aluminum foil, which is folded into a packet and autoclaved at 121°C for 15 minutes. The sterile discs or squares are inoculated by immersing into drops of a heavy yeast suspension prepared in 5% skim milk. The packets are dried in a desiccator for 2–3 weeks at 4°C and then placed in an airtight container and stored at 4°C. Cultures are recovered by aseptically removing a piece of filter paper and inoculating the appropriate solid or liquid medium. Yeast maintained in this manner have been viable after 2–3 years.

Liquid-drying

L-drying refers to liquid-state drying in such a way as to prevent freezing. Drying occurs under vacuum at temperatures generally no colder than 5–10°C. The method has been improved and simplified by Malik for the preservation of sensitive microorganisms that are damaged by freezing or freeze-drying. The technique has been used to successfully preserve some yeasts and filamentous fungal spores.

Lyophilization

The equipment, cryoprotectants, and theoretical aspects of freeze-drying have already been discussed. Fungi are grown under optimum conditions using media that will produce maximum sporulation. Mature spores are harvested by flooding the agar cultures with approximately 2 ml of a 20% sterile skim milk solution. The skim milk must be cold when used, and thus it is stored at 2–8°C until required. The spores are gently scraped from the surface of the culture to yield a suspension containing at least 10^6 spores/mL. The spore suspension is pipetted back into the tube containing the remaining milk and mixed thoroughly. If more than one tube is used, repeat the procedure and pool the suspension in one tube to yield a concentration of at least 10^6 spores/mL. The spores of cultures grown in broth medium are centrifuged and resuspended in sterile 20% skim solution.

Approximately 0.2 mL of the spore suspension is dispensed into sterile vials for freeze-drying. The interval between harvesting the spores and dispensing must be minimized because many spores begin to germinate when suspended in liquid. Spores should not be in the skim milk for more than 2 h before being processed. Filled vials should be refrigerated while awaiting further processing. The fungi are freeze-dried using one of the ATCC methods already described. Lyophilized spores can be stored at 2–8°C. Extended shelf life can be achieved by storing the vials at –70°C in a mechanical freezer.

Freeze-dried fungal cultures are rehydrated with 0.5–0.9 mL of sterile water, then transferred to tubes containing 6 mL sterile water and allowed to soak for 50–60 min before transferring to solid medium. At the ATCC, yeasts are processed in the same manner as bacteria for the component freeze-drying method. Cultures are grown on the appropriate solid medium and suspended in sterile 20% skim milk or reagent 18 to a concentration of at least 10^6 cells/mL. The suspension is mixed thoroughly, and 0.2 mL is dispensed into each vial. The cultures are freezed-dried using the component freeze- drying method described earlier.

Cryopreservation

The preservation of fungi at ultralow temperatures of liquid nitrogen (–196°C for liquid phase and –150°C for vapor phase) is presently considered the best method of storage. The method can be applied to both sporulating and nonsporulating strains, as well as those fungi that have been genetically altered or harbor plasmids. The freezing rates and cryoprotective agents described for bacteria are also applicable to the freezing of fungi. Spores or mycelial fragments are harvested by flooding slants or plates with 10% glycerol or 5% DMSO and gently scraping the surface of the cultures. The fungal suspension is dispensed in 0.5-mL amounts into sterile plastic freezing vials and frozen at a controlled rate.

Fungi that produce sticky mycelia or whose mycelia grow embedded in the agar can be prepared for freezing by cutting agar plugs containing new growth (hyphyl tips) with a sterile cork borer (5 mm). Three or four plugs are place into each sterile plastic vial with 0.4 mL of 10% glycerol and frozen at a slow controlled rate. Cultures grown in broth are fragmented in a sterile Waring blender in a biological hood and suspended in equal parts of 20% glycerol and growth medium or equal parts of 10% DMSO and growth medium to give a final concentration of 10% glycerol (vol/vol) or 5% (vol/vol) DMSO, respectively. Some strains must be concentrated by centrifugation in order to obtain sufficient material for freezing.

Pathogens must not be macerated in a mechanical blender because of the hazard of aerosol dispersion. Some cultures, such as *Agaricus* strains, can consist of seeds, grains, or pollen and be frozen without the presence of a cryoprotectant. Slime molds can be preserved by freeze-drying or freezing. Spores, microcysts, and spherules are preserved by a controlled freeze in 10% glycerol and stored in the liquid-nitrogen vapor phase. Thawing of the frozen culture should be done rapidly in a 37°C water bath with moderate agitation. Once the ice melts, the culture can be transferred to a suitable medium.

Yeast cells have been successfully preserved by controlled freezing to liquid-nitrogen temperature (–196°C liquid, –150°C vapor phase). The yeast cultures are handled in the same manner as bacteria and grown as described in the freeze-drying section for yeast. Cell suspensions are prepared in 10% glycerol to a concentration of at least 10^6 cells/mL. The yeast are frozen under controlled conditions as described in the bacterial section. Several reports have described the loss of plasmid expression or the induction of respiratory-deficient mutants in yeast as a result of freeze-drying or improper freezing. Nierman and Feldblyum reported that when plasmid-containing *Saccharomyces cerevisiae* were preserved by freezing in liquid nitrogen, no loss of viability was observed, and the fraction of plasmid-containing cells was the same before and after freezing. The preservation of plasmid-containing yeast is superior in liquid nitrogen with a controlled freezing rate than by freeze-drying.

PRESERVATION OF CELL CULTURES

Storage in liquid nitrogen is currently the best method of preserving cultured cells. Cryopreservation methods for a variety of cell lines are well developed and used extensively for culture storage. The general events that occur during freezing and the effect of the cooling rate, cryoprotective agents, and a controlled rate of cooling have been discussed in the bacterial section. Cell suspensions are prepared in the same manner used for routine subcultivation; trypsin is added if necessary to produce a uniform suspension. The cell suspension is centrifuged at approximately 100 *g* for 10 min, and the resultant pellet is resuspended in an appropriate amount of fresh medium containing a cryoprotectant. Normally, 5–10% DMSO or glycerol is employed. The freeze medium should be prepared just prior to use by mixing fresh growth medium and the cryoprotective agent and kept at room temperature. The remaining cryoprotectant is discarded because oxidative contaminants can accumulate. A cell suspension containing approximately 10^6–10^7 cells/mL is satisfactory. The cell suspension is dispensed in 1-mL amounts into each glass vial or plastic ampule and frozen at a controlled rate of 1°C/min to about –50°C. The vials are immediately transferred into liquid nitrogen for storage in the liquid or vapor phase. Rapid thawing is essential for recovery of frozen cells, thus the frozen vials are immersed directly into a 37°C water bath and agitated until the suspension is completely thawed. The thawed cell suspension is transferred to 10 mL of sterile growth medium and centrifuged at 100 *g* for 10 min. The supernatant containing the cryoprotectant is discarded, and the cell pellet is resuspended in fresh medium and propagated using standard procedures.

Standard Operating Procedure

It is important that all preservation methods include standard operating procedures for the determination of culture purity, viability, retention of morphological and physiological characteristics, production of desirable products, and other key attributes. Purity checks must be conducted before and after processing microorganisms for freeze-drying and microorganisms and cell lines for freezing. Generally, the cell suspension is diluted and inoculated into suitable medium and incubated at optimum temperature. After growth appears, the cultures are examined for purity. It is also important to conduct pre- and post-preservation viability checks for freeze-drying and freezing to determine the effectiveness of the process. In addition, periodic viability assays should be performed to ascertain the shelf life of the culture. Cell lines should also be examined for *Mycoplasma* contamination, because it has been estimated that about 10% of cell lines are contaminated. It is also important to conduct tests to verify cell-line identity; several rapid techniques have been developed that use DNA probes to identify the cell lines.

Role of Documentation

Documentation plays an important role in the maintenance and preservation of cultures. Important data include the identity of the culture, isolation source and methods, the individual who isolated it, geographical location, morphological and physiological characteristics, maintenance data, and other salient features. If the culture was obtained from another investigator, the documentation should include the name of the investigator, culture history if available, and date of acquisition. The greater the amount of data recorded and regularly updated, the greater the value of the culture.

4

FOOD BIOTECHNOLOGY

Bread is one of the most common and low-cost foods and is associated with traditions in many different countries. Bread is also closely related to *biotechnology*, which is synonymous with *high technology*. The term that relates these two subjects is *enzyme*. Consumers have certain criteria for bread, including appearance, freshness, taste, flavor, variety, and a consistent quality. It is a great challenge for the baking industry to fulfill these criteria for several reasons. Firstly, the main ingredient of the bread—flour—varies due to varied wheat qualities, weather, and milling technology, although millers attempt to blend the wheat sorts to produce a good and consistent baking flour quality. It often proves difficult to satisfy both high-quality and low-cost standards at the same time. Secondly, because bread traditions differ, the baking industry needs various qualities of ingredients and procedures. For instance, an English toast bread with fine crumb structure and very soft texture is not popular with the French, who want baguettes with crispy crusts, large holes, and good chewiness in the crumb. Thirdly, healthier products are becoming more popular due to more intensive cultural exchanges and changes in consumer preferences.

Some new variety breads can be made by simple adjustment of the formulation or baking procedure. However, others require the bakers to develop new techniques. Therefore, both millers and bakers need agents or process aids such as chemical oxidants, emulsifiers, and enzymes to standardize the quality of the products and diversify the product range. For decades enzymes such as malt and microbial α-amylases have been used for bread making. Due to the changes in the baking industry and the demand for more varied and natural products, enzymes have gained more and more importance in bread formulations. Through new and rapid developments in biotechnology, a number of new enzymes have recently been made available to the baking industry. One example is pure xylanase, with single activity instead of traditional hemicellulase preparations, which improves the dough machinability.

A lipase has a gluten-strengthening effect that results in more stable dough and improved bread quality, and a maltogenic α-amylase has a unique antistaling effect. This article reviews the effects of these enzymes. It also presents synergistic effects when these enzymes are combined with either each other or traditional enzymes such as fungal α-amylase.

ENZYMES IN BREAD MAKING

Wheat and thus wheat flour contain endogenous and indigenous enzymes, mainly amylases. However, the level of amylase activity varies from one type of wheat to another. The amount of α-amylases in most sound, ungerminated wheat or rye flours is negligible. Therefore, most bread flours must be supplemented with α-amylases, added in the form of malt flour or fungal enzymes. Many methods are available for the determination of amylase activities, as reviewed by Kruger et al. Other methods such as falling number (FN) and Brabender amylograph are used by the baking industry and millers to determine the amylase content correlating to bread making. Although FN is excellent for measuring the activity of cereal amylases including added malt flour, it is not suitable for measuring the activity of fungal α-amylases. Because fungal α-amylases are generally less thermostable, they are inactivated at temperatures near 65°C. Therefore, fungal α-amylases cannot be detected by the standard FN method, which is conducted at 100°C.

Perten et al. reported a modified FN method for measuring the fungal α-amylase activity in flour. However, in practice, the method is less suitable for routine analysis because it is necessary to define the measuring conditions for each type of flour (because the indigenous and endogenous amylases vary in different flours) to determine the modified FN with added fungal α-amylases. Fungal α-amylases act on the damaged starch, which varies depending on wheat sort and milling condition. Generally, flour made from hard winter wheat contains more damaged starch than soft wheat. The α-amylases widely used in the baking industry can hydrolyze amylose and amylopectin to release soluble intermediate-size dextrins of DP2–DP12. α-Amylases provide fermentable sugar, which results in an increased volume, better crust color, and improved flavor.

Due to hydrolysis of the damaged starch, a suitable dosage of α-amylases results in a desirable dough softening. However, extensive degradation of the damaged starch due to an overdose of α-amylases leads to sticky dough. Fungal glucoamylases are a possibility but less common as baking enzymes. Literature on glucoamylases for baking is scarce. The volume and crumb structure improve with increasing dosage of fungal α-amylases. Although a high dosage can provide a larger volume increase, the dough would be too sticky to work with. The optimum dosage is thus defined as the dosage with maximum reachable volume without a sticky dough. For the examples, the optimum dosage for both flours is 15 FAU/kg flour.

ANTISTALING

Bread staling is responsible for significant financial losses. An estimated 3–5% of all baked goods produced in the United States are discarded every year, representing a value of $1 billion. A maltogenic α-amylase has been found to have unique antistaling effects. It can degrade both amylose and amylopectin at the gelatinization temperature and produces mainly α-maltoses and a small amount of dextrins of DP1–DP12. A study of the mechanism of antistaling effects of various amylases compared with monoglycerides showed that this maltogenic α-amylase not only has a substantially improved antistaling effect compared with fungal α-amylases but also improves the elasticity of the bread crumb, which is an important factor that influences the palatability of bread. Both crumb softness and crumb elasticity are important characteristics for the description of crumb freshness perceived by consumers. Softness indicates the force needed to compress the crumbs, and elasticity indicates the resiliency or resistance given by the crumb while being pressed.

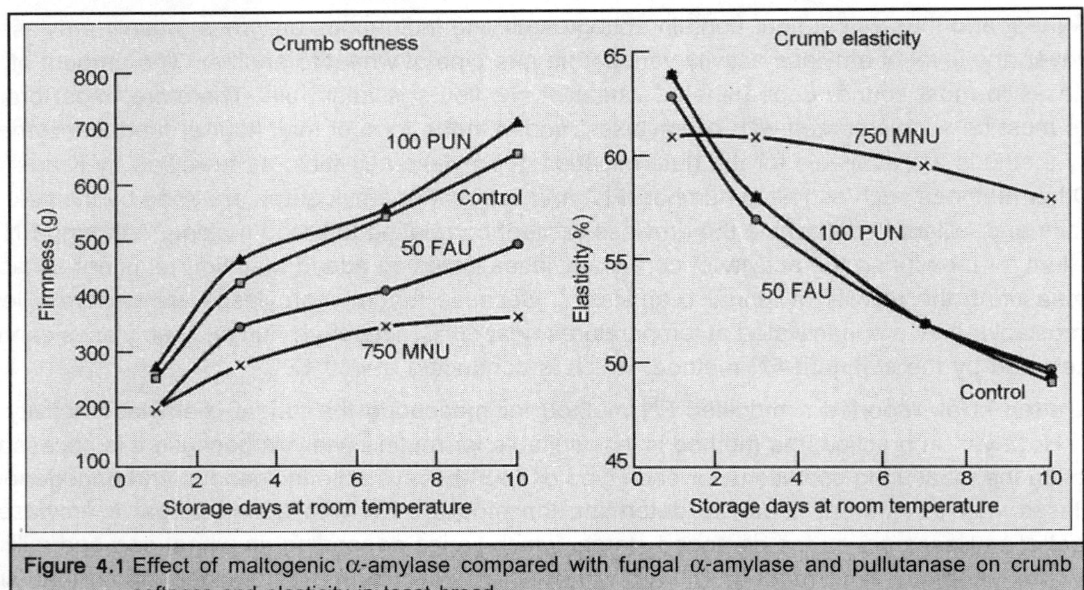

Figure 4.1 Effect of maltogenic α-amylase compared with fungal α-amylase and pullutanase on crumb softness and elasticity in toast bread.

The two texture characteristics may not necessarily correspond with each other; that is, the softest bread may not necessarily have the most elastic crumb texture or vice versa. Bread staling results in reduction of crumb elasticity and increase of crumb firmness. Commercial amylases used in the baking industry are generally α-amylases that specifically hydrolyze the α-1,4 glycosidic linkages of amylose and amylopectin molecules in starch. Most known products are cereal enzymes derived from barley malt and the fungal α-amylase from *Aspergillus oryzae*. These amylases are effective in partially hydrolyzing damaged starch and are often added to flour as flour correction to develop desirable properties, such as oven spring and brown color to the crust. These amylases have limited antistaling effect due to their limited thermostability and are for the most part inactivated before the onset of starch gelatinization during baking.

It was reported by Si and Simonsen that fungal α-amylase, although it reduces the crumb firmness, has no effect on reducing starch retrogradation. Pullulanase, which specifically hydrolyzes the α-1,6 glycosidic linkages of amylopectin, does not have a positive effect and may even have a negative effect on antistaling. The maltogenic α-amylase is a unique α-amylase that produces a significantly softer bread and maintains a high level of crumb elasticity during storage. Addition of maltogenic α-amylase in sponge and dough breads can prolong shelf life for at least 4 days longer compared with 0.5% powdered distilled monoglycerides. The crumb softness of the loaf with maltogenic α-amylase at day 7 was as soft as the loaf with 0.5% distilled monoglycerides at day 3, whereas the elasticity of the loaf with maltogenic α-amylase at day 7 is even higher than the loaf with distilled monoglycerides at day 3. Maltogenic α-amylase has a thermostability between that of fungal α-amylase and thermostable bacterial α-amylase. Therefore, it is able to hydrolyze the glycosidic linkages during the gelatinization of starch during the baking process, but it does not excessively degrade the starch because it is inactivated during the later stage of baking.

Another major advantage of maltogenic α-amylase is its tolerance of overdosing during the bread-making process in the bakery. Too much fungal α-amylase can cause sticky dough or over-browning of the crust. The bacterial α-amylase can easily be overdosed. Its pure endo action excessively degrades the starch during baking, causing collapse of the bread immediately after baking and sticky crumb during retail storage. Maltogenic α-amylase does not affect the dough rheological property because of its low activity at a temperature less than 35°C. It is highly active only at the temperature during starch gelatinization and does not excessively degrade the starch but mainly produces small, soluble dextrins. The overdosing risk is thus much lower than with the other two types of amylases. The bacterial α-amylases from *Bacillus subtilis* are able to inhibit staling by hydrolyzing glycosidic linkages within the amorphous areas of gelatinized starch. However, because of the high degree of thermostability, the enzymes can persist throughout baking and produce an excessive level of soluble dextrins.

As a result the final product is often unacceptable, with a gummy or even sticky crumb texture that causes problems during slicing and retail storage. The bacterial α-amylase gave the softest crumb but also a very gummy crumb with extremely low elasticity. The excessive degradation of starch during baking and storage creates a sticky crumb. Soy β-amylase has been reported to have an antistaling effect. Our results, based on both sponge and dough and straight dough procedures, show that 3,000 BANU/kg flour, which is a quite high dosage, has a similar antistaling effect as 0.5% powdered distilled monoglycerides. The acid fungal α-amylase at a dosage of 100–200 AFAU/kg flour has an effect similar to 0.5% distilled monoglycerides. To understand the mode of action of this maltogenic α-amylase on bread staling, the effects of enzymes on the starch retrogradation were measured by differential scanning colorimetry (DSC).

The endotherm area indicates the retrogradation of wheat starch. The endotherm area per gram starch gel containing monoglycerides was lower than the control, indicating its effect on retarding starch retrogradation. The endotherm area per gram of the starch treated with maltogenic α-amylase was significantly lower than with the other samples; the difference increased after 4 days of storage and was even more pronounced after 9 days. This indicated a significantly slower rate of starch retrogradation. The fungal α-amylase had no effect on retarding starch retrogradation. The results indicate that an enzyme can reduce the crumb firmness by means of volume or crumb structure improvement, but it has no effect on retarding the retrogradation of starch during storage. Pullulanase had no effect on retarding starch retrogradation. This means that reducing the number of branches of amylopectin does not reduce the rate of bread staling and may even increase the staling rate, as reported earlier.

HEMICELLULASES IN BAKING

Xylanase and pentosanase, also being called hemicellulases, have long been used as dough-conditioning enzymes, especially in European-type bread. Pentosanases improve dough machinability and oven spring, resulting in a volume increase. Through recombinant gene technology, some xylanases made from genetically modified organisms (GMO) are coming on the market. The benefit of using a xylanase instead of a traditional pentosanase preparation is that there are far fewer side activities in the xylanase product. Consequently, a lower dosage is needed to achieve the same effect with less risk of possible interference from side activities. The enzyme quality is also more consistent. Pentosanases and xylanases can be overdosed due to over-degradation of wheat pentosans, thereby destroying the water-binding capability of the wheat pentosans. The result of overdosing is dough stickiness. A suitable dosage of these enzymes results in a desirable softening of dough, thereby improving the machinability. An optimum dosage is therefore defined as the dosage that gives the maximum improvement of the bread properties without causing dough stickiness.

The optimum dosage of pentosanases or xylanases varies according to the flour. The optimum dosage for type 1 flour is 120–200 and for type 2 flour is 80 fungal xylanase units (FXU). The true mechanism of xylanase in bread making has not been clearly elucidated, although a number of publications have attempted different approaches. The composition of pentosans varies depending on the flour type. The interaction between pentosans and gluten plays an important role that has not yet been elucidated. Most commercially available enzyme preparations used for studies consist of several enzyme activities including amylase, protease, and several hemicellulases. Works by Hamer indicate that the use of pentosanases increased gluten coagulation in a diluted dough system. Si and Goddik reported that a good baking xylanase increases the gluten strength measured by dynamic rheological methods on a Bohlin rheometer VOR.

By measuring the gluten extracted from a dough system, the good baking xylanase increased storage modulus G′ and G″. At the same time the gluten became more elastic because the phase angle δ decreased, whereas the two other xylanases, which did not show good baking performance, had no significant effect on gluten strengthening. The effect of a commercially available baking xylanase on the rheological properties of gluten extracted from a wheat flour dough measured on Bohlin rheometer VOR. At the dosages that give good baking performance, 50–100 FXU/kg flour, the gluten was strengthened and exhibited a more elastic property. These improved rheological properties of the gluten explain the positive effect of the xylanase on bread making (i.e. increased oven spring, resulting in larger volume and improved dough stability). At an overdosed level of 400 FXU/kg flour, which provided a very sticky dough, the G′ is at the same level as the control and the phase angle shows a more viscous property of the gluten.

Different xylanase or pentosanase preparations have different effects on arabinoxylans in terms of their scission points and reaction products; therefore, they have different effects on bread making. Although many analytical methods are available for assaying pentosanases or xylanases using a wide range of substrates, it is still not possible to have an analytical method of which the enzyme units can correlate to the baking performance when comparing different types of xylanases or pentosanases. It has been reported that no correlation to the baking performance was found when different enzymes from different sources were submitted to the same standardized assay and baking tests using purified substrate. One possible explanation is that once the wheat pentosans, both soluble and insoluble, were extracted from the flour, the mechanism of the interactions between wheat pentosans and gluten could no longer be assayed. Xylanase influences the gluten, as discussed earlier. A possible mechanism of pentosanases mentioned by Hamer is that the enzyme can offset the negative effects of insoluble pentosans present in the flour, because insoluble pentosans are regarded as having a negative effect on loaf volume and crumb structure. A study conducted by Jakobsen and Si using four different pure xylanases concluded that the best enzyme in terms of baking performance is the xylanase that has a certain level of activities toward both soluble and insoluble wheat arabinoxylans in a dough system. A xylanase having the major activity toward the insoluble arabinoxylans gave a dough too sticky to be accepted.

IMPORTANCE OF LIPASE

The use of lipase for bread making was almost unknown just a few years ago, although it was reported as early as 1968 that the use of certain lipase preparations in a bread dough significantly retards the tendency of bread to become stale. Until recently, some 1,3-specific lipases were found to have a good dough-conditioning effect, in terms of improved dough rheological properties, increased dough stability upon over-fermentation, increased oven spring resulting in a larger volume, and improved crumb structure in dough systems without added shortening. Like most other enzymes, the effect of lipase also depends on the type of flour and the baking formulation. All three types of flour are European baking flours with similar characteristics in terms of lipids, gluten content, and water absorption. For Meneba flour the optimum dosage was 2,000 LU/kg flour, and for Manitoba and Baguepi flours it was 1,000 LU/kg flour. An addition of 500–2,000 LU/kg flour results in a volume increase of 20–30%. At too high a dosage of lipase the dough becomes dry and stiff with a reduced volume. Whereas lipase significantly improves bread quality in fat-free formulations or those with added oil, it has no beneficial effect when the formulation contains added hydrogenated shortening in terms of volume increase.

Because of its effect on volume increase and especially on the improved, more uniform crumb structure, the lipase crumb softness during storage. As discussed earlier, larger volume and more uniform crumb structure contribute to a softer bread. Lipase reduces the crumb firmness compared with the control. It can provide a crumb softness as good as 3–6% shortening. However, it has no significant effect on improving crumb elasticity. The mode of action of the lipase is still unknown. The postulation about producing monoglycerides in situ cannot explain the effects of the lipase demonstrated earlier. Monoglycerides are known to have an antistaling effect but have limited dough-conditioning effects compared with lipase. Furthermore, the contents of total lipids in most wheat flours is in the range of 1–1.5%. Regarding the limited hydrolysis degree and limited quantity of triglycerides of saturated fatty acids in dough systems, an insufficient amount of monoglycerides of saturated fatty acids is produced to exhibit the effects described earlier.

Besides, the monoglycerides produced by this 1,3-specific lipase would be the second-position monoglycerides. It was reported that first- and third-position monoglycerides are more able to form complexes with starch and thus having a retarding effect on starch staling. Addition of lipase to a flour dough does not change the rheological properties of the dough measured by both Farinograph and Extensograph with optimum water absorption. This is a good characteristic for bread making, because most bakers do not like major changes in the dough system. However, when we measured the rheological property of the gluten treated by the lipase using the dynamic rheological method, we found that the reason for the good dough-conditioning effect of this lipase is the increase of gluten strength. The gluten taken from lipase-treated wheat flour dough is significantly stronger, with an increased G′ and more elasticity, a lower δ than the control. An overdosing of lipase results in too strong of a gluten complex with a very high G′, which gives a dough that is too stiff and a smaller volume increase, as mentioned earlier.

The gluten treated with lipase has a high G filt and increased relaxation time, indicating an increase of cross-links in the gluten networks measured by the dynamic stress relaxation method. It is not yet possible to elucidate the nature of these extra cross-links. The knowledge of the chemistry of wheat lipids in bread making is still fragmentary. It was assumed that glycolipids, through hydrogen bonds and hydrophobic interactions, form a linkage between gliadin and glutenin. Bekes et al. reported that there is a strong positive correlation between the lipid-mediated aggregates and loaf volume. However, overaggregates of the gluten may be responsible for the deterioration of bread-making quality as reported by Weegels and Hamer. A possible explanation of why this lipase has the positive effective is that it degrades some of the lipids, namely triglycerides, thereby reducing the overaggregation and forming a more stable gluten network. A recent study conducted at Lund University indicated that lipase increased the thermostability of the reversed hexagonal phase of the liquid–crystalline phase during heating up to 100°C. This observation is suggested as a possibility for one of the mechanisms of lipase in bread making. Because lipases are gaining more and more attention in the baking industry, studies in the near future should give us a better understanding of the mode of action of lipase for bread making.

OXIDATIVE ENZYMES IN BREAD MAKING

Oxidants, such as ascorbic acid, bromate, and azodicarbonamide (ADA), are widely used for bread making and have been well studied and reported on. However, the true mechanism of the oxidants in bread making has not been established, although a number of hypotheses were offered by numerous studies. It is generally known that bromate is a slow-acting oxidant and becomes active at high temperatures. In relation to baking, it has the maximum effect in the later stages of proofing and in the early stages of baking, whereas ascorbic acid and other oxidants are fast-acting oxidants and having the maximum effect during mixing and proofing. Increasing demands by consumers for more natural products with fewer chemicals and especially the possible risks with bromate in food have created the need for bromate replacements. Therefore, oxidases are gaining more and more attention within the baking industry.

Although glucose oxidase for bread making has been known since 1957 and lipoxygenase, lysyl oxidase, sulfhydryl oxidase, peroxidase, laccase, and transglutaminase have been reported to have good oxidizing effects, the knowledge about oxidases for bread making is scarce and only glucose oxidase is currently available commercially. Glucose oxidase has good oxidizing effects that result in a stronger dough. It can be used to replace oxidants such as bromate and ascorbic acid in some baking formulations and procedures. In other formulations it is an excellent dough strengthener along with ascorbic acid. Examples of applications of glucose oxidase for bread making are given elsewhere in this volume.

SYNERGISM

Using enzyme combinations for bread making is not new. It is well known that hemicellulase or xylanase used in combination with fungal α-amylase has synergistic effects. A high dosage of a pure xylanase may give some volume increase, but, the dough with this dosage of xylanase would be too sticky to be handled in practice. When the xylanase is combined with even a very small amount of fungal α-amylase, a lower dosage of xylanase with the α-amylase provides a larger volume increase and better overall score without the dough stickiness problem. The combination of xylanase and fungal amylase still has its limitations because of the dough stickiness easily caused by these enzymes. Because of its effect on gluten strengthening, lipase improves dough stability against over-fermentation. Without any enzyme, the baguette had poor stability on over-fermentation.

With the addition of α-amylase alone or in combination with xylanase, the volume and overall scores of the bread improved. When α-amylase, xylanase, and lipase were used together, the overall scores were further improved. The dough exhibited especially good stability on over-fermentation, which resulted in significantly better baguettes. Because lipase does not make the dough sticky and significantly improves dough stability and crumb structure, the synergistic effects between xylanase or amylase and lipase provide many possibilities for improved bread quality. A dosage of 3.3 g Fungamyl Super MA consisting of fungal *a*-amylase and xylanase gave satisfactory dough consistency but a limited volume increase. A doubling of this dosage gave a larger volume increase but a dough too sticky to be handled. When combining the low dosage of Fungamyl Super MA with the lipase, the volume increase was significantly higher without any dough stickiness. The combination of lipase with α-amylase or xylanase can provide a fine, silky, and uniform crumb structure to loaves based on a straight dough process.

Improving Quality

As mentioned earlier, the maltogenic α-amylase is a true antistaling enzyme that affects neither bread volume nor crumb structure. Therefore, it is more feasible to use this enzyme in combination with enzymes such as fungal α-amylase, xylanase, and lipase to ensure the other bread quality parameters such as volume, dough stability, and crumb structure. Most enzymes or other bread-improving ingredients including emulsifiers can improve the crumb softness due to the effects on bread volume and/or crumb structure, but they have a limited effect on crumb elasticity. The exception is the maltogenic α-amylase. Although the fungal α-amylase and xylanase or that in combination with thermostable bacteria α-amylase can reduce the crumb firmness during storage, the bread with the maltogenic α-amylase was significantly softer and had more elastic crumb than the other samples. The preceding example also illustrates that the crumb firmness and elasticity may not necessarily correspond with each other.

Many ingredients can reduce crumb firmness without affecting crumb elasticity, such as the bacterial α-amylase shown earlier. Our studies have also found that emulsifiers such as sodium stearoyl lactylate (SSL) may also have a negative effect on crumb elasticity. The maltogenic α-amylase has the synergistic effect together with lipase, xylanase, and fungal α-amylase: 30 ppm of the maltogenic α-amylase together with the other enzymes gave a softer crumb during storage than using 45 or 75 ppm of the maltogenic α-amylase alone. The combination of the four enzymes (fungal α-amylase, xylanase, lipase, and maltogenic α-amylase) gave the softest crumb throughout the entire storage period of 9 days. The softness of the bread with these four enzymes at day 9 was at the same level as the loaf with 0.5% distilled monoglycerides at day 3. However, because only the maltogenic α-amylase has an effect on improving crumb elasticity, the more maltogenic α- amylase added, the more elastic is the bread crumb.

Strengthening the Dough

Glucose oxidase has good oxidizing effects that result in stronger dough. It can be used to replace oxidants such as bromate and ADA in some baking formulations and procedures. In other formulations such as with ascorbic acid, it is an excellent dough strengthener. Use of glucose oxidase creates a dry and strong dough. A high dosage of fungal α-amylase gives the dough extensibility. The combination of these two enzymes can therefore achieve a synergistic effect. When the two enzymes are used together with a smaller amount of ascorbic acid, the dough is not only very stable but also absorbs 1–2% more water, resulting in a greater volume increase and a crispier crust. Glucose oxidase combined with fungal α-amylase can replace the bromate in some bread formulations. This procedure requires a very stable dough because after two series of 50 min of proofing the dough is turned around before baking. A dough without strong dough stability will collapse when turned.

The basic formulation contains both ascorbic acid and K-bromate. By adding glucose oxidase and fungal α-amylase instead of bromate, the final bread has a much improved appearance and a volume increase of approximately 40%. In the previous sections each of the most commonly used enzymes for bread making has been reviewed, because of the rapid development of biochemistry and enzyme technology and the increasing interest in enzyme applications for the baking industry, more and improved enzyme products will be introduced in the near future. There will also be more reports and studies that will bring us a better understanding of the mechanism of baking enzymes. In turn this understanding will lead to further improved products. An exciting future for baking enzymes lies ahead.

5

BIOCATALYTIC PRODUCTION OF FLAVOR

Flavor and aroma chemicals are exceptionally bioactive molecules that exert their very characteristic taste and smell effects even when present at very low concentrations. This is because of their highly selective interactions with receptors in the mouth and nose. It is this high activity, in combination with the usually very pleasant, evocative, and identity-conferring properties that make flavor and aroma chemicals so valuable. In particular their tastes and aromas are very characteristic of particular foods, beverages, flowers, and so forth. This is made possible because of the many combinations and permutations of different flavor and aroma chemical molecules, with different characteristic mixtures found in each particular natural source. Hence complex mixtures of flavor chemicals are generally much more valuable because of their more rounded tastes. These organoleptic characteristics are generally very precisely determined by the exact molecular structures of the chemicals involved.

As a result, even slight modifications, such as another hydroxyl group, a *trans* rather than a *cis* double bond, or an *R* rather than an *S* isomer, can completely change or abolish a molecule's taste and smell properties, both in terms of its character and identity, and also the threshold concentration at which it can be detected. Many flavor and aroma chemicals are produced naturally in foods by the action of endogenous enzymes and/or by naturally occurring microorganisms such as those contributing to the tastes and smell of bread, cheese, wine, beer, and many oriental products such as soy. The need to make processed foods with taste and smell characteristics redolent of traditional products creates a derived demand for natural flavors, which in turn, creates a need for processes to make them. This is particularly the case for many highly processed foods that lose many of their volatile or heat-unstable flavor chemicals during processing.

It is very appropriate to use modern enzyme and microbial technologies to manufacture flavor and fragrance materials, particularly as biocatalysis techniques provide the opportunity to carry out new reactions, and especially more selective reactions not possible with conventional organic chemical synthesis. Use of biocatalysis methods is especially advantageous for the production of flavors because, when carried out rigorously, such techniques allow them to be described as "natural," and so command the premium prices conferred on them as a result of the insistent consumer demand for natural or "green" food ingredients. This advantage is, however, less strong for aroma chemicals because the consumer demand for natural fragrance materials is, so far, less well established.

Biocatalysis processes are also very useful when alternative methods such as chemical synthesis or agricultural production are either not possible or not cost-effective. Perhaps the biggest challenge is in developing processes that can reliably and reproducibly deliver products at cost-effective prices. The normal range of isolated enzyme, fermentation, and bioconversion processes are used. However, biocatalytic processes for flavors and fragrances are different from those in other areas of biotechnology in that they generally do not utilize genetic engineering techniques. This is because the products of genetic engineering don't fully meet the stringent consumer criteria for natural products. However, in all other respects this article on the use of bioprocessing to make flavors has been written in the context of, and with reference to, other successful bioprocesses used in the drug, agrochemical, food, and chemicals industries.

Ultimately the goal is products that are of good quality, are cost-effective and safe to use, and that meet customer demands. So far this combination of market pull, the consumer demand for natural flavors, and the technical push of enzyme and microbial technology has been successful in bringing a wide range of natural flavours to market, produced by enzyme, bioconversion, and fermentation techniques—whichever is most effective for each particular flavor material. Notable exceptions include a microbially produced natural vanilla flavor, and indeed any product using plant cell culture as a manufacturing technology. The world market for flavor sales has been estimated at $4.5 billion in 1994; this figure excludes the very considerable in-house production and use of flavors by some major food and beverage manufacturers. Europe still constitutes 40% of these sales (ca. $1.8 billion per year), followed by the United States, with sales in Asia having expanded most rapidly and with eastern European markets offering potential for the future.

The major uses of flavors are in soft drinks (29%), followed by savory flavors such as meat and snack products (26%), and then applications in many other categories of products such as alcoholic beverages. Major commercial trends in the flavor industry include globalization, consolidation, compliance, coping with the pressure of flavor costs, and customer service. Globalization trends are especially aimed at gaining a commercial presence in emerging markets such as China. Consolidation is proceeding apace as companies join together and are acquired so as to create the critical mass in research and development, purchasing, sales, and so forth in order to compete worldwide and against larger competitors. Globalization and consolidation trends combine as the smaller local companies with particular skills and niche markets are purchased by multinational flavor companies. This often results in a rationalization of manufacturing and fewer but much larger flavor-producing plants. Compliance with regulatory, legislative, and customer requirements is increasing rapidly, and the resources that must be devoted to compliance issues are increasing much more rapidly than the growth in sales revenues. This is despite growth in flavor sales of 6–7% per year over the last decade, which is significantly higher than for the economy in general.

Due to the pressure on large food and beverage manufactures to reduce the contribution of flavor ingredients to the overall cost of their final products, and increasing competition between different flavor manufacturers, there has been an interesting trend to provide a more valuable service, for instance, in the form of more highly formulated flavors that are easier for the end-manufacturer to use. Other trends are to focus on the absolutely key "character-conferring" flavor chemicals, especially when they are sufficiently powerful to allow use at only low concentrations in the final product; to search for generic manufacturing processes whereby a small range of different flavors, such as C6 alcohols and aldehydes, can all be made using the same manufacturing plant; and to anticipate emerging consumer demands, such as for ethnic food influences or particular types of healthy ingredients. Despite recent consolidation, the flavor ingredient supply market is still very fragmented, with the top 10 flavor companies having together still only 54% of the 1994 market of $4.5 billion in sales, and with the 5 leading companies, IFF, Quest, Givaudan, Haarmann & Reimer, and Firmenich, having 14, 10, 9, and 7% market shares, respectively. This means that even these market leader companies have turnovers that are much smaller than their biggest customers such as Nestle, Unilever, and Sara Lee, who are among the largest companies in the world.

Another indication of economic importance is that European Community estimates show that eventually the impact of biotechnology in food and related businesses could be greater than in pharmaceuticals. Any successful product requires a strong market demand and also a cost-effective manufacturing process to meet that demand. This article concentrates on the latter, but it is important to touch on some aspects of market demand first. Any bioprocess to make a natural flavor or aroma chemical exists only because of the market need for that flavor or aroma chemical. In turn the need for the flavor or aroma chemical arises as a result of a derived demand from the consumer's choice of products that contain the flavor or aroma chemical in a fully formulated and often branded product.

The current world consumption of flavor and fragrance products is summarized, and Hausler and Munch report more than 65% of all flavoring ingredients used commercially in the United States are labeled as natural and have a food market potential exceeding $9 billion. This estimates the worldwide market for flavors alone at approaching $4 billion per year, and the total for all flavor, aroma, and essential oil ingredients at almost $10 billion per year. Another illustration is the high percentage of naturally flavored foods sold in Europe and the United States that contain natural flavor chemicals. The economic value of biocatalytic processes is not just in terms of the materials produced, but also in the very much higher economic value of the flavored end products, which include branded consumer goods that depend on availability of flavors of the right quality and cost.

WORK DONE

Flavors have a long history. The first products involving the bioproduction of flavors included beer, wine, and bread; yogurt made using lactobacilli; cheeses matured using a variety of microorganism including fungi. Also in Asia soy, miso, tempeh, and other products made using *Aspergillus oryzae* and other strains. Very often the success of these processes depends on the use of fast-growing anaerobes that accumulate products that discourage the growth of contaminants, such as ethanol or lactic acid, and also supplementation with additives such as hops and spices.

This first phase of products was refined over many millennia. Many of the traditional flavors and flavored products still set the pattern for modern products because of deeply ingrained consumer preferences. For instance, beer and cheese were first made at least 6,000 years ago, that is some 250 generations ago. So there has been plenty of time for humans to develop inate flavor preferences, as well as preferences for other qualities such as color, clarity, foaming behavior, and mouthfeel properties.

Some of these additional properties are the direct result of microbial biochemistry, for instance, glycerol is important in conferring mouthfeel, and it is produced from dihyroxyacetone by yeast. It was not until the nineteenth century and the rise of a technological culture that second-phase products arrived in the form of much more chemically defined flavor products such as monosodium glutamate, citric acid, lactic acid, acetic acid in the form of improved vinegar producing processes, and also the products of a range of microbial enzymes. These processes required strain selection, development of special growth media, and the maintenance of hygienic or sterile conditions. In 1837 Liebig and Wohler identified the first flavor compound, benzaldehyde.

The isolation and synthesis of vanillin by Tiemann, Haarmann, and Reimer in 1874–1876 began the modern flavor industry, and microbially produced flavors were first described by Omelianski in 1923. Also, once the technology for producing enzymes and cells had become widespread, the use of these enzymes and cells as biocatalysts in biotransformation processes became possible. Some other indications of the high sensory effects of flavors and the impact that biomanufacturing has already had are the estimates that only about 400 aroma chemicals are manufactured in quantities of greater than 1 tonne per year, and only 68 materials are reported to be used at a rate greater than 3 tonne per year, but that some 50–100 flavor materials are reported to be made by bioprocesses. Several key considerations need to be recognized because they strongly influence the type of processes used.

GENERAL PROPERTIES

Raw materials often contain endogenous enzyme activities and are often physically and chemically complex and diverse. They include oils and fats, proteins, and carbohydrates. Many are biopolymers that are usually produced from substainable agricultural sources, rather than petrochemicals, and contain many chiral molecules. Products have the following characteristics:

1. Have a wide range of value, from £ several 000/kg for some flavors to less than £100/tonne for animal feed (compare the prices of citric acid and vitamin B_{12}).
2. Product value depends on consumer satisfaction (flavor, mouthfeel, etc.), which can vary with age, culture, and so forth.
3. Genetic engineering, especially genetically modified organisms, have poor consumer acceptability.
4. Flavor/aroma chemicals with comparatively similar chemical structures usually have quite different characteristics and intensities. For example, both di- acetyl (2,3-butadione), which gives a buttery taste to dairy products, and hop-α-iso acids, which gives beer its bitter taste, are both vicinal diketones.

5. It is not possible to record taste and smell sensations in the same way as sight and hearing sensations, and so the flavor and aroma quality of a material cannot be accurately assessed on-line, or even easily off-line. In a process, only the proportions of the various chemicals that contribute to the overall taste and smell can be analyzed. True flavor quality can be properly assessed only by an expert flavorist or trained taste panel, or on a gross basis by the reaction of consumers to a particular flavored product. However, some progress toward a taste–aroma sensor is being made, with olfactory neurones being shown to respond to physiological levels of volatiles in culture media and a bovine binding protein being coupled to a resin. A functional odor receptor has been expressed in the noses of rats and shown to exhibit specificity toward just four C7–C10 aliphatic aldehydes and with the greatest response being to octanal. This will hopefully lead to a proper working knowledge of how the 500–1,000 odor receptors can recognize the thousands of distinct aromas, probably by acting in combinational mechanisms, with any one receptor responding to several different aroma chemicals and any one such aroma chemical capable of triggering several receptors.
6. Flavor and aroma chemicals are often very potent materials, well suited to small-scale processing, and are usually diluted considerably when added to the final product. Their flavor value is defined as the concentration of flavor used, divided by its flavor threshold (that is, the concentration at which it can be just perceived). Because of the potent activities of flavors, care must be taken in multipurpose processing operations that produce a range of flavor products to ensure that there is no cross-contamination of one particular flavor by residues of another flavor produced in the preceding run.
7. The actual perceived flavor of a molecule depends to some degree on the physical form in which it is eaten, thus the same flavor molecule formulated in products based on fat or oil, carbohydrate, water (as in carbonated beverage), and ethanol (as in an alcoholic beverage) can have quite different tastes, for instance due to solubility effects and the influence of mouth-feel and other factors on taste perception.
8. Often one single flavor chemical will substantially define the characteristic of a particular flavor or aroma chemical, such as benzaldehyde for almond taste and vanillin for vanilla. These character-impact flavor chemicals can often command particularly high prices because of their high flavor value.
9. Even quite closely related molecules can have very different flavor sensations, for instance geraniol (flower aroma) can be microbially converted into 6-methyl heptene-2-one, which is a component of tomato flavor.
10. The natural or nature-identical status of flavors is important and deserves a more detailed consideration because of the impact of such definitions on the design and operation of flavor bioprocesses.
11. Flavor and aroma chemicals must meet both industry and government standards for safety such as those provided by the International Organization of the Flavor Industry and the U.S. Food and Drug Administration, respectively. Increasingly, kosher and halal standards are also being used as criteria of good quality. For a more detailed discussion of the issues involved see work by Manley.

12. Very often different isomeric forms of the same molecule have quite different flavor and aroma sensations. For instance (*S*)-(+)- and (*R*)-(–)-linalool are described as sweet and lavender, respectively, *R* and (–)-*trans*-α-ionone are violet-fruity and woody, (*S*)-(+)- and (*R*)-(–)-carvone are caraway and spearmint, (+)- and (–)-*cis*-rose oxide are sweet and fruity, (+)- and (–)-nootkatone are grapefruit and woody, (+)- and (–)-limonene are orange and turpentine, and (+)- and (–)-*p*-menthene-8-thiol are fruity and grapefruit tasting, respectively. In addition marked differences in flavor threshold often occur. Thus the (+)- and (–)-nootkatones have odor thresholds of 0.6–1.0 ppm and 400–800 ppm, respectively, and the (*R*) and (*S*)-isomers of δ-decalactone have odor thresholds of 1.5 and 5.6 ppb, respectively.

Natural and nature-identical flavors are defined as follows. These definitions obviously very much influence the types of processing operations that can be used.

Flavor in nature

These are materials obtained by appropriate physical processes (including distillation and solvent extraction) or enzymatic or microbiological processes from material of vegetable or animal origin, either in the raw state or after processing for human consumption by traditional food-preparation processes (including drying, torrefaction, and fermentation).

Identical substances

These are materials obtained by chemical synthesis or isolated by chemical processes and are chemically identical to a substance naturally present in material of vegetable or animal origin.

Characters

In order to succeed, a product made by any process needs to meet market demand factors such as the following:

1. The declining availability and quality of many traditional materials, especially when they cannot meet the demand of modern high-volume-, high-quality-specification mass-market products, for instance, some plant spices and colors
2. Changes in the technologies used for cooking and food processing (e.g., microwaved foods, extruded products, and liposome delivery of cosmetic ingredients)
3. Increasing size of low-cost mass markets, especially in branded and own-brand consumer products, detergents, air fresheners, washing powders, and so on
4. Changes in consumer demands, for example, healthy eating trends that include preferences for low saturated fat, salt, sugar, azo dyes, high fiber, and vitamins
5. Increased legislative concerns that amplify the drive toward safety and healthiness

These factors create opportunities for technical innovations such as the following:

1. Making available alternative raw materials
2. New reaction and processing techniques, as well as processes adapted to use new raw materials

3. Improved analytical techniques including those useful for process monitoring and control, and improved formulation for better performance (e.g., encapsulation and knowledge of synergy effects)

If technical innovations are successful, the flavor will then be used in formulated products recommended by flavorists on the basis of availability and cost of raw materials, the organoleptic value of the product, and the product's physical properties that affect ease of formulation and so on. A more detailed business analysis will include a consideration of the following factors:

1. Applications
2. Customer needs and product segmentation
3. Volume of demand and likely longevity of demand, price, and especially likely profit margins
4. Specifications for different grades of product
5. Competitors, both current and prospective
6. Raw materials prices and availabilities
7. Processing costs
8. Patent position, including freedom to patent and non- infringements position
9. Likely time and investment required to reach first sales
10. Return on investment, especially allowing for risk factors, regulatory and safety considerations, and strategic benefits such as other uses for the technology
11. Manufacturing capacity required, including downstream processing and any special requirements such as continuous operation

The dynamic nature of the industry is illustrated by recent changes. The Tastemaker company has recently been acquired by Hoffman LaRoche, Quest International has now been sold by Unilever to ICI, Bayer has integrated Haarmann and Reimer and Florasynth, and Danisco has purchased both Borthwicks and Becks, is an interesting example of a major food ingredients company expanding its flavor businesses, rather than the more usual diversification of flavor companies into food ingredients that will move them into the top dozen flavor companies.

Making Flavors

The necessary bioprocesses to make flavor and aroma chemicals fall into four broad categories: (i) traditional processes, (ii) processes for low-intensity flavors, (iii) processes for high-intensity flavors, and (iv) production of taste enhancers.

Traditional processes

Traditional processes are used for products such as beer, wine, cheese, yogurt, and soy sauce. These generally involve microbial action on chemically complex plant- or sometimes animal-derived raw materials, with the formation of relatively low concentrations of flavor chemicals that may very well have good preservative properties as well. These can be described mostly as flavored foods produced by naturally occurring microorganisms and/or enzymes derived from them. For instance, in the beer-brewing process, flavor is derived both from the ingredients used, such as hops and malt, and from the processing: the action of the yeasts and also the heating that forms the butter-tasting α-iso acids from precursors present in the hops. The traditional brewing process is typical of more recently developed bioprocesses that consist of three main stages; work-up of raw materials, the microbial reaction step, and product processing, which puts it into the form required by the customer. This last step may involve formulation with other flavor materials and/or bulking agents. In many cases traditional processes have been industrialized and adapted. For instance, intense cheese flavors can be manufactured by the hydrolysis of casein with proprietary proteases and peptidases combined with the hydrolysis of butter fat by lipases and esterases. *Aspergillus niger* and *Rhizomuor miehei* are the preferred sources for lipases for cheese making. Other bioprocesses for the dairy flavor chemicals diacetyl, acetaldehyde, and various lactones, as well as for accelerated milk souring, have been developed.

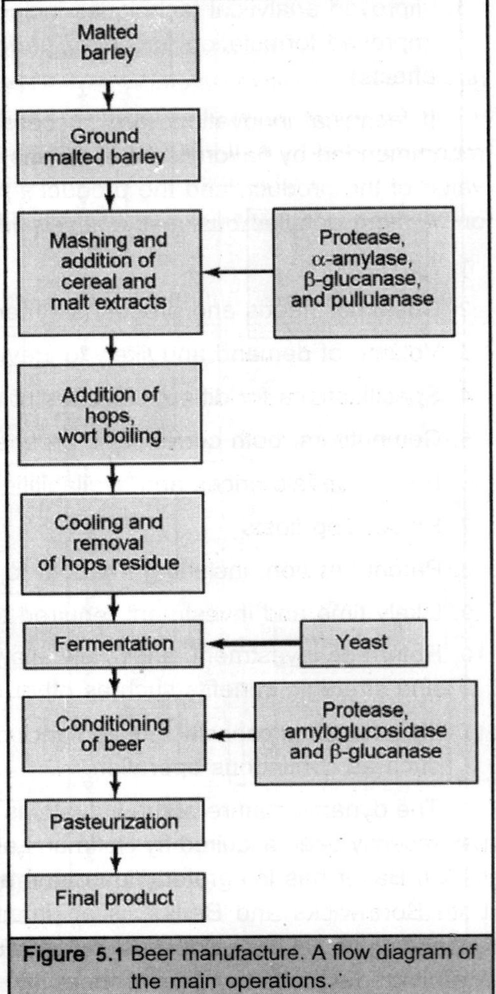

Figure 5.1 Beer manufacture. A flow diagram of the main operations.

Low-intensity flavors

Manufacturing has already been established for a range of low-intensity flavoring materials such as high-fructose corn syrups (10 billion tons/year produced worldwide), citric and acetic acids (acidulants), protein hydrolysates, and cheeses and cheese flavors. The bioprocesses that have been developed for their large-scale manufacture are well described elsewhere: citric acid, milk souring, protein hydrolysates. More recently processes involving the sequential use of enzymes to modified cheese flavors have been developed.

High-intensity flavor chemicals

More recent developments include the production of high-intensity flavors, such as methylketones (blue cheese flavor) and *c*-decalactone (peach and other fruit flavors), in the form of pure natural chemicals by fermentation and bioconversion processes.

Taste enhancers

Taste enhancers include materials such as monosodium glutamate (MSG) and mononucleotide flavor enhancers such as guanosine monophosphate (GMP) and inosine monophosphate (IMP), which are also produced as pure chemicals.

Scaling Up a Process

It is important to appreciate that scaling up a process to meet cost and product specification targets can be at least as challenging as the original discovery research necessary to invent the process. Much lower flavor concentrations need to be achieved for a flavor chemical when it is actually created in the food, in the way that flavor chemicals are produced in traditional products such as beer, cheese, and bread, rather than as a pure chemical, which is then added to the food or beverage at the required level. This does, of course, presume that any microorganisms involved are completely safe and acceptable, and also that flavor production will occur under the very different conditions existing in the food. Where processed foods are concerned, production in as pure and concentrated a form as possible is the most economic. The flavor can then be diluted as required for each end-use product and also combined with other flavors and flavor chemicals to give improved flavor quality.

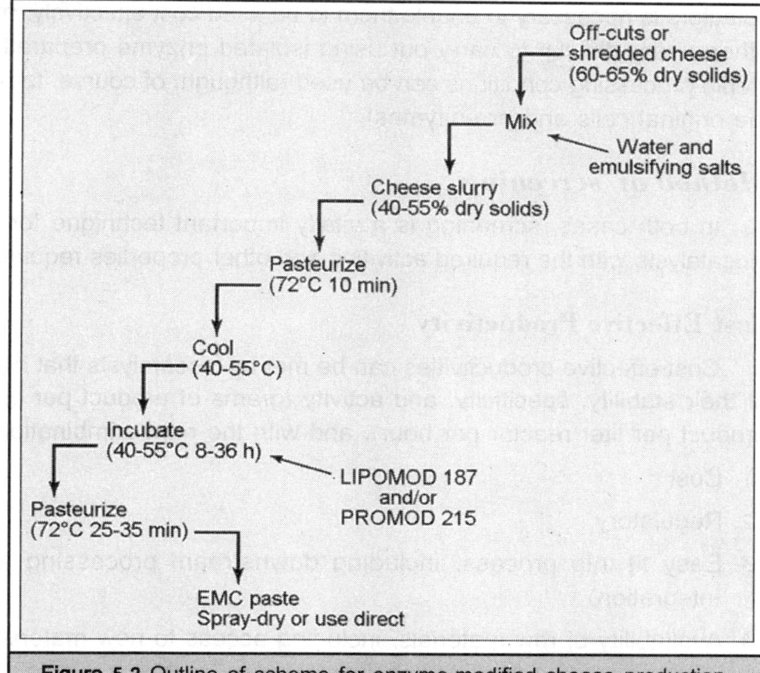

Figure 5.2 Outline of scheme for enzyme-modified cheese production.

Some indication of the relationship between the concentration at which a flavor is produced and the concentration at which it is used in the final food or beverage product is given by the production-to-use ratio, which can be illustrated by the following. If a material is produced at a concentration of 10 g L^{-1} and then used at 0.5 g L^{-1} (500 mg L^{-1}) in the product, or is present in a natural food or beverage at a concentration of 0.5 g L^{-1}, then its production-to-use ratio will be 20:1. Key variables Include the type of biocatalyst to be used, its performance, and in particular the medium used to carry out the reaction. The key features of the main methods are summarized as follows:

Multistep reactions

Fermentation is very suitable for multistep reactions. When using unstable enzymes and when cofactor regeneration is needed, it requires expensive fermenters and associated equipment. Operation is relatively expensive, because of the requirement for sterile conditions. The product is usually dilute and heavily contaminated with cells and by-products, and product quality can be variable due to batch-to-batch variations.

Role of extracellular enzymes

This method is especially suitable when extracellular stable enzymes are available, when single-step reactions are required that do not require cofactors (up to 10^3- to 10^4-fold regeneration of cofactors is necessary to enable them to be used cost-effectively, making synthetic reactions, among others, very difficult to carry out using isolated enzyme preparations), and when hygienic but not sterile processing conditions can be used (although, of course, fermentation is necessary to produce the original cells and/or enzymes).

Method of screening

In both cases, screening is a vitally important technique for discovering new and proprietary biocatalysts with the required activities and other properties required to make the needed molecules.

Cost Effective Productivity

Cost-effective productivities can be met by biocatalysts that have the required activities in terms of their stability, specificity, and activity (grams of product per gram of cells per hour, or gram of product per liter reactor per hour), and with the right combination of the following characteristics:

1. Cost
2. Regulatory
3. Easy fit into process, including downstream processing and process tolerance (process integration)
4. Availability of raw materials, including access to new materials
5. Easy scale-up
6. Broad specificity
7. Activity on concentrated reactants
8. Equilibrium point (high degree of conversion)
9. Good stability
10. High selectivity (few side products)
11. Sterile conditions *not* required (or fermentation not susceptible to contamination)
12. Ease of strain improvement by mutation or genetic engineering
13. Rapid growth rates and nonfastidious nutritional requirements
14. Robust and easily reproducible performance

The complexities involved in selecting biocatalysts with the optimal performance is illustrated by the cascade of enzymes used to create cheese flavors, and by a comparison of the rates of reaction of some common lipases when used either to hydrolyze triglycerides, or in the reverse reaction, to synthesis a flavor ester. Similarly there is a very large variation in the ability of different lipases to make a range of esters. These differences in reactivity are less surprising. It is noted that the catalytic structure of various lipases are different. Thus, whereas the catalytic triad in the active site of *Homo sapiens* and *Mucor miehei* lipases is Asp, His, Ser, in *Humicola lanuginosa* it is Asp, His, Try, and in *Geotrichum candidum* it is Glu, His, Ser. Fermentation medium requirements include the following important considerations when developing a production medium that will be as inexpensive and reliable as possible:

1. Selective advantages for the production strain, allowing fast growth, high yields, and no remaining by-products (a big challenge, not least because most wild-type strains are surface growers, and considerable effort may be necessary to get them to adapt to efficient production in submerged fermentations)
2. Defined and balanced medium (no excess of any medium component)
3. No by-products, color, odor, or substrates, remaining at the end of biotransformation
4. Substrate(s) stable, soluble, and of constant quality Reduced risk of contamination due to medium composition
5. Product not degraded by use as C-or N-source for cell growth in any circumstances
6. Oxygen requirements (this can become rate-limiting for rapidly growing cultures; provision of air for growth is expensive because of filtration costs to ensure sterility and mixing power costs necessary to ensure good gas transfer throughout the fermenter)
7. Little or no foaming problem
8. Short lag period

Overall the bioprocess will be quantified and costed on the basis of the following seven factors:

1. Specific growth rate
2. Biomass concentration (g dry wt/L)
3. Oxygen transfer rates and power consumption rates (usually turbine impellors are used so as to minimize power input costs and maximize mixing)
4. Grams of product produced per liter of reactor per hour
5. Concentration of product achieved (g/L)

Yields are very dependant on the type of metabolite being formed by the cells. Estimates of rates vary from 100 to 500 nmol/g cells^{-1} h^{-1} for amino acid formation, to 1 to 5 nmol/g cells^{-1} for vitamin and coenzyme synthesis. Degree of conversion of precursor(s) into product (%), and purity of product (especially with respect to other chemically similar materials that could make its purification difficult) are also important. This can be also described as the selectivity of the reaction: grams of product required per gram of all products formed. Problems in bioprocessing are well illustrated by the terpenes. Whereas terpenes are very common flavor and aroma materials with a very wide range of tastes, smells, and applications, production to date is entirely by extraction from plant raw materials. It appears that no true bioproduction process for a terpene has yet been established. This is probably because monoterpenes have low water solubilities, combined with high cytotoxicities, together with high volatilities and often poor chemical stabilities.

By contrast, sesquiterpenes can be biotransformed much more successfully. For instance, the processes for producing Ambrox and sclareolide are described elsewhere in this article. Of course, each product and process can be expected to have some individual features. These include the optimization of fermentation media to maximize productivity and minimize cost, and to allow for particular features, for instance, often the product concentration achieved can vary with the specific growth rate of the fermentation. Some indication of the challenges encountered in developing a bioprocess are illustrated by the examples given in this article.

Commercial Importance

In addition to the scientific factors just discussed, the following technicocommercial factors also have a big influence in many instances. Flavors and fragrances are manufactured only when required by end-user companies, who are usually the manufacturers of consumer end products, that is, they are subject to a derived demand. This is accomplished following the issuing of a competitive brief to a number of companies. Thus, the requirement for the individual flavor chemicals and materials required to create the flavor is subject to a derived demand, and as the actual flavor is usually tailored specifically to the needs of one end-user company, the amounts of any one flavor chemical or material required can vary considerably from customer to customer. Compounding and formulation of flavor chemicals is very often the last step in the manufacturing process and is a comparatively simple operation that adds significant value because of the improved complexity and character that can be created. This step depends on the creative input of skilled and experienced flavorists who create the formulations that dictate the types and amounts of flavor chemicals that are used.

High flavor/aroma impact and character- or identity-conferring chemicals are very important because of their high value, even when they are expensive to make. Control of organoleptic purity involves far more than just maintenance of chemical purity, as a trace impurity with a high flavor impact present at instrumentally undetectable levels can still have a very detrimental effect on flavor and aroma quality. This factor is very important when considering the batch-to-batch variations in quality because very few flavor chemicals are required in large enough volumes to justify production by continuous processes. Control of organoleptic quality also extends to the storage of ingredient chemicals and formulated flavors because cross-contamination can occur from ingredient to ingredient (vanillin is especially prone to this) and also from materials leached out of containers and packaging.

Economies of scale of manufacture are difficult to achieve for all but the largest-volume flavor chemicals. Instead, the small scale of manufacture creates the opportunity for generic manufacture using the same multipurpose equipment to make a range of ingredient chemicals on a campaign basis in direct response to market demands. This requires careful scheduling so as to properly match demand with supply, and scrupulous cleaning of equipment between runs. Patent protection is important, especially for new flavor materials, but the enforcement of process patents is sometimes difficult, which is a particular problem for key producer microbial strains. Because of the individual nature of the flavor chemicals, including extraction from natural raw materials and complex formulation, special manufacturing know-how is often involved that cannot be protected by patents. The very extensive process validation requirements for biopharmaceutical processes are not generally required for flavor chemicals. The standards that are required are consistent with best practice in the food and fine chemicals industries, such as food hygiene standards and the sterility necessary to perform uncontaminated fermentions.

Manufacture of high-volume flavor and aroma chemicals is increasingly becoming attractive to chemical companies, rather than just flavour and fragrance companies, and internationalization is occurring fast, as indicated by the entry of Chinese manufacturers. Because of the volatile nature of flavor and aroma chemicals, distillation is a dominant downstream processing operation in their manufacture. Usually, sophisticated equipment is used, controlled by computers, and upgraded to minimize oxidation, overheating, and so forth that can so easily produce off-flavors. Solvent extraction is common, and new methods are gaining credibility, for example, the use of supercritical carbon dioxide as an extraction technique; it has a polarity similar to that of hexane, and because of its volatility, leaves no residue in the product. Formulation of flavors often begins with spray-drying, often onto a carrier such as maltodextrins, which also serve to stabilize the flavor. Raw materials are usually purchased from a great variety of sources, rather than produced in-house.

Sometimes the availability of a raw material is a key driving force in a company's decision to produce a flavor. One example is that a good supply of by-product yeast is very useful for making savory flavors. Variabilities in the supply, price, and quality of natural raw materials, such as vanilla beans, can frequently pose big problems. In other cases it is the availability of the natural precursor molecule for a bioprocess that is the problem that can prevent the commercialization of that process. Examples include the 11-hydroxypalmitic acid required to make δ-decalactone, the methionine required for methional and related flavor chemicals, and also precursors for raspberry ketone. Effluent disposal from manufacturing plants can create some novel problems because of the objectionable smells that can be created for neighbors.

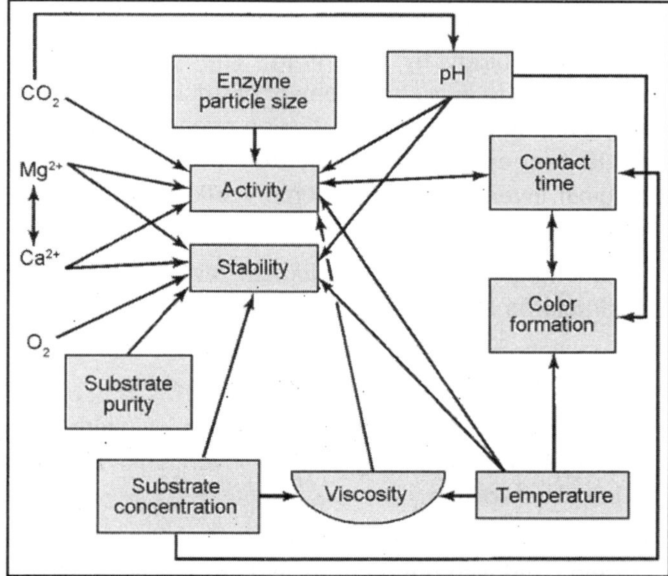

Figure 5.3 The interactions between the various process variables that can influence the productivity of glucose isomerase when used for the industrial production of high-fructose syrups from hydrolyzed starch.

One widely adopted process to combat this problem is to route all exit gases through a bed of supported microorganisms that metabolize much of the objectionable materials, thereby valorizing the exit gases to acceptable emission levels. In any process, its performance, and therefore the cost and quantity of the product, is dependent on a complex interrelationship between a large number of process variables, as illustrated for the use of glucose isomerase (GI) to make high-fructose corn syrup bulk sweeteners. This is a splendid example of the high degree of process integration that must be achieved in order for a mature bioprocess to continue to be successful. The process involves a semicontinuous process with three enzyme reaction steps together with raw material processing, downstreaming, and sophisticated chromatographic separation of glucose and fructose sugars. Since GI is now a mature product, new versions of this enzyme demonstrate the big improvements necessary for second-generation processes to succeed. GI from *Streptomyces murinus* has about a 60-fold greater productivity in terms of tons of product produced/kg enzyme, which results in a 30-fold lower cost of product.

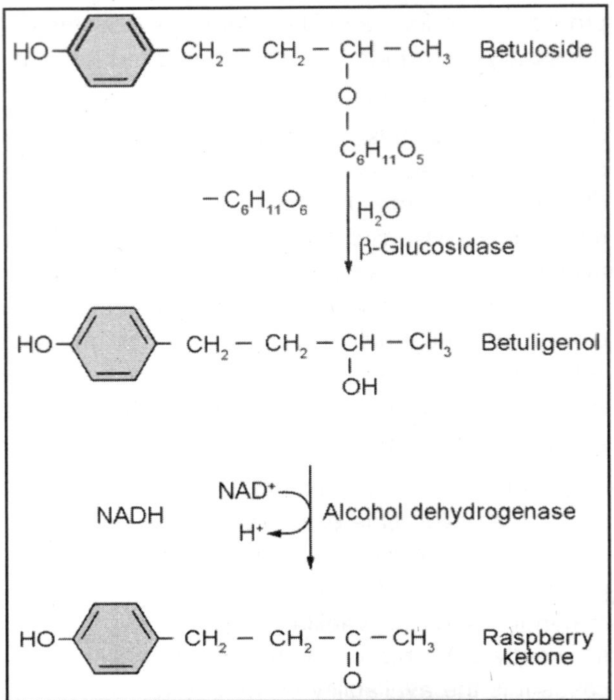

HO–C_6H_4–$CH_2 - CH_2 - CH(O{-}C_6H_{11}O_5) - CH_3$ Betuloside

$\downarrow$ H_2O, β-Glucosidase ($-C_6H_{11}O_6$)

HO–C_6H_4–$CH_2 - CH_2 - CH(OH) - CH_3$ Betuligenol

$\downarrow$ Alcohol dehydrogenase (NAD^+ → NADH + H^+)

HO–C_6H_4–$CH_2 - CH_2 - C(=O) - CH_3$ Raspberry ketone

Figure 5.4 Scheme for making raspberry ketone flavor chemical from the plant-derived precursor betuloside.

Patenting

Another important factor is the protection of the technology by patenting. The key objectives are to avoid infringement of existing patents and to gain patent protection oneself. In this respect biochemical and microbiological inventions create some particular problems:

1. Functionally identical biological materials (e.g., enzymes) can have significantly different structures and amino acid sequences.
2. Functionally identical biological materials can be obtained from very different biological sources (e.g., enzymes with the same substrate specificity from different microbial sources).
3. The requirement to deposit samples of microorganisms is a substantial over disclosure as compared to patenting of other inventions. This often leads to some processes being kept as in-house secrets.
4. Biological materials, especially living organisms, are too complex to be described in sufficient detail and with enough precision to easily satisfy the requirement of patent law. For instance, existing microbial classifications are not always directly applicable to newly isolated microorganisms.
5. Varying criteria and definitions are used to distinguish biocatalysts, for example, substrate and inhibitor specificity and immunochemical cross-reactivity in the case of enzymes, and nutrient requirements and fatty acid composition and so on for microorganisms.

Types of Bioflavor Processes

Depending on the complexity of the biochemical reactions required to produce the desired products, bioflavor processes can be categorized along the lines of the following examples:

Involvement of Single-enzyme

Examples of single-enzyme reactions include glutaminase, 3′–5′ nucleotide diesterase (e.g., from *Penicillium citrinium*), adenylic deaminase (e.g., from *Brevibacterium ammoniagenes*), glycosidases for the release of terpene flavors from grape and other flavor sources, lipases and esterases for flavor ester synthesis and for resolution of racemic mixtures (such as obtaining *l*-menthol from *dl*-menthol).

Involvement of Mixed-enzymes

Examples of mixed-enzyme reactions include use of mixed proteases for producing protein hydrolysate flavors; β-glucosidase and nitrile lyase to make benzaldehyde from amygdalin; lipoxygenase, hydro- peroxide lyase, and oxidoreductase to make, for instance, *cis*-3-hexenol from linoleic acid via *cis*-3-hexenal), and proteases, peptidases, lipases, and esterases to make enzyme-modified cheese flavors (EMCs). Recently, mixed- enzyme preparations have become widely used in the extraction of fruit flavors.

Factor Affecting the production of fruit flavors

Recently it has been found that very many fruit flavor chemicals such as the 13-norisoprenoids and terpenes (e.g., linalool and geraniol) occur in plants as glycosides. Geraniol was the first glycosidically bound aroma chemical to be identified as such in rose extracts, with a structure of geranyl-xylose-glucose, and the C13-norisoprenoids (e.g., ionones, raspberry ketones, and others such as *trans*-3- hexenol are present in the fruit of origin, predominantly in glycosidially bound forms). Usually the proportions of these glycosidically bound flavor chemicals to free flavor chemicals are in the ratio of 2:1 to 5:1. This is possibly because the glycosides are more water soluble, more resistant to acid hydrolysis, and effectively detoxify these terpenes, phenols, and similar compounds. The glycosides have similar structures with the aglycone bound first to glucose and then to a another sugar such as α-L-arabinose, α-L-rhamnose, β-D-xylose, β-D-apiose, or another β-D-glucose unit, depending on the plant source.

Hydrolysis of such glycosides liberates the aglycose, thereby enhancing the flavor content of a conventional fruit product and even allowing the generation of flavor from waste materials such as peel and stems. However, the enzymes required to achieve these advantages must have some special characteristic such as good activity at low pH (about pH 3.5), resistance to product inhibition by glucose, and even resistance to moderate concentrations of ethanol in the case of fermented products. Most importantly, flavor chemicals are released, and flavor is enhanced only when two enzyme activities are present: a β-glucosidase and also another glycosidase such as α-L-arabinosidase or α-L-rhamnosidase, depending on the plant material being treated. Because of these requirements, glycosidically bound flavors are released only rarely in existing processes, such as fermentation, because yeast β-glucosidase is insufficiently active.

However, one process that absolutely depends on endogenous glycosidase action is in flowers, where it appears that aroma materials are stored as glycosidases and then, upon the flower opening, glycosidases are produced that hydrolyze the glycosides, releasing the aroma chemicals that create the fragrance of the flower. Use of such special glycosidase combinations have now been widely adapted, for instance, in grapefruit juice extraction, and the cloning of the appropriate enzymes into suitable host strains is underway. An example is the *A. niger* enzymes used by Gist-brocades. A very good example of the improved yields of flavor that can be achieved by the use of glycosidase is the improved yields of vanilla flavor that can be obtained by treating vanilla beans with emulsin, which is a mixed enzyme preparation especially rich in β-glucosidase that is obtained as a crude extract of almond meal.

Role of single microbial strains

Single microbial strains are used to produce the following: methylketones with spores of *A. niger*, γ-decalactone with *S. cerevisiae* strains, δ-decalactone with *Cladosporiun suaveolens*, diacetyl with *Streptococcus cremoris* and *Streptococcus diacetylactis*, soured milk for margarine production with *S. cremoris*, methylbutyric acids with *Acetobacter acetii*, and pyrazines with *Corynebacterium glutamicum*.

Role of combined use of several microorganisms

A combination of several microorganisms is used for such processes as the Unilever two-stage continuous process for manufacturing soured milk, which makes use of *Lactobacillus* and *Streptococcus* strains, and the accelerated process for soy sauce manufacture using *Pediococcus halophilus*, *Saccharomyces rouxii*, and *Torulopsis/Candida versatilis*. Also several microbial strains are used to achieve a synergistic effect in cheese fermentations. For instance, propionic acid is an important flavor component of Emmental and some other Swiss cheeses. It can be produced by first forming lactic acid with strains such as *Lactobacillus lactis* or *Streptococcus thermophillus* and then metabolizing the lactic acid into propionic acid by *Propionibacter* strains. Using this two-stage method, concentrations of propionic acid of 40 g L^{-1} can be made. Propionic acid is also useful to form flavor esters such as ethyl, benzyl, citronelyl, and geranyl propionates, as is butyric acid, also produced by fermentation in concentrations of 20–30 g L^{-1}, which is used to make ethyl, isobutyl, and amyl butyrate flavor esters.

Interaction between Enzymes and Microorganisms

Production of savory flavor

Production of savory flavors using proteases, lipases, ribonucleases, and polysaccharidases to break down plant raw materials, and then microorganisms to convert the products of hydrolysis into flavor chemicals are used in processes such as the Biosol process. Improved flavors using this approach could be possible by genetically engineering yeast strains to contain higher concentrations of the savory flavor enhancers GMP and IMP, and also of the flavorsome octapeptide "beefy meaty peptide."

Production of fruit and dairy flavor

The traditional use of the acetic acid bacterium *A. acetii* to make vinegar by oxidizing the ethanol in wine into acetic acid has been redeployed by modern flavor biotechnologists. This time *A. acetii* is used for the bioconversion of isobutyric alcohols present in fusil oil into isobutyric acids. Thus both the 2- and 3-methylbutanols are oxidized into 2- and 3-methylbutyric acids. This oxidation takes place with very little racemization. The overall reaction requires two sequential reactions: a dehydrogenation into the aldehyde intermediate than can tautomerise easily, and then oxidation to the acid. These isobutyric acids are good flavor chemicals in their own right, but in addition their value and range of uses can be increased and expanded by esterification either by using lipases or esterases "in reverse," or by azeotropic distillation.

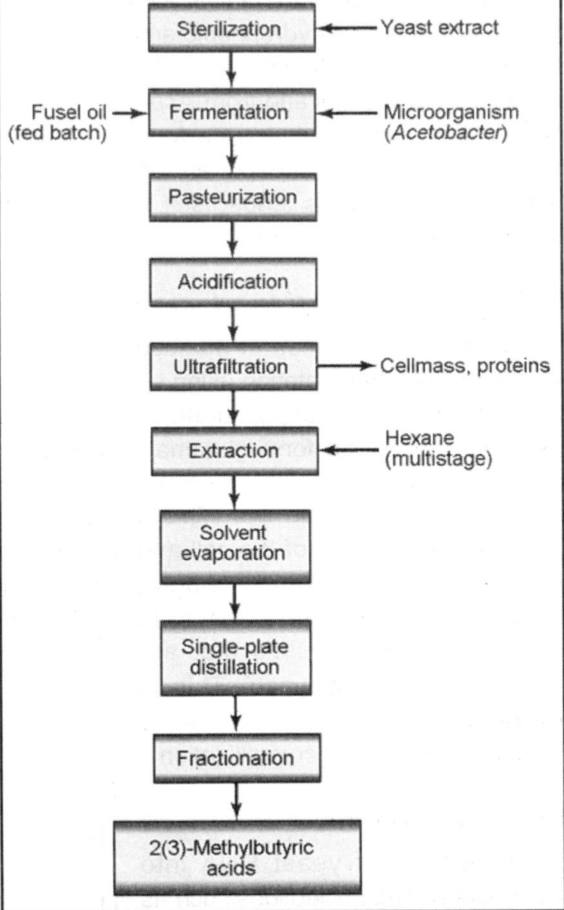

Figure 5.5 Large-scale process for the manufacture of 2- and 3-methylbutyric acids for flavor chemicals and flavor precursors, using fusel oil as the source of the 2- and 3-methylbutanol precursors.

Enzymes of plants

Plant-derived biocatalysts include lipoxygenase for the production of β-ionone, *cis*-3-hexenol, and so forth. Soy lioxygenase produces 13-hydroperoxides that break down to C6 aldehydes, whereas tomato lipoxygenase has a different substrate specificity and produces 9-hydroperoxides from linolenic acid, rather than the 13-hydroperoxides, with the 9-hydro-peroxides breaking down to form C9 aldehydes such as *cis*-3-nonenal. By contrast, Fusarium lipoxygenase forms 9- and 13-hydroperoxides. *Morchella* sp. are used for 1-octen-3-ol production and hydroperoxide lyase obtained from plant sources such as green pepper. So, in general there would appear to be a historical trend from traditional fermented products to the use of isolated enzymes, submerged fermentation, and now bioconversion methods. Obviously the complexity of the biocatalytic system used has a direct effect on the ease of operation and costs of the process, and so a more technically sophisticated process can be justified only when the resulting product is sufficiently valuable, both in terms of profit margins created and markets that can be gained, to justify its use.

Bioprocesses

Isolated enzymes are used to make a variety of flavor products. Different mixtures of microorganisms and plant and enzyme proteases are used to produce protein hydrolysates, depending on the source of the protein, the process used, and the product needed, especially because different end-use applications require different degrees of hydrolysis of the proteins. A particular challenge is to minimize the bitter taste of the hydrolysate, which is caused by peptides that have N-terminal hydrophobic amino acids. This can now be achieved using specially developed peptidases, especially aminopeptidases, which are usually obtained from lactic acid bacteria such as *Leuconostoc lactis* and from *Aspergillus oryzae*. These were discovered as enzymes from some dairy starter cultures that were found to have debittering activity and are being cloned to make them available more cheaply and in greater quantities. Some carboxypeptidases are also used. Note that the process lends itself to the creating of a range of savory products, depending on whether the material is dried, concentrated, or supplied as a dilute liquor.

Industrial production of monosodium glutamate

Over 500,000 tonnes of monosodium glutamate the taste enhancer is made worldwide each year, mostly by fermentation using specially selected strains of *Corynebacterium glutamicum*. These strains have a low α-ketoglutarate dehydrogenase activity that prevents the formation of succinate, and so they instead form glutamate by reductive amidation of the α-ketoglutarate. In addition, successful and high-yielding fermentations depend on the use of media that have a very low biotin content (less than 5 $\mu g\ L^{-1}$). This makes the cell walls of the microorganism leaky so that the glutamate passes out of the cell and accumulates in the medium. After fermentation the cells are removed, and the glutamate is precipitated (by the addition of HCl) as α-form crystals. These are heated in water to convert them to β-form rod-shaped crystals that are easier to filter and wash. This is followed by neutralization, decolorization, and recrystallization.

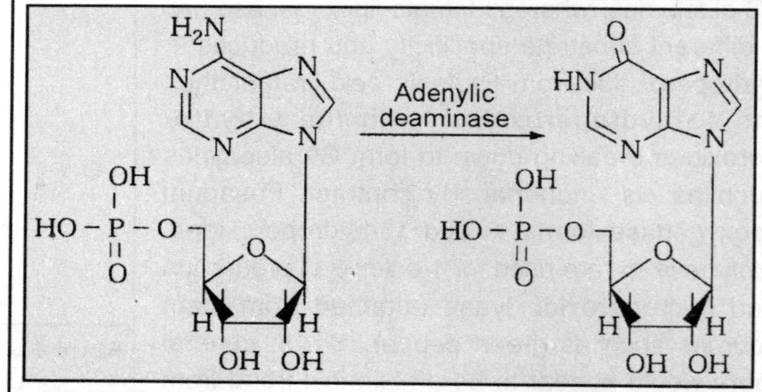

Figure 5.6 Conversion of 5'-AMP to 5'-IMP by the action of adenylic deaminase.

Yeast RNA flavor

Hydrolysis of yeast RNA into flavor-potentiating nucleotides such as guanosine-5′-monophosphate is carried out by processes similar to the following. Yeast is grown such that it has a high RNA content. It is then harvested in late log phase with a hot alkaline solution. Since inosine monophosphate (IMP) is a potent flavor enhancer, the activity of the hydrolysate is frequently enhanced by treatment with an adenylic deaminase, which converts 5′-AMP into 5′-IMP.

Biological extraction of plant materials

Plant materials are a traditional source of flavoring materials. Recently it has been found that many plant flavor chemicals, especially terpenols such as linalol, nerol, and geraniol, occur glycosidically bound to the plant tissue. As a result, specialized glycosidase enzyme preparations have been developed to augment traditional processing by enabling increased rates of extraction and higher yields of products. This approach has been increasingly employed in wine making;

enzyme preparations such as those from *A. niger* are added to the must. Curiously, endogenous plant glycosidases and the glycosidases of wine and beer yeasts are barely active against these glycosidically bound flavors, which explains the low yields obtained during traditional extraction and fermentation.

This is because the glycosidases present in plants and yeast tend to be active at a moderate pH (around pH 5), are inhibited by glucose and ethanol, and have only a narrow aglycone substrate specificity. Such specialized enzymes have now been made available by enzyme supplier companies such as Gist-brocades. One good example of extraction is the preparation of L-rhamnose, which is the key precursor molecule for the very important flavor Furaneol (2,5,-dimethyl-4-hydroxy-(2*H*)-furan-3-one), which is prepared by heating rhamnose with an amine source such as proline. L-Rhamnose can be isolated by the enzyme hydrolysis of glycosides such as naringin, neohesperidin, and rutin present in citrus peel and other plant sources. These molecules are flavanone glycosides with rhamnose, in the terminal position.

Enzymes are preferred that exclusively or, more usually, preferentially hydrolyse rhamnose, while minimizing the release of other sugars. This is to avoid an off-taste that develops during the reaction of glucose with the proline. Therefore any glucose released is removed by fermentation to ethanol and CO_2 by yeast, or into 5-ketogluconic acid, with its removal by precipitation as its calcium salt. Taking this approach a stage further, by coupling hydrolysis of a precursor glycoside with its bioconversion into a useful flavor chemical, raspberry ketone (4-(4′-hydroxyphenyl)-butan-2-one) has been synthesized.

Enzyme treatment can also be used to reduce an undesirable flavor of a plant juice, rather than to enhance a pleasant taste. For instance, the bitterness of citrus juices is due to the flavanones limonin, naringin, and neohesperidin, which are often present in fruit juice as glycosides. Bitterness is a particular problem for grapefruit, in which naringin predominates and has a taste threshold of just 20 ppm. Commercially available naringinase, which is a mixture of glucosidase and rhamnosidase, acts sequentially by converting the naringin to prunin and to then the much less bitter naringenin, has been proved in trials to provide an effective solution to this problem.

Natural benzaldehyde

Another source of an enzymatically treated plant-derived flavor is natural benzaldehyde. A number of companies have operated a bioprocess to make natural benzaldehyde, based on the combined use of β-glucosidase and mandelonitrile lyase, which are both conveniently present in almond meal (emulsin). This is used to treat the mandelonitrile (a nitrile glycoside) present in certain plant materials. A preferred industrial source is the cherry stones remaining after processing to remove the cherry flesh.

However, the liberation of poisonous hydrogen cyanide by the lyase creates safety problems and has led to a search for alternative processes. L-Phenylalanine is a suitable precursor, especially since it is now available cheaply from a fermentation process developed to make phenylalanine as a component of the high-intensity sweetner Aspartame. The basidiomycete *Ischnoderma benzoinum,* combined with effective in situ product recovery, produces 1 g benzaldehyde/L, together with some 3-phenylpropanol.

Characteristic flavor of blue cheese

Methylketone provides the characteristic flavor of blue cheeses such as Roquefort, Camembert, and Stilton and consists of mixtures of C5 to C11 2-alkanones, especially 2-pentanone, 2-heptanone, 2-nonanone, and 2-undecanone. This is because only medium-chain-length fatty acids are converted into methyl ketones. This may be because only such medium-chain fatty acids, and not longer-chain fatty acids, can enter the cells as free acids and seem to cause significant morphological changes to the cells that take them up. Such flavor chemicals are commercially important in the manufacture of consumer products, such as salad dressings, soups, and pizza toppings, in which a cheese flavor can predominate, and also as a component of other flavors, especially savory flavors.

Methylketones are manufactured using the spores of microorganisms, such as *Penicillium roquefortii*, that still retain the required enzyme activities despite being dormant. This makes the production-scale use of spores very easy, as once grown they are easy to store until required and then require no growth media. They are eminently suitable for use in a logistically convenient and cost-effective biotransformation process. This process is much simpler than a conventional fermentation because of, for instance, very moderate oxygen transfer requirements. Using this method, lipolyzed fats or oils can be transformed into cheese flavors with 10 times the methylketone content of normal blue cheeses. The details of the biochemical pathway whereby microorganisms convert fatty acids into methylketones have been elucidated and used as the basis for large-scale fermentation processes, although the detailed physiologically control processes are still not properly understood.

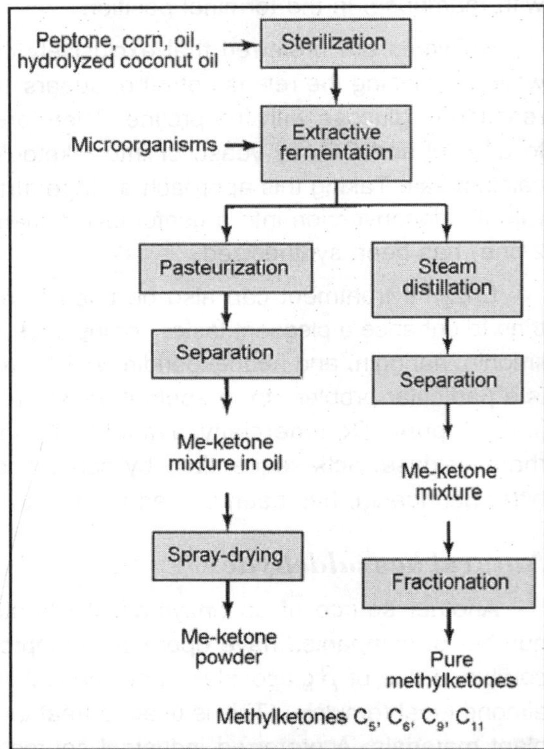

Figure 5.7 The unit operations involved in manufacturing processes for the methyl ketone blue cheese flavoring materials.

Commercially successful microbial processes for making methylketones have been used for some time by companies such as Unilever and Bush Boake Allen. The way microorganisms such as *P. roquefortii* produce methyl ketones from medium-chain fatty acids is of considerable scientific interest because it resembles conventional β-oxidation until the final step, at which a decarboxylase produces the methyl ketone, replacing the thioesterase found in β-oxidation. Methylketone formation is favored under conditions in which cell growth is constrained. Spores are more active in producing methyl ketones than mycelium; use of immobilized cells have been described, and solid-state fermentations are often used. For instance, Bush Boake Allen's process involves mixing cellulose powder with coconut oil, which contains a high concentration of octanoic acid, and inoculating with *A. niger* spores, which converts the octanoic acid into 2-heptanone, which has strong blue cheese aroma and flavor. The yield is reported to be about 40%, and 2-undecanone to be 60% of the total products. By comparison, using spores of *P. roqueforti* in solid-state fermentation, Larroche and Gross achieved product concentrations of 75 g 2-heptanone L^{-1}, with a yield of almost 80%. This appears to be the only manufacturing process for flavor chemicals that employs solid-state fermentation, despite the advantageous features of this technique. This may be because the "culture" of applied microbiology at the manufacturing level is dominated by the stirred-tank approach, developed with such success for antibiotic production, except for traditional processes such as cheese ripening and silage making.

Commercialy important flavors

The γ- and δ-decalactones are commercially very important flavor chemicals for which successful industrial bioconversion processes have been established. This is despite the fact that de novo production by fermentation of glucose and other carbon sources is very low yielding. The yields of lactones obtained by this process are greatly enhanced by acidification during product recovery. Only about 25% of the 4-hydroxydecanoic acid is lactonized during the fermentation at pH 6.0–7.0, but full yields are reported when extraction by steam distillation and solvent extraction are performed at pH 2. By-product formation can also be an important consideration affecting the yield of a product and the ease of its isolation. For example, 3-hydroxy-γ-decalactone continues to be formed even after production of the desired product, γ-decalactone, has ceased. It is associated with the fermentation becoming more aerobic, although some 3,4-dihydroxydecanoic acid can be cyclized into 3-hydroxy-γ-decalactone, which has no positive flavor value, or it can be dehydrated into a mixture of 2-decen-4-olide and some 3-decen-4-olide, which together make up some 5% of the microbial metabolites. Upon reduction by yeast, these metabolites give δ-decalactone and the *R*-γ-decalactone, respectively. In addition, minor amounts of other lactones are also formed from hydroxylated fatty acid intermediates in the β-oxidation pathway; 3-dodecen-6-olide is formed from the C12 fatty acid intermediate.

Yields of the desired γ-decalactone product can be maximized by terminating the fermentation at the right point, or the 3,4-dihydroxy-γ-decalactone can be converted into 3,4-unsaturated-γ-decalactones by distillation and then converted into some additional γ-decalactone by a stereoselective *S. cerevisiae* reduction. In all of the processes developed, yeasts are used as the biocatalyst. This is in part because yeast allows fatty acids to diffuse freely across their cell membranes, making them readily available for reaction. By contrast most bacteria find fatty acids to be toxic, particularly in the undissociated forms that prevail at low pH, and cellular uptake is deliberately limited and requires specific uptake proteins.

It should also be remembered that the process from the microorganisms' viewpoint consists of diffusion of the precursor molecule across the cell membrane and into the cytosol, thence into the peroxisome, where reaction takes place, with subsequent movement of the lactone out of the peroxisome and then out of the cell. Hence, the concentration of fatty acid precursor supplied is important: too little and it is all consumed to meet microbial growth and maintenance requirements, too much and the excessive fatty acid concentration is toxic, which also reduces yields of the desired product. Considerable research has been devoted to finding the best raw material for γ-decalactone production. Castor oil is a cheap and readily available source of *R*-ricinoleic acid, but plant sources of oleic acid have also been converted into ricinoleic acid by microbial hydration. In addition, alkyl ricinoleate has been shown to have the advantage of reducing the amount of foaming that takes place in the fermenter.

Even more ingenuity has been displayed in developing sources of the precursors and biosynthetic pathways to δ-decalactone. Bioconversations have also been carried out using water-immiscible extracting solvents separated from the aqueous phase by a semipermeable membrane to allow selective passage of the δ-decalactone into extraction solvent. These routes to δ-decalactone are also of interest because *S. cerevisiae* is used to carry out two different reactions: reduction of the 2-decen-5-olide and lactonization of the 11-hydroxypalmitic acid. Jalap resin also contains some 3,11-dihydroxymyristic acid, and this is converted by *S. cerevisiae* during the process into its corresponding lactone δ-decalactone.

Dairy flavor

Shukunobe and Takato described the production of diacetyl, useful as a component of dairy flavors, from citric acid using *Streptococcus cremoris* and *S. diacetylactis,* achieving yields of 14 g L^{-1} of product. A feature of this process was the requirement of oxidizing agents for the effective recovery of the diacetyl by distillation.

Flavor of mushroom

Schindler and Seipenbusch reported an industrial process for making 1-octen-3-ol, which is a character impact flavor chemical of mushrooms, by fermenting mycelia of morchella and then adding linoleic acid. Under high shear conditions, about 2.5 g product/kg dry w mycelia was formed, presumably by lipoxygenase action on the linoleic acid.

Esterified flavor

2-Methylbutyric acid and 3-methylbutyric acid are useful flavor components, especially when esterified. They are manufactured from their corresponding 2- and 3-methylbutanols, which are extracted from fusel oils by oxidation using *A. acetii* (vinegar bacteria) by the process.

Leafy green and fruit taste flavor

As an example of a more recently developed bioprocess, Firmenich claims to have developed a flexible and viable industrial-scale biotechnological process for the production of natural C6 aldehydes and alcohols. These include *trans*-2-hexenol, which has a leafy green and fruity taste, and *cis*-3-hexen-1-al, which has a grasslike green and fresh taste. The current world market for these materials is estimated to be high, at up to $50M per year. Another stimulus to develop a bioprocess is that although demand for C6 "green" flavors is growing, the traditional sources such as mint oil continue to be in short supply compared with demand. However, there are no good microbial sources of the key hydroperoxide lyase enzymes required; plant sources have hitherto been too low in activity, and cloning of these plant enzyme genes to provide more active enzyme sources is difficult because of the reluctance of consumers to accept gene modification as natural. The Firmenich process simulates the metabolic path whereby these flavor molecules are made in plants: from polyunsaturated fatty acids by the sequential action of lipoxygenase, hydroperoxide lyase, and oxidoreductase enzymes.

Firmenich uses soy bean lipoxygenase to make the hydroperoxide fatty acid derivatives, with the reaction carried out in aqueous medium at pH 9.5, and supplying pure oxygen to ensure that the reaction is completed within 1 h. The next step, cleavage of the hydroperoxides, was carried out at pH 8.5 using a plant extract specially selected to have a high lyase activity. For instance, for producing *cis*-3- and *trans*-2-hexenals, green pepper and cucumber extracts had the highest activities. If desired these flavor aldehydes could be converted to their corresponding alcohols, such as *trans*-2-hexenol, using the alcohol dehydrogenase activity of *S. cerevisiae*. Using this process, typical results include the production of *cis*-3-hexenol at 4.2 g L^{-1}, with a yield of 42% based on the fatty acid supplied, and *trans*-2-hexenal at 1.8 g L^{-1} at a yield of 24%. These values compare with typical concentrations of these flavor chemicals in fruits and vegetables of only about 2.1 mg kg^{-1} and 6.2 mg kg^{-1}, respectively; in this respect, the bioprocesses are almost 2,000- and 200- fold more effective, respectively. However, this method is not so easily applicable to *cis*-3-hexenol because, although *trans*-2-hexenal and *cis*-3-hexenal are initially produced as a mixture, the *cis*-3-hexenal is then subject to rapid nonenzymic conversion into *trans*-2-hexenal, and then the former is easily reduced by the yeast first into *trans*-2-hexenol and then into *n*-hexanol. In addition, *trans*-2-hexenal can react with acetaldehyde to form 4- octen-2,3-diol. Therefore, in order to make *cis*-3-hexenol, the 13-hydroperoxide, plant lyase, and yeast must be incubated simultaneously so that *cis*-3-hexenal is immediately reduced into *cis*-3-hexenol without any isomerization. A similar approach is used in the manufacture of 1-octen-3-ol and other related C8 alcohols and aldehydes that are very characteristic of mushroom flavors from *Agaricus bisporus* and related species. This fungus has a lipoxygenase with a special substrate specificity that converts linoleic acid into the 10-hydroperoxide, which is then cleaned to give the 1-octen-3-ol. This flavor is manufactured using unwanted mushroom stems as the source of the biocatalyst.

Use of lipoxygenase

β-Ionone production is a good example of an emerging bioprocess that also uses lipoxygenase, but requires the development of novel reaction conditions. Both α- and β-ionones are widely used in the perfume and flavor industries, and are probably produced in nature by oxidation of carotenes and fatty acids. In this process soya lipoxygenase is used to convert carotenes, from carrots or algae, in a very concentrated reaction together with fatty acids. The reaction is carried out in a kneading trough under pressure and with air renewal. Products include α- and β-ionones and aldehydes such as hexanal and 2,4-decadienal, produced presumably from the carotene and fatty acids, respectively.

Vanillin from pseudomonas

An example of a new and emerging approach, but using a different technical approach is the Harmaan and Reimer process for vanillin, the key flavor constituent of vanilla. Screening allowed the isolation of a novel *Pseudomonas* strain capable of converting eugenol, which is readily extracted from clary sage, into a number of products, several of which could be isolated in good yields, the choice of product being dictated by the fermentation conditions employed. From a biochemical perspective, there is an interesting parallel between phenylpropenoid degradation, such as of ferulic acid and fatty acid β-oxidation. They have characterized the pathway of ferulic acid degradation in a strain of a soil bacterium, *Pseudomonas fluorescens*, which they isolated by its ability to grow on ferulic acid as sole carbon source. A key feature of the pathway is the hydration and retroaldol cleavage of feruloyl-CoA to produce vanillin and acetyl-CoA, catalyzed by a newly discovered enzyme, (+)hydroxycinnamoyl-CoA hydratase/lyase. Comparison of the amino acid sequence of this enzyme with other known enzymes reveals that it is a member of the enoyl-CoA hydratase/ isomerase super- family.

Flavor of roasted food

Pyrazines contribute substantially to the characteristic tastes of roasted and toasted foods and also contribute to many cheese, vegetable, nut, and fermented food flavors. *Bacillus subtilis* and *Brevibacterium linens* have been used to produce 2,5-dimethyl-pyrazine and tetramethylpyrazine at 4 g L^{-1} and 1 g L^{-1}, respectively, by fermentation of soy beans or flour enriched with the precursors L-threonine, acetoin, and ammonia and then supplied in a fed-batch mode, followed by extraction and distillation. The highest reported yield of a pyrazine as produced by a bioprocess is 12.5 g L^{-1} of 2-isobutyl-3-methoxypyrazine, produced by a mutant strain of *Pseudomonas perolens*. This pyrazine has a characteristic bell pepper flavor. The report provides a good example of strain-to-strain variations in flavor production by microorganisms, as the parent strain of *P. perolens* produced only 46 mg L^{-1} of this pyrazine when grown on the same medium.

Flavor from iris rhizomes

Irones are degradation products of carotenoids that are valuable because of their very high flavor impact and low flavor thresholds. Traditionally they have been isolated from stored iris rhizomes, but now accelerated microbial processes have been developed, for instance, by Givaudan-Roure using microbial strains isolated from rhizomes such as *Serratia liquifaciens*. Using this method, 1 g mixed irones/kg rhizome could be obtained after 8 days, compared with 0.4 g/kg after 3 years for the traditional process (a 340-fold increase in productivity), and with the proportions of *trans-α-*, *cis-α-*, *cis-γ-*, and β-irones being very similar in both products.

Production of terpenes

Despite terpenes comprising a class of molecules that is very important in flavor and aroma products, very little real progress toward developing robust industrial processes appears to have been made. This is in part because of the water insolubility and cytotoxic characteristics of terpenes due to their effect on cell membranes. Toxicity seems to be related to their log *P* (partition coefficient) values, and terpenoids are often more toxic than hydrocarbon terpenes. However, a couple of promising attempts to develop productive process have been made. (*R*)-(+)-limonene used at 20 g/L has been converted on a 70-L scale into 900 g of pure 1*S*,2*S*4(*R*)-*p*-menth-8-ene-1,2-diol, with very few side products, using *Corynespora cassiicola*. So as to avoid toxicity problems, the limonene was fed continuously; glucose, which is required to maintain enzyme activities, was also fed continuously in response to the CO_2 level in the exit gases (an indicator of cellular respiration rates). Using a second microorganism, *Penicillium digitatum,* the (*R*)-(+)-limonene component of racemic limonene was transformed by the same workers into α-terpineol in a 46% yield. Another promising terpene biotransformation is of the sesquiterpene patchoulol into 10-hydroxypatchoulol, a key intermediate to nor-patchoulenol, which is the major odiferous component of patchouli oil; 2–4 g L^{-1} of the 10-hydroxypatchoulol was obtained using strains of *Pithomyces* obtained by a selective screening exercise.

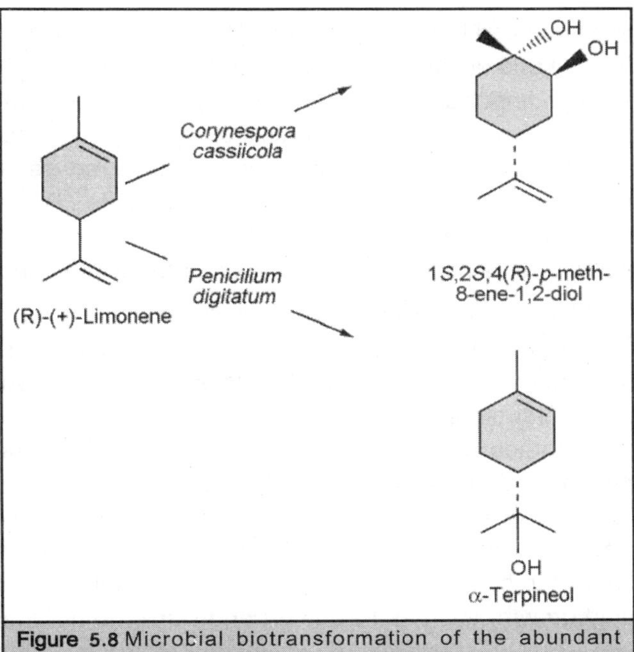

Figure 5.8 Microbial biotransformation of the abundant monoterpene (R)-(+)-limonene.

Another promising terpene biotransformation is the use of immobilized 3-β-hydroxysteroid dehydrogenase from *Cellulomonas turubata* to convert L-menthone into L-menthol. In this reaction, NADH was recycled by oxidation of L-methylisobutyl carbinol by the same enzyme. Although this dehydrogenase was active only on L-menthone at 0.3% of its activity toward its natural steroid substrate, it was stable, with a half-life for the immobilized enzyme of 25 days, and produced 25 g L^{-1} of L-menthol product. Reactor productivity was improved by continuous in situ product removal by adding solvent to extract the L-menthol, with selective passage of solvent out of the reactor through a hydrophobic membrane. This process is potentially very useful for industrial use because immature peppermint plants produce high yields of an essential oil that consists mostly of L-menthone, only 40% of which is converted into the L-menthol and is normally harvested, which opens up the possibility of a higher-yielding process involving harvesting the immature peppermint oil and then producing L-menthol by the sort of dehydrogenase reaction described earlier.

Production of soy sauce

Traditional soy sauce takes 6 months to make, which adds to the costs of the process. Osaki et al. succeeded in making soy sauce of acceptable flavor with a system of sequential immobilized cell columns. Best results were obtained using *P. halophilus, Zygosaccharomyces rouxii,* and *Candida versatilis* cells immobilized on ceramic supports and run in series, which required just a 6-day reaction period.

Mixture of dairy flavors

The variety of different yogurt, cheese, and other flavors is well known, each being due to a different mixture of flavor chemicals, most of which are produced by microbial action. These include proprionic and butyric acids, peptides and amino acids, δ- and γ-lactones, methylketones, pyrazines, diacetyl and acetoin, acetaldehyde, phenylethanol, methanediol, dimethylsulfide and even phenol. Another important factor is control of the degree of bitterness caused by peptides containing hydrophobic amino acids by use of the most appropriate mixture of proteases and peptidases. The original objective of bioprocesses was to reduce costs by accelerating the natural fermentation and thereby reducing storage times and costs. The next objective has now become a quest for improved flavor quality, and even the generation of novel flavors. Since the actual flavor development process remains the same, the goal is to develop superior starter cultures.

Microbial improvement of cheese making is a challenging task for several interrelated reasons. First, mixed cultures are common, so there is often sequential growth of different strains, with growth of the second strain taking place after the pH has been changed and using metabolites produced by the first strain. Then there is the solid nature of the cheese as the fermentation medium, with both surface growth of strains and growth within the cheese taking place. Consequently both microbial growth and flavor production can be very slow compared with submerged fermentation. The type of microbial growth that takes place also seems to influence the flavor of the cheese produced. For surface-ripened cheeses, such as Camembert and Brie, flavor development is dominated by proteolysis and the resulting peptides formed, whereas for cheeses ripened by internal mold growth, such as Roquefort and Danish Blue, the flavor is dominated by methyl ketones and fatty acids, respectively. A well-established innovation is the so-called enzyme modified cheese flavors available from a wide range of suppliers. These are made by treating cheeses with selected lipases and proteases and result in well-balanced cheese flavors with intensities up to 20 times those of the conventional mature cheeses.

Meat flavor

Gluconobacter suboxydans is used to manufacture 5-ketogluconic acid from glucose as the key precursor for the important meat flavor 4-hydroxy-5-methyl-2,3-(2*H*)-furanone. The precursor is easily recovered from the bioconversion as its calcium salt and is then converted into the furanone by heating. A related sulfur-containing furanone is formed by reacting in the presence of cysteine. The process has been developed by Quest International. It can be seen as a more advanced form of yeast flavor built up from traditional yeast extracts. By varying the types of enzyme and microorganisms used, the flavor profile of the product can be modified to best match the needs of particular applications. Additional flavor character and range can also be created by a subsequent reaction flavor step.

BIOTECHNOLOGICAL PRODUCTION OF BIOFLAVORS

The public acceptance of genetically engineered products in the food industry is not very well developed. Also there is considerable opposition to being able to assign the label "natural" to flavors produced by genetic engineering. Hence, there are still not very many good examples of successful processes using this technology. However, there are many instances of well-researched processes that are waiting in the wings to be used once consumer perceptions have changed. What is quite well established is the use of enzymes produced on an industrial scale using engineered producer strains in which the gene for the required enzyme has been cloned into a host strain with good fermentation characteristics. These include lipases, and chymosin for milk clotting. A comprehensive review by Muheim et al. also describes the use of genetic engineering technology to make debittering proteases, to produce a flavorsome beefy meaty octapeptide, and to make *cis*-3-hexenol and methional. One very effective use of genetic engineering is to remove the off-flavor caused in beers by the accumulation of diacetyl and thereby reduce the time required for the slow metabolism of diacetyl by the maturing yeast, which converts it into acetoin.

The diacetyl is formed from α-acetolactate, and it can either be prevented from forming, or be rapidly removed before it can be converted into di- acetyl. In order to prevent its formation, acetolactate decarboxylase (ALDC) can be cloned into the yeast. Yeasts transformed with the *Acetobacter aceti* ssp. xylinum ALDC gene have been used for beer production at the pilot scale and produce much less diacetyl during fermentation than their untransformed parents, resulting in a significant saving in maturation time and therefore production costs. Diacetyl formation can be reduced by modifying genes in the amino acid biosynthetic pathway that lead to α-acetolactate, especially the IL V2 gene that codes for acetolactate synthetase, followed by mating of these strains to generate hybrids that produce both little diacetyl and good quality beer. Another success is the development of *B. subtilis* strains that produce the flavor enhancer IMP in enhanced yields because of an absence of IMP dehydrogenase that is responsible for its metabolism into guanosine synthesis. Alternatively guanosine production can be enhanced by increasing the number of gene copies of this enzyme.

BIOCHEMICAL ENGINEERING

Biochemical engineering has the key role of transforming promising laboratory discoveries into cost-effective and reproducible large-scale manufacturing processes in a timely fashion and with minimal expenditure. This usually involves real-time analysis of pH and dissolved oxygen and carbon dioxide levels and sophisticated data logging and analysis as well as a careful consideration of the chemical and physical properties of the materials involved, for instance, the separation characteristics of biological materials are very often pH dependent. Also of critical importance is maintaining good sterile or hygenic conditions (e.g., by using cleaning-in-place methods), product specifications and cost targets, by-products and their disposal costs, the intended scale of production and any economies of scale possible, effects on the sizing of process equipment, operating regime (e.g., batch vs. continuous reactors, distillation vs. solvent extraction of product), and safety and good management practice constraints including process validation requirements. Of particular consideration for flavor and aroma products is that they often consist of a mixture of different chemicals, many of which are necessary to give the product its real value.

As well as downstream processing, upstream processing is also important. This is because so many flavor and aroma products are manufactured from natural raw materials extracted from plants, so upstream processing operations for such intrinsically variable raw materials also pose special challenges, especially for successful process integration. The main problems that are usually associated with bioprocessing are the low concentrations of product(s) that are formed, and consequently the need to remove large amounts of water, and the instability of these biochemical products with respect to heat and other extreme conditions that may occur during their purification, concentration, and recovery. Consequently the costs of downstream processing can be greater than those of the fermentation or bioconversion unit operation and can be as great as 50% of the overall processing costs. Because the value of a flavor or aroma chemical is so dependent on its taste and smell qualities, the effects on these of trace impurities carried over from fermentation media and other contaminants can be very significant. For example, in a process to produce δ-decalactone flavor, considerable efforts were made to purify the 11-hydroxypalmitic acid precursor to an acceptable purity so as not to compromise the flavor quality of the product. This involved extraction of the botanical source (sweet potato residue or jalap roots) followed by hydrolysis and purification.

Obviously it is the performance of the overall process that determines the success of the process and the product it produces. Therefore the integration of all the individual unit operations, whether bioproduction steps or product isolation operations, is vitally important. Also important is the reproducibility of the process; an important factor in many processes for natural flavors is the variability in the quality and prices of raw materials. As compared with chemical processing, precursor and product inhibition effects often dominate bioreactions, necessitating specialized techniques including fed-batch reactions, continuous supply of precursors, use of slow-release forms of products, two-phase reactions (in which the second phase or resin acts as a product reservoir for product sequestered out of the aqueous phase), loop reactors (in which product is recovered by an in-line resin column or membrane extraction unit), and the recovery of product by its adsorption from the exit gases of a reactor.

The potential for such sophisticated bioengineering is illustrated by the good productivities already achieved for some nonflavor products such as L-tertiary leucine and *N*-acetylneuraminic acid, which have productivities of 638 and 470 g product (L day)$^{-1}$, respectively, using membrane reactors in which the product is continuously removed and the cofactor is regenerated in situ, allowing the enzymes to work at as high a rate as possible. On occasions the maximum permitted amounts of residue solvents are defined, and multiple extractions are required to obtain good yields. It may also be necessary to use more sophisticated extraction equipment, such as rotating disc extractors. In general the solvents used should be nonvolatile, not prone to emulsion formation, and immiscible with water to facilitate solvent recovery and reuse and minimize environmental problems, and cells should be dried prior to solvent extraction so as to prevent emulsion formation with residual water. In any consideration of the use of solvents, the importance of good quality water is vital. An early illustration of this in the history of biotechnology is that the best beers were brewed using the special, high-quality water obtained at Burton-on-Trent, U.K. The subsequent introduction of ion exchange for the purification of water for brewing allowed the production of superior quality beers to become widespread. A good account of the biochemical engineering aspects of process development is given by Chisti and Moo-Young. These include the following unit operations.

Chromatographical Technique

Chromatography is very often expensive and difficult to fit into processes, and so it is used only for high-value products. Reverse-phase chromatography is very useful because the desired material can be absorbed straight from the aqueous phase in which it has been produced. Chromatography is usually used after clarification to remove any impurities that would foul the column, such as proteins released by any cell lysis, and then product is recovered by washing with ethanol, in which form the flavor can often be used directly. An example is the malt flavor produced by fermentation using *Streptococcus lactis* var. maltigenes, which (following ultrafiltration to remove cell mass and protein remaining from the fermentation media) can either be subject to reverse osmosis and then spray- dried with a carrier to make a powder form of the flavor, or steam distilled and the reverse-phase chromatography carried out with ethanol as the eluent to make a liquid flavor.

Liberation of Intracellular Products

Cell disruption is necessary for the liberation of intracellular products. This imposes two constraints on the process. First, the amount of product that can be made is limited to that which can be accumulated within the cell, whether the intracellular product is an enzyme or a small molecule. Second, cell disruption liberates most of the contents of the cell together with the desired product, and so considerable purification is often required. It is also interesting to note that cell disruption is a unit operation that has had to be developed de novo in response to the need from bioprocesses. Therefore, unlike many other operations such as filtration, centrifugation, and crystallization, cell disruption has no equivalent in conventional chemical engineering.

Mechanical Separation

Centrifugation is required variously to recover cells, for the clarification of liquids prior to extraction, chromatography etc., for breaking emulsions (such as are often formed by cell disruption), or for washing cells prior to another operation such as disruption or immobilization.

TRACE CONTAMINATION

Many other biological factors are also important in the production and processing of flavor chemicals. Trace contaminant strains can introduce traces of undesirable flavor component that make the product less valuable or even unusable. The BOD of fermentor wastes needs to be minimized. Exit gases from fermentation need to be sterilized often by microfiltration. Noxious smells should be removed prior to emission into the local environment, although flavor factories are often sited with respect to prevailing winds so that emissions are less likely to reach major sites of population. In biotransformation processes the precursors will ideally be appreciably water soluble and with minimal cytotoxicity. As in any bioprocess, the media will be formulated so as to achieve a number of aims, including minimization of the lag phase, the inoculum size required, and the duration of the fermentation.

Other objectives are to minimize power input costs and to maximize mixing so as to maximize the utilization of expensive filtered air. Perhaps the newest innovation in flavor recovery technology are the use of in situ solvent extraction, and of perevaporation, which has been demonstrated to be effective for the in situ recovery of benzaldehyde from fermentation broth using *Bjerkandera adusta*. Another continuous in situ product recovery method is the continuous addition of an extracting solvent to the reaction vessel, with its selective removal together with the product using a hydrophobic membrane. In addition per-evaporation has been used to recover 6-pentyl-α-pyrone from *Trichoderma viride* cultures, obtaining an 85% yield of product and a 20-fold enrichment after 2 weeks of fermentation.

DIVERSIFYING ACTIVITIES

In comparing various processes, the productivity of the biocatalyst is very important. Production of flavors by plant cells is a very desirable method that still requires major advances to make it both practicable and cost-effective. These advances may be developed for pharmaceutical products and then transferred to flavor processes. For instance, Phyton has made big advances in developing plant cell fermentations to make paclitaxel, the active ingredient in Bristol-Myers Squibb's taxol anticancer drug. Therefore, it is interesting to note that, by comparison with the microbial systems already described, it is reported that the rates of capsaicin biosynthesis in ripening pepper plants is ca. 0.5 mg g dry wt^{-1} day^{-1}, and in pepper plant cell culture it is ca. 0.1 mg g dry wt^{-1} day^{-1}, which is equivalent to a turnover rate of 0.00006 mmol g^{-1} h^{-1}. To make progress it is usually necessary to elucidate the biochemical pathway the plant uses to make the flavor. For instance, the biosynthesis of Furaneol by strawberries (2,5-dimethyl-4-hydroxy-(2*H*)-furan-3-one) has been recently described, as has the biosynthesis of raspberry ketone by coupling coumaroyl-CoA and malonyl-CoA, followed by reduction, by raspberry fruit. If sufficient research and development resources are available, biocatalysts with very high productivities can be developed, and these have created vast industries. For example, consider the commodity products made by biocatalysts.

Because flavor and aroma chemicals are much higher-value products, the bioprocesses can afford to be less productive but still need to achieve good productivities because the volume of demand is not as great as for other products. For instance ca. 10^7 tons/year of high fructose corn syrup are made, but processes for flavors and aroma chemicals do not benefit from such big economies of scale. The concentration at which a product can be produced in an enzyme or microbial reaction is a good guide to how cost-effectively it can be manufactured. High reaction productivity must be complemented by effective downstream processing and by good process intensity in each unit operation of the manufacturing process. This is because product concentration reflects the size of equipment, especially of reactors, and the hydraulic throughput rates that are required to make a given amount of product, and thus influences both capital costs and costs of mixing, drying, pumping, and so forth. Also the product concentration indicates the volumes of fluid that will have to be processed to obtain the required amount of product and the cost of concentrating and isolating the product.

The relationship between production concentration and sale price is well established for a wide range of bioproducts ranging from very high value biopharmaceuticals to low-cost food ingredients such as citric acid. Thus, if a flavor material can be produced only at quite low concentrations, it is very likely that the only market opportunity for it will be in prestige and/or high-value niche markets, if they exist. A major constraint on the product concentrations that can be achieved is product inhibition. Therefore, a determination of inhibition constants is as important as determining the rate of reaction parameters, and work on developing effective in situ product recovery methods that will remove product as it is made are very important. For flavor molecules, such in situ product recovery techniques can include the recovery of volatile flavors from the exit gases of bioreactors. Product inhibition is, of course, exacerbated if it is difficult for the product to diffuse out of the cell in which it has been produced.

ECONOMIC VALUES

Once a flavor has been produced it must always be remembered that its value will vary with its applications. A graphic example is provided by vanilla flavor when used in low-fat ice creams. Reformulation of the flavor is required because the less oil soluble flavor chemicals are less well perceived in the low-fat products and so need to be present in relatively higher concentration than does the more oil soluble flavor component in traditional high-fat ice creams. Big advantages can be obtained if positive applications results can be obtained, for example, in creating reductions in fat and salt contents of products, improved flavor stability and controlled release, emulsion stabilisation, as well in as the growing area of health benefits (nutraceuticals). Changes can also occur during food processing and cooking, especially when extreme conditions are used. For instance, so-called reaction flavors can be formed from flavor components during heating, particularly via the Mailard reaction that occurs between reducing sugars and the amino groups of amino acids. Condensate flavors can also arise upon cooling, especially when local high concentrations of reactants occur. An example is the reaction of acetaldehyde and hydrogen sulfide to form 3,5-dimethyl-1,2,4-trithiolene.

Commercial Steps

The necessary tests and procedures for product approval and registration can be both lengthy and expensive, and different countries often have different requirements. The most difficult task is to gain approval for a new chemical; approval for new processes to make existing materials with a proven record of safety is easier to gain. The ideal is "generally regarded as safe" approval, which indicates that long-term and significant consumption in the diet results in no evidence of any adverse effects.

Consumer Acceptability

However novel the underlying science and however efficient the performance of the manufacturing process, the product will be successful only if it achieves both regulatory approval and consumer acceptability, and indeed, generates a marked consumer preference because of the perceived value of the products to individual consumers and interest groups. This is a difficult area not least because of the marked difference between lay and expert perceptions of risk, with the consumer tending to put more weight on the risk of new uncontrolled factors and the effects directly on themselves and their families, rather than relying on the statistical risk factors based on broad demographics.

In particular individual consumers fear loss of control, particularly to organizations that they find difficult to trust because of their vested interests that may act counter to their own interests, such as companies' profit motive. So in this respect there is a very big difference between the use of bioprocesses for flavor and for pharmaceuticals. Whereas, both types of product have very high biological activities and are discovered and then developed along similar lines, the consumer needs and decisions for their uses are quite different. There is a strong "cure" imperative for drugs, which are usually used only an occasionally, whereas the use of food flavors is much more an everyday pleasure-based activity, leading to an ethical concern expressed as a preference for natural-flavored products. For instance, Abraham et al. reports that about 70% of the food flavors used in Germany in 1990 were natural.

CONCLUDING REMARKS

Despite biopharmaceuticals being widely regarded as the cutting edge of modern biotechnology, a surprisingly high proportion of current flavors and food ingredients are manufactured using bioprocesses. These include not only refined and developed versions of traditional processes, such as for cheese, yogurt, and protein hydrolysates, but also many new and sophisticated bioprocesses for key chemically defined materials such as the flavor chemicals methyl ketones, and γ-decalactone, and flavor enhancers such as GMP and IMP. This article has attempted to explain the technical basis for these products as used in industry and indicate exciting future developments in this field that are based on the pull of market demand forces but still constrained by cost, patents, labeling, and regulatory considerations. However, a comprehensive account is not yet possible because full details are not available concerning the bioprocesses now known to be operating. In many cases it is even uncertain at what scale processes are operating. In conclusion, some important commercial factors that must be considered including the following:

1. Significant market demand (especially for label-friendly natural ingredients and for products with superior performance such as fat reduction, health benefits, or improved convenience)
2. Substantial profit margins
3. Economies of scale of production
4. Availability of low-cost raw materials
5. Patentable (proprietary) technologies
6. Weak competition
7. A range of products with different market uses from a single process
8. Ease of regulatory approval

As regards cost, Welsh has suggested a relationship between selling price and market size. This shows that although the highest-priced products, up to $1,250 per kg, appear superficially most attractive, the largest markets are for the lower-priced products, with an approximate market size of $20 million, as compared with only $3 million and $1.25 million for the higher-priced products. These values of natural flavors do not compare unfavourably with the price of gold, currently about $12,000 per kilogram.

Much exciting research is taking place, and the pace of change in the food and beverage industries is intense, which is creating many new opportunities for bioflavors, especially considering the sustained demand from consumers for products containing natural ingredients. However, these opportunities must be balanced against constraints, including the availability of natural raw materials in sufficient quantity, quality and low price, the difficulty in developing bioprocesses that give high conversions of concentrated substrate streams into products with no by-products, and the difficulties in easily concentrating and purifying the required bioproducts. Perhaps in this last respect more efforts should be made to obtain the benefits of biomolecular selectivity for product isolation and recovery. The objective of bioprocesses in the flavor and fragrance industry is to produce the highest quality product at the minimum processing costs. I hope this chapter illustrates the complex interrelationships between what is needed (consumer demands), how to make it (the development of bioprocesses), and also the important conditioning factors, such as what it is possible to patent, safe operation, and the need to meet other demands, such as for naturalness. Perhaps this chapter can in some small way help others in this exciting area of technology to succeed in developing more new and improved bioprocesses for flavors.

6

INDUSTRIAL PRODUCTION OF WINE

During the past 20 years there have been considerable developments in winemaking techniques, with the introduction of processes and machinery borrowed from other sectors of the food industry. These developments have affected all phases of wine production, wine being the end product of a technological chain that includes the preparation and treatment of the must, fermentation, aging, and bottling. Thus, technological progress has improved the quality of the wine produced. Wine quality is measured by finesse, intensity, and originality in taste and smell and by microbiological and physicochemical stability. Among these, during the past few years particular attention has been paid to enzyme treatments and to the use of selected fermentation starters (both free and immobilized). Some of these processes have already been applied to production.

BIOTECHNOLOGICAL PROCESS

Wine making is a biotechnological process in which enzymes play a fundamental role and in which the use of exogenous enzyme preparations makes it possible to overcome the problem of insufficient activity of endogenous enzymes in the grapes. The enzymic treatments have a multiple purpose in that the changes made to the grapes, the must, and the wine have an influence on the processes of clarification and filtration, the processes of extraction of color and aroma, and the stability of the wine. It must, however, be remembered that the use of enzymes in the wine industry, apart from pectolytic enzymes, remains extremely limited for a series of reasons that can be summarized as follows:

1. The traditionalism of the wine industry, which is still anchored to classical methods
2. Limitations of enzymic activity connected with the composition of the must and the wine (low pH, ethanol concentration, presence of tannins, and a fairly low working temperature compared to the optimum for enzyme preparations)
3. Legal restrictions

However, during the past few years, research has suggested the use of enzymes that catalyze the transformation of glucan (glucanase), proteins (peptidase), and polyphenols (laccase, peroxidase, phenolase, and tannase), all of which are compounds that condition the stability and organoleptic characteristics of wines. Particular attention has been paid to glycosidases that are able to free odorous compounds by hydrolyzing terpene glycosides.

Enzymes of Pectic Substances

Pectolytic enzymes have been widely used for many years, because they act on the pectins. Pectins are complexes of acids (homogalacturonan and rhamnogalacturonan) and neutral polysaccharides (arabinan, galactan, and arabinogalactan) that are found in the must in colloid form. They act as protective colloids, causing problems in the clarification and stabilization.

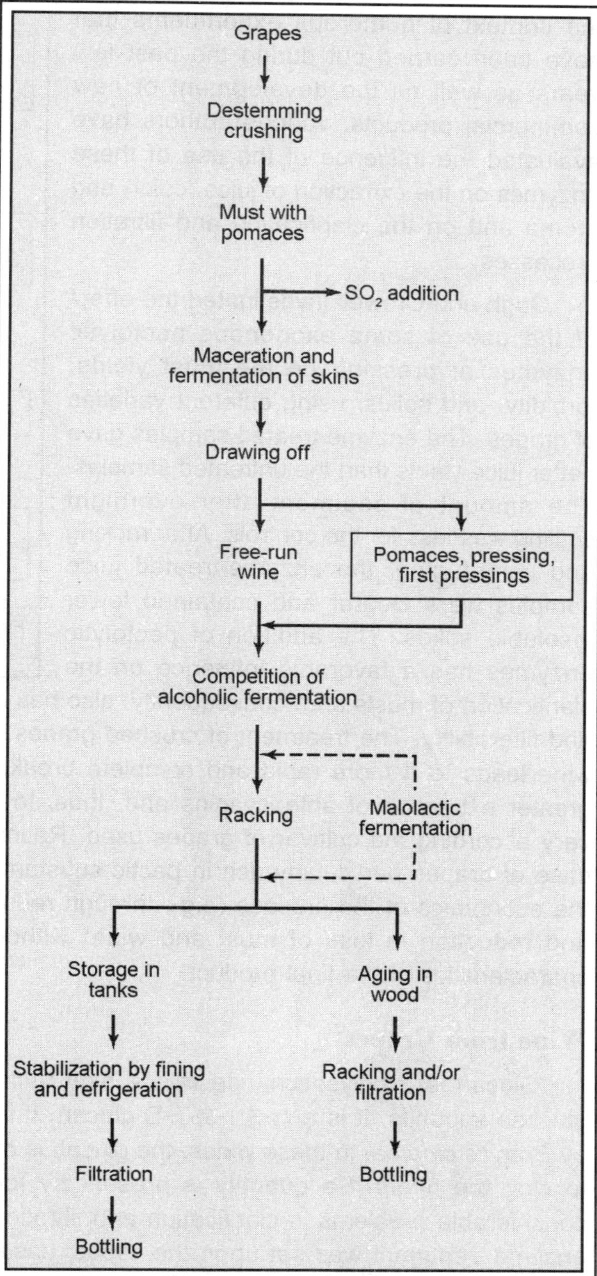

Figure 6.1 Scheme of a process for red wine preparation (fermentation with skins on).

Enzymes that can act on the pectic substances are usually classified into two groups: de-esterifiers (pectinmethylesterase) and depolymerases that catalyze both hydrolysis (polygalacturonase) and β-elimination reactions (pectin lyase and pectate lyase). Purified enzymic preparations from cultures of microorganisms normally contain polygalacturonase and lyases associated with hemicellulase and cellulase. However, the presence of pectin methylesterase must not be too great in order to limit the liberation of methanol, which is found esterified with the carboxylic groups of the galacturonic acid in the pectic molecule. Pectolytic enzymes are used at vintage time in the preparation of both white and red wines. Italian law permits the use of pectolytic enzymes in the processing of pressed grapes, the must to be used for producing white wine, musts and wines obtained using the thermovinification system, and wines after they have been separated from the pomaces. In the context of numerous experiments that have been carried out during the past few years as well as the development of new commercial products, various authors have evaluated the influence of the use of these enzymes on the extraction of juice, color, and aroma and on the clarification and filtration processes.

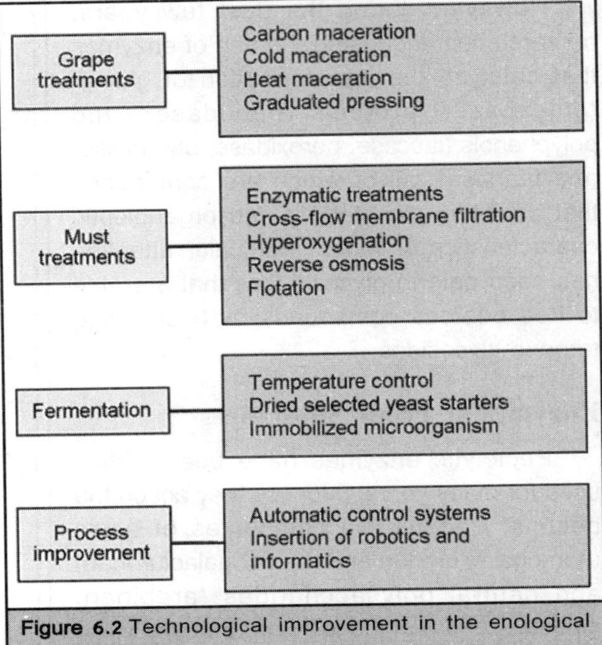

Figure 6.2 Technological improvement in the enological industry.

Ough and Crowell investigated the effect of the use of some exogenous pectolytic enzymes at pressing on the must yields, turbidity, and solids, using different varieties of grapes. The enzyme-treated samples gave better juice yields than the untreated samples. The amount of sediment after overnight settling was less for the controls. After racking and centrifuging, the enzyme-treated juice samples were clearer and contained fewer insoluble solids. The addition of pectolytic enzymes has a favorable influence on the clarification of musts and, consequently, also has a positive effect on such parameters as viscosity and filterability. The treatment of crushed grapes with pectolytic enzymes in the production of red wine leads to a more rapid and complete breakdown of the cell wall of the skin. This leads to greater extraction of anthocyanins and, thus, to an increase in color, although the results may vary according the cultivar of grapes used. Rational use of pectolytic enzymes, especially in the case of grapes particularly rich in pectic substances (e.g., muscat) makes it possible to improve the economics of the process (e.g., through reduction in time required, better use of equipment, and reduction in loss of must and wine) without negatively altering or modifying the quality characteristics of the final product.

Wine from Grapes

Glucan is a polysaccharide with a high molecular weight (M_r 10^5-10^6), consisting entirely of glucose subunits. It is a (1-3:1-6)-β-D-glucan. It is found in must and wines from grapes infected by *Botrytis cinerea*. In these wines, the glucan is of great technological importance because it tends to clog the filters if a quantity is present by forming a glutinous layer on them. This causes considerable problems in clarification and filtration. In order to solve this problem of clogging, an enzyme treatment was set up in the 1980s, based on β-glucanase, which hydrolyzes the glucan excreted by the *Botrytis cinerea*. Experiments carried out by Villettaz and Wucherpfenning and Dietrich to evaluate the efficacy of β-glucanase have demonstrated the technological and economic importance of this enzyme in improving the filterability of wines produced from grapes affected by *Botrytis*. β-glucanase has been included in the practices and treatments permitted by the EEC, although the conditions for use have yet to be defined.

Treatment with Proteolytic Enzymes

In winemaking, treatments are carried out to lower the protein content, because in must and wine, proteins are found in colloidal dispersion, which can become insoluble, forming haze and precipitates. When this occurs during processing it causes delays, but if it occurs when the wine is already on sale it causes economic loss. This phenomenon occurs predominantly in white wines and only very rarely in red wines. Red wines contain tannins, which are believed to react with the proteins to form insoluble tannoprotein compounds. During fermentation and aging, this leads to a drastic reduction in the protein levels. Although the proteins responsible for the instability have not yet been well characterized, it seems that the protein fractions of lower MW (12,600 and 20,000 to 30,000) and lower P_i (4.1 to 5.8) and glycoproteins are the major and most important fractions contributing to protein instability in wines. Treatments with proteolytic enzymes have been tried as an alternative to using traditional finings such as bentonite. These enzymes cause hydrolysis of the proteins into amino acids or short-chain peptides that are completely soluble in the wine; however, intervention with exogenous enzymes is not free from problems.

Many of the traditional proteases (papain, pepsin, and trypsin) are not suitable for use in wine; they require high temperatures that are deleterious to wine quality, and they do not have their optimal pH in the range of 3 to 4, which is ideal for wines. For these reasons, studies have mostly been directed toward the use of acid proteases of fungal origin in free and immobilized forms; however, so far the results have not always been positive. Modra and co-workers assessed the action of commercial peptidases on proteins isolated from muscat grapes. This treatment modified the HPLC elution profile of a high relative molecular weight fraction containing protein. The most generally discernible change was a decrease in the M_r 23 .000 peak, with a concomitant increase in the 10.000 M_r peak, and a loss of resolution in both. The change in protein profile varied among the peptidases, but none of the enzymes gave complete proteolysis after 7 days.

The tests carried out on muscat must confirmed that the peptidases modified the proteins; however, their efficacy was not sufficient to make the wine stable, and treatment with bentonite was required. Woiwodow et al. tested the action of an acid protease from *Aspergillus niger* immobilized on Sylochrom C-80 on the proteins of white table wines. Experiments were carried out with different quantities of enzyme (2 to 10 g/L) and with different times of contact (2 to 48 h) at a temperature of 23 to 25°C. Complete stability of the wine was reached after prolonged treatment (48 h) with very high quantities of immobilized enzyme (10 g/L). In other tests, electrophoresis of the proteins isolated from the wine before and after treatment showed a diminution in both number and intensity of protein fractions, but the wine was not completely stable.

Bakalinsky and Boulton studied the use of food-grade acid protease (from *A. niger*) covalently linked to agarose spheres of 100 μm. Four types of wine were treated at temperatures of 30 and 37°C in a recirculating flow packed-bed reactor containing 226 mg of dry weight conjugate. At the end of the treatment, the wines were subjected to heat-stability testing. The results showed that the enzyme, although active in all of the wines tested, achieved protein stability only in Riesling wine. The Riesling was the only wine whose pH was adjusted to 3.0 for treatment, whereas the other wines were treated at their natural pH levels. The pH adjustment may have been the cause of the successful stabilization. That immobilized protease was active in the wine but failed to have any significant effect on the protein stability may indicate that the proteases do not hydrolyze the protein fractions responsible for the instability or that the hydrolysis occurs along with the formation of products that are still unstable. If the potentially unstable wine proteins were complexed with polyphenolic compounds, one would expect them to be more resistant to enzymic hydrolysis. Currently, the use of free or immobilized proteases to stabilize wine requires more information on the nature of the proteins, their interactions with the other components of wine, and the conditions that cause them to become insoluble. Furthermore, sensory evaluation of the treated wine is recommended to assess possible detrimental alterations in flavor caused by the probable increase in peptide content. Small hydrophobic peptides, in particular, are recognized as having a bitter flavor.

Phenolic Components in Wine

Phenols in wine predominantly arise from the grapes and are primarily flavonoid. They are crucial components in wines because of their basic importance to the color and flavor. Flavonoids (catechins, proanthocyanidins) are also responsible for discoloration, turbidity, and flavor changes in wines. Review of the importance of phenolic compounds in grapes and wine has recently been published. Chemical and enzymic oxidation of phenolic compounds leads to the formation of highly reactive intermediates (quinones) that interact with amino acids to reduce sugars and alcohol and give rise to chromophores and volatile substances. This results in a loss of freshness, anywhere from yellowing to browning of the wine, and turbidity.

Table 6.1 Phenolic enzymes

Oxidases
Peroxidase
Donor (catecol) + H_2O_2, oxidized donor (quinone) + $2H_2O$
o-Diphenoloxidase (catecholoxidase, polyphenoloxidase, phenolase, tyrosinase, cresolase)
2 *o*-Diphenol + O_2; 2 *o*-quinone + $2H_2O$
p-Diphenoloxidase (laccase)
2*p*-Diphenol + O_2; 2*p*-quinone + 2 H_2O
Hydrolases
Tannin acylhydrolase (tannase)
Galloylglucose, gallic acid + glucose
Anthocyanase
Anthocyanidin 3 -monoglycoside, anthocianidin + hexose

The removal of polyphenols is a common practice in winemaking processes to prevent discoloration and to obtain stabilization against oxidation, especially in white and rose wines. Wine technology tends toward the adoption of systems for extraction of the must that limit the solubilization of oxidizable substances and toward the use of finings (gelatin and casein), adsorbants (active carbon), and polyvinyl polypyrrolidone (PVPP). The use of these additives in wine production is aimed at reduction of the level of polyphenols and compounds that catalyze their transformation, particularly enzymes and heavy metals. To limit the amount of treatment with chemical-physical agents, it has been suggested that enzymes active against polyphenols (oxydases, hydrolases, and transferases) be used in the prefermentative phase.

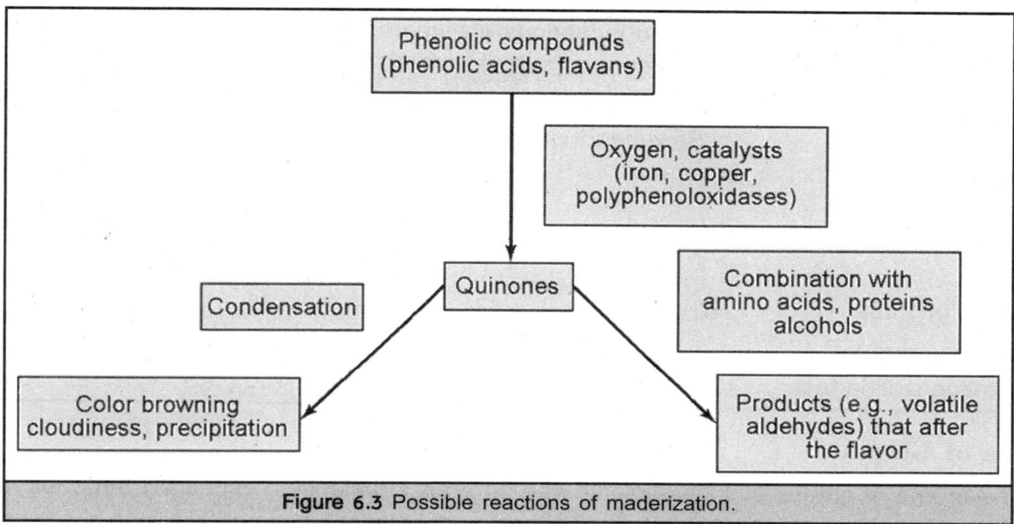

Figure 6.3 Possible reactions of maderization.

The aim of this is to set off a process of enzymic oxidation of polyphenols in the must under controlled conditions, so as to destabilize them and accelerate the process of polymerization and flocculation. The possibility of eliminating or achieving reactivity loss of the polyphenols responsible for wine instability has been tested on the bench and on a pilot scale using enzymic preparations containing, respectively, tannase (from *Aspergillus* sp.), phenolase (prepared as an extract or acetoin powder from *Agaricus campestris*), and laccase (from *Polyporus versicolor* in a liquid culture, partially purified). These different enzyme treatments were compared with each other, and some were then compared with conventional systems of stabilization. The polyphenol concentration in wines prepared from the same, but differently processed, pinot grapes showed that laccase was more effective than the other enzymes tested and produced a wine with a more stable color.

The possibility of using laccase was checked on a pilot scale with the must from Trebbiano grapes. Laccase treatment was compared with clarification carried out with casein, active carbon, and bentonite, with or without sulphur dioxide. The use of the enzyme was shown to be highly effective, preferable or practically identical to the results from traditional processing. The wines treated with laccase had different and better organoleptic maderization-resistant features. In the case of muscat and Riesling, the action of laccase (8,300 units/L) was compared with that of fining agents (gelatin /SiO_2). The results obtained with muscat must without SO_2 and Riesling must with SO_2 (150 ppm) shows that the enzymic treatment is highly effective, preferable, or practically identical to traditional processing, especially when using must without sulphur dioxide. These interesting results allow us to conclude that lac-case prefermentative treatments make it possible to eliminate instability in white wines caused by oxidizable polyphenols. Obviously, the process of winemaking must include physical (e.g., ultrafiltration) or chemical (e.g., addition of sulphur dioxide, deproteinization) systems that make it possible to eliminate the enzyme. The enzymic treatment coupled with a filtration system (filtration by diatomaceous earth, PVPP, or silica of ultrafiltration) is probably a technique that gives sufficient stability for white wines.

Table 6.2 Polyphenol content and color increases found in maderization test of pinot wines obtained by various enzyme treatments

	No enzyme	*Laccase*	*Phenolase*	*Tannase*
Polyphenol (mg/L)				
Total	339	310	319	333
Nontannic	267	261	284	284
Color				
E_{420mm} (x 1,000)	92	82	78	64
Increase due to maderization test	121	95	160	156

Aroma of Grapes

The aroma of grapes includes volatile free odorous substances, especially terpenes (linalool, geraniol, nerol, citronellol, α-terpineol, linalool oxide) that play a fundamental role in giving character to certain cultivars, for example, muscat. In addition, grapes also include compounds called aroma precursors, which are glycosides mainly of linalool, nerol, and geraniol (for example 6-*O*-α-L-arabinofuranosyl-β-D-glucopyranosides, 6-*O*-α-L-rhamnopyranosyl-β-D-glucopyranosides and β-D-apiofuranosyl-β-D-glucopyranosides.

Table 6.3 Comparison of untreated and use of fining or laccase enzyme on total phenolics and catechins of muscat and riesling must

	Untreated	*Fining*	*Laccase*
Muscat (raw)			
Total phenolics	232	184	90
Catechins	43	56	30
Color E_{420mm} (x 1,000)	226	99	89
Riesling (150 ppm SO_2 added)			
Total phenolics	303	282	286
Catechins	29	28	29
Color E_{420mm} (x 1,000)	83	53	60

The ratio between bound and free monoterpenols ranges from 1 to 5 in juice grape cultivars of muscat and Riesling and can be as high as 15 in the Gewurtztraminer variety. From a technological point of view, the existence of such bound forms is interesting because these glycosides, nonvolatile precursors, can give flavoring aglycones when hydrolytic reactions take place. High-temperature acid hydrolysis can be used, but it may lead to extensive rearrangements of the terpenols. A study of the transformations of the terpenic compounds has in fact shown that, as the pH of the aqueous medium diminishes, the hydration and cyclization of linalool increase. Enzymic methods are attractive for increasing the concentration of free flavorants with minimal changes in the natural monoterpene composition. In the first stage, the intersugar linkage is cleaved by arabinofuranosidase, rhamnopyranosidase, or apiofuranosidase (depending on the structure of aglycon moiety), and the corresponding monoterpenyl-β-D-glucosides are released. In the second stage, the liberation of monoterpenols takes place after action of a β-glucosidase on the previous monoterpenyl β-D-glucosides. In the search for an enzymatic compound suitable for the hydrolysis of these precursors, different commercial products (pectinase, cellulase, hemicellulase, lypase, and protease) have been tested.

Some of the preparations tested have contemporaneously good α-arabinofuranosidase, α-rhamnopyranosidase, and β-glucosidase action, but their efficacy in the must and wine is a function of numerous factors. The efficiency of hydrolysis of monoterpenyl β-D-glucosides by β-glucosidases was found to be dependent on the structure of the aglycon and the origin of the enzyme. Moreover, the efficiency of β-glucosidase is influenced by the concentration of glucose and gluconolactone, and in sweet wines, the hydrolysis is interrupted at the stage of the monoglucoside without freeing the odorous monomers. In the case of dry wines, on the other hand, there is a considerable increase of most of the free terpenes. Considering that the glycosidases used so far have not given the results hoped for, and that this is principally because of the properties of β-glucosidase, the hydrolytic activities of various plant and fungal β-glucosidases toward terpene glycosides are being studied. The goal of these studies is to find an enzyme with large aglycon specificity and good stability to low pH as well as tolerance to glucose and ethanol.

CHARACTERISTICS OF ALCOHOLIC FERMENTATION

Of the two fermentation processes, alcoholic fermentation causes the transformation of sugars into ethyl alcohol and CO_2, whereas malolactic fermentation causes the transformation of malic acid into lactic acid and CO_2. However, both processes produce compounds (organic acid, alcohols, ester, and carbonilic compounds) that influence or determine the organoleptic characteristics of wines and so play a fundamental role in fixing wine quality. For this reason, the microorganisms responsible for these transformations, yeast and bacteria, have for some time been the object of great interest. The work of Pasteur (1886) was followed by numerous studies of the microflora of grapes and of the factors that regulate the equilibria between the various species and their metabolism. In fact, a great variety of microorganisms are present in grapes and musts, and the yeasts responsible for alcoholic fermentation are but a small part of them.

The fermentative yeasts (particularly *Saccharomyces cerevisiae* and related species) generally predominate because the chemical composition of the must is more favorable for their development than for that of other eumycetes and schizomycetes. However, it is difficult to control alcoholic fermentation that develops naturally, because certain parameters (such as temperature, degree of oxygenation, degree of clarity of the must, and phenomena of microbial antagonism) can modify the natural selection and therefore condition the development and completion of the process. This alters the ratio between the principal (ethanol and CO_2) and secondary (acetic acid, succinic acid, acetoin, 2,3-butanediol, acetaldehyde, glycerol) products.

Production of White Wine

To overcome the problems mentioned earlier, for many years it has been standard practice, especially in the production of white wine, to use dry pure cultures of enologically suitable species (in particular *Saccharomyces cerevisiae* and all its physiological variants) with known properties. This ensures a quicker start to alcoholic fermentation (thanks to the massive addition of yeast), prevents the development of a typical microflora, and therefore eliminates the uncertainty and variation that occur in spontaneous alcoholic fermentation. Selected culture has therefore been the subject of many investigations that have considered the desirable characteristics of wine yeasts and their influence on the composition of wine and its organoleptic characteristics. Numerous properties of yeasts have been studied: some are always desirable, and others are always undesirable. During the past few years, the wine industry has looked with interest at yeast genetics and strain development programs in order to obtain modified yeasts suitable for winemaking technologies that do not impair the flavor and bouquet of the final product.

Table 6.4 Some Characteristics of *Saccharomyces cerevisiae* affecting the winemaking process

Desirable
Rapid initiation of fermentation
Efficient conversion of grape sugar to alcohol
Ability to conduct even fermentation
Ethanol tolerance
Ability to settle rapidly at the end of fermentation to aid clarification
Fermentation at low temperatures such as 10-14°C
Retention of viability during storage
Sulfur dioxide resistance (sulfur dioxide is normally used in winemaking)
Production of desirable fermentation bouquet and reproducible production of the correct levels of flavor and aroma compounds
Killer factor or resistance to killer toxins
Film-forming capacity
Undesirable
Production of sulfite, hydrogen sulfide, or mercaptan
High production of acetaldehyde, acetic acid (volatile acidity), ethyl acetate, and higher alcohols foaming capacity
Production of urea, which can result in the formation of ethyl carbamate

Yeast geneticists have carried out programs of genetic improvement to obtain new yeast strains that possess a high fermentation rate and high ethanol production as well as other relevant winemaking characteristics. This program has used both classical genetics (e.g., hybridization through conjugation between spores or aploid cells) and more recent biotechnological methods, such as somatic hybridization through fusion of protoplasts induced chemically or physically and transformation through the introduction of fragment of extraneous DNA using molecular vectors. Thus far, it has been possible to eliminate foaming capacity of strains thought to be highly suitable for inducing alcoholic fermentation. Certain wine yeast strains produce proteins that are responsible for the foaming associated with fermentation. Excessive foaming during wine fermentation is an undesirable characteristic of wine yeast strains.

The formation of a froth head can result in the loss of grape juice or reduce the capacity of plant equipment, because part of the fermentation vessel may have to be reserved to prevent the froth from spilling over. Foaminess also detracts from the visual appeal of the product. Furthermore, it has been possible to effect genetic marking to differentiate the selected strains from those of the indigenous microflora (always present in musts) and to give the killer phenotype to a strain to increase its competitiveness with indigenous flora. The killer phenomenon is related to the ability of some yeasts to secrete toxic proteins that are lethal for so-called sensitive strains. Particular attention is being given to the following possibilities: (i) introduction of the flocculent characteristic into yeast strains to facilitate the separation of cells from wine at the end of fermentation; (ii) introducing malolactic enzymic activity into yeasts with the object of effecting alcoholic fermentation and malolactic fermentation with a single microorganism (yeast), a modification that would yield obvious technological advantages; (iii) obtaining modified strains for the production of the aroma components (esters, fatty acids, and alcohols); and (iv) obtaining strains that do not produce H_2S, a component responsible for organoleptic alterations in the wine.

Achievements are still modest in the field of genetic improvement of enological strains because of various factors: (i) the complex genetic background of industrial strains (ploidy, homothallism, poor sporulation, etc.); (ii) the impossibility of describing the genetic basis of most enological characteristics and, thus, of elaborating simple screening tests for them; and (iii) the presence of indigenous microflora in the must that, unlike other substrata, cannot be subjected to sterilization without altering the wine characteristics.

Optimization of New Process

Another line of research that has developed in the field of enology is the search for and optimization of new processes that are able to meet the technological and organizational needs of the industry (greater productivity, reduction in the time needed for the process, reduction in capital and running costs). In this field, we can mention the use of yeasts immobilized in natural nontoxic polymers or contained in a microporous membrane and of the cell-recycle batch fermentation process (CRBF) for the preparation of white wines and sparkling wines. The advantages that are sought are summarized in this section.

Champenois method

It would be desirable to eliminate the phase of remuage, with a considerable savings in labor and time, thanks to instant sedimentation of the immobilized yeasts by gravity. Sparkling wine produced by the champenois method is characterized by fermentation of wine in the bottle, to which sugars, yeasts and other fermenting agents have been added. The bottles thus prepared are stacked horizontally in thermoregulated environment (12 to 18°C) where the secondary fermentation is carried out. Once fermentation has terminated and the wine has been in contact with the yeast for at least a year, the bottles are turned frequently and inclined (an operation referred to as remuage), so that the yeast sediment slides toward the bottle neck and settles on the stopper. The sediment is then eliminated by degorgement. Remuage is both long and arduous in that it requires a large amount of labor over a long period of time (1 to 4 months) and the occupation of large wine cellars. By using immobilized yeasts, it is possible to move directly from the stacks to degorgement with a saving in production costs of about 80% when compared with the traditional method.

Charmat process

The possibility of using a high cellular concentration and thus of obtaining a very high rate of reaction so as to achieve a continuous process is another desiderable innovation. This would allow a reduction in the dimension and number of pressure tanks, with a diminution in the costs of running and equipment. In the bulk process for sparkling wine production, the Charmat method, sugars, other nutrients, and yeasts are added to the base wine, and the second fermentation occurs in large stainless steel sealed tanks. The sparkling wine is then filtered, fined, and bottled. This process is the cheapest for producing sparkling wines, but it still requires a considerable number of expensive tanks.

Improvement in wine production

The possibility of improving the rate and efficiency of transformation by increasing the concentration of yeast would be an advantage in general wine production. In this way, a linear development of the fermentative process is achieved, excluding the yeast induction phase and exponential growth. This reduces the time required by the fermentative process and the final level of the secondary products of fermentation (acetaldehyde, acetic acid, products of catabolism of the amino acids, etc.) that accumulate in the first phase of fermentation when the biomass is formed. The use of yeasts entrapped in calcium alginate and in other materials has been investigated by Cantarelli; Duteurtre et al.; Lemonnier and Duteurtre; Loureiro; Zamorani et al.; and researchers at our Institute. This research has made it possible to observe the following:

1. Of the various compounds used for entrapment—agar, agarose, calcium alginate, barium alginate, κ-carrageenan, gelatin, pumice stone, porous porcelain, controlled porous glass, and dyalysis membranes—agar and calcium alginate gel are the most suitable for fermentation.
2. The retention power of the immobilizing matrix is inversely linked to the fermentative capacity of the yeast and to the alginate/yeast ratio.
3. The presence of structural constituents of alginate (uronic acids) in the wine may be ignored up to bead content equal to 14 g/bottle.
4. The phosphate and tartaric, malic, and citric acids present in wine do not have a disintegrating effect on the structure of the alginate.

5. The enrichment of cells that proliferate in nucleic acids, reserve polysaccharides (trehalose and glycogen), and structural polysaccharides (mannan and glucan) is not reflected in a variation in the content of the wine.
6. When the concentrations of alginate and cells in beads are equal, then the smaller the inoculum at draught, the greater the cellular release; however, this release can be avoided by using nonproliferating yeasts.

Table 6.5 Analytical characteristics of sparkling wine produced by free yeast cells and yeast cells immobilized in 3% alginate

	Free yeast	*Immobilized yeast*
Ethanol (% vol)	12.29	12.29
Sugar (g/L)	2.3	1.7
Ashes (g/L)	1.48	1.48
Ashes alkaline (mequiv/L)	11.0	11.0
pH	3.25	3.21
Titratable acidity (g/L)	6.79	6.93
Volatile acidity (g/L)	0.36	0.34
Tartaric acid (g/L)	1.73	1.82
Malic acid (g/L)	1.47	1.47
Lactic acid (g/L)	2.10	2.06
Citric acid (g/L)	0.20	0.24
Succinic acid (g/L)	0.40	0.44
Pyruvic acid (mg/L)	47.5	49.7
α-Ketoglutaric acid (mg/L)	58	63
Acetaldehyde (mg/L)	71	68
Glycerol (g/L)	5.90	5.60
Color (410 nm)	0.073	0.072
Gauge pressure CO_2 at 20°C (atm)	8.0	8.3
Calcium (mg/L)	66.5	65.5

As for the evaluation of the influence of the alginate immobilization technique on the fermentative process, the composition, and the organoleptic characteristics of the bottled sparkling wine, Fumi et al. observed that the use of immobilized cells does not cause any significant change in the second fermentation process.

The main components (ethanol, organic acids, total nitrogen, and higher alcohols) are not appreciably different in sparkling wines obtained by immobilized yeast when compared with traditional sparkling wines, but some differences are found as regards certain amino acids and some aroma components. As shown by tasting, however, these differences do not affect the organoleptic characteristics. Sensory evaluation of sparkling wines obtained with immobilized yeasts versus a control was performed using a duo-trio test and black glasses.

Table 6.6 Free amino acids and aroma components of sparkling wines produced by free and immobilized yeast in 3% alginate

	Free yeast	*Immobilized yeast*
Amino Acids		
Histidine (mg/L)	38.60	36.80
Asparagine (mg/L)	167.76	253.62
Arginine (mg/L)	542.11	612.24
Serine (mg/L)	67.83	65.88
Aspartic + glutamic acid (mg/L)	152.39	147.50
Glycine + threonine (mg/L)	88.17	84.16
Alanine (mg/L)	362.0	359.5
γ-Aminobutyric acid (mg/L)	151.2	145.8
Proline (mg/L)	227.03	220.97
Valine (mg/L)	22.75	23.08
Methionine (mg/L)	trace	trace
Tryptophan (mg/L)	trace	trace
Phenylalanine (mg/L)	17.70	18.38
Isoleucine (mg/L)	31.52	31.67
Ornithine (mg/L)	18.58	19.54
Lysine (mg/L)	12.53	14.26
Tyrosine (mg/L)	trace	trace
Total nitrogen (mg/L)	534	528
Esters		
Ethyl butyrate (mg/L)	0.38	0.27
Ethyl caproate (mg/L)	0.53	0.52
Ethyl pyruvate (mg/L)	0.56	0.43
Ethyl lactate (mg/L)	143	113
Diethyl succinate (mg/L)	0.63	0.60
Isobutyl acetate (mg/L)	0.13	0.08
Isoamyl acetate (mg/L)	1.15	0.82
Alcohols		
Methyl alcohol (mL/100 mL of EtOH)	0.03	0.03
1-Propanol(mg/L)	29	31

	Free yeast	Immobilized yeast
1-Butanol (mg/L)	1.62	1.23
2-Methyl-1 -propanol (mg/L)	55	52
2-Methyl-1-butanol(mg/L)	24	19
3-Methyl-1-butanol(mg/L)	100	94
Hexanol (mg/L)	1.46	1.33
Benzyl alcohol (mg/L)	trace	trace
β-Phenylethanol (mg/L)	17.1	16.7
Acids		
3-Methylbutanois acid (mg/L)	0.55	0.57
Caproic acid (mg/L)	3.70	0.57
Caprylic acid (mg/L)	5.63	4.98

Divies and Duteurtre conducted comparative tests with free and calcium alginate immobilized yeasts for bottled sparkling wine production, and they also noted that there are no variations in the principal chemical or physical characteristics. Experiments to avoid cell release were performed by coating alginate/yeast beads with a protective cell-free alginate gel layer. These studies were carried out by Duteurtre, Fumi et al., and Loureiro for bottled sparkling wine production. A complete bottle fermentation was achieved without cell release, independent of the different conditions adopted—concentration of the alginate, time of gelification, and $CaCl_2$ concentration. However, Ors et al. noted that the double layer must have a minimum critical thickness, equal to 0.2 mm, in order to avoid cellular release. In an assay where agar cylinders covered with a protective layer of agar without cells were used, equally satisfactory results were not achieved. Although the problem of cell release from the immobilization matrix during fermentation may be solved,

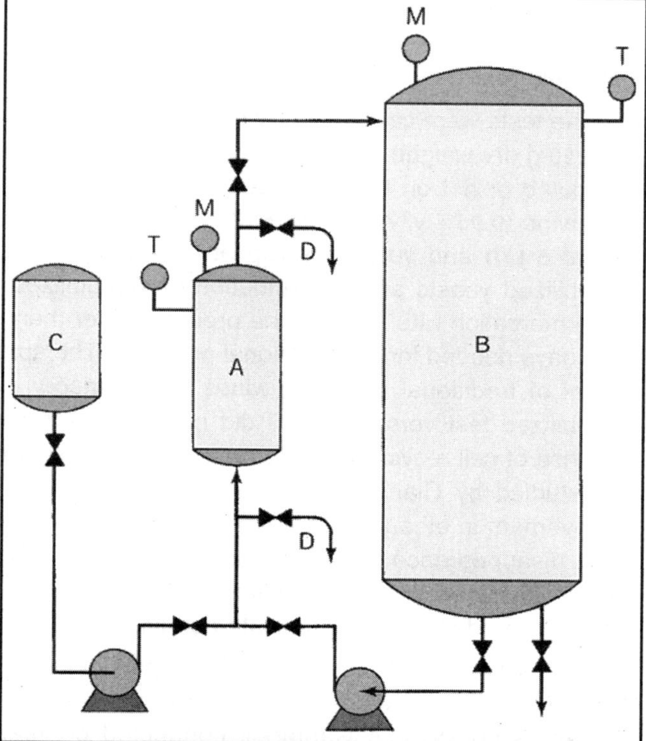

Figure 6.4 Schematic of the system used in the production of sparkling wine in tanks with immobilized yeast; A, bioreactor with immobilized yeast; B, sealed tank; C, syrup tank; D, sampling taps; M, manometers; T, thermometers.

there are other difficulties to overcome for the industrial application of calcium alginate immobilization. The possibility of supplying wineries with ready-to-use immobilized cells would be a significant step for the implementation of this technique. For this, Loureiro has performed freeze-drying experiments of cells immobilized by gel occlusion.

These tests with different polysaccharide gels showed that it was possible to freeze-dry without changing bead shape using either κ-carrageenan or κ-carrageenan plus calcium alginate. However, cell release always occurred, and further studies will be necessary to optimize the thickness and rigidity of the protective bead layer for controlling cell release. Lemmonier and Duteurtre developed a device consisting of a small cartridge containing two microporous membranes, one hydrophilic and the other hydrophobic, which is filled with yeast and placed in the neck of the bottle. The two membranes hold the yeasts captive, enabling a free and total exchange between the yeasts and the wine. When the cap is opened, the cartridge is automatically ejected. The first results showed that production of sparkling wine with immobilized yeasts occurs in a manner similar to that using traditional methods with free cells. Some differences have been noted regarding protein and polysaccharides. The use of a continuous fermentative system with a calcium alginate immobilized yeast bioreactor was tested by Fumi et al. for the production of sparkling wines in tanks. The general idea was to run the second fermentation outside the principal stainless steel tank using equipment that could easily be moved from one tank to another.

The tests were carried out by placing 2 L of 1.5% alginate and 8% yeast beads (yeast inoculum was 250 g dry weight) or 3 L of 1.5% alginate and 10% yeast (yeast inoculum was 500 g dry weight) distributed or not on five distanced supports of pierced stainless steel, in a 10-L bioreactor. The base wine (9.95% v/v ethanol, 5.61 g/L titratable acidity, pH 3.04) was continuously supplied at a rate of 5 L/h and sucrose syrup at 0.14 L/h. The results showed that a high concentration of immobilized yeasts and a distribution of immobilized yeast mass in the bioreactor greatly affect the fermentation rate, making the process faster than the traditional one (78 h, as compared to 10 to 15 days needed for the traditional process). The sparkling wines have a composition comparable to that of traditional sparkling wines, and sensory evaluation (expert panel of eight tasters) of immobilized test versus control did not show significant differences at a level of $p < 0.05$. The influence of cell-recycle batch fermentation process on sparkling wines produced in pressure tanks was studied by Ciani et al. They noted a reduction in the time of fermentation as well as an improvement in ethanol productivity and yield. This was linked to the increased inoculum and also to the disappearance of the lag phase, which lasted approximately 150 h in batch. The increase in ethanol yield is probably correlated with a resting cell state highlighted by a lower percentage of new cells produced in steady-state conditions.

The analytical characteristics of the sparkling wine obtained were not significantly different from those of sparkling wine produced by traditional batch fermentation. As far as the fermentative process of must is concerned, Cantarelli conducted comparative tests with free and calcium alginate immobilized yeasts. A fluidized or packed-bed reactor consisting of a 200-L tubular fermenter with immobilized yeast beads at various concentrations (maximum 8 g/L) was used for wine production. The results showed that immobilized yeast, compared with free yeast, had almost linear fermentation kinetics without a lag or exponential phase. In addition, organoleptic and chemiometric evaluation of analytical data both showed that there were no significant differences in by-products. Therefore, the final by-product content of wine produced by a fluid-bed fermenter was lower than by a packed one. Batch fermentation tests on Riesling grape juice were carried out by Fumi et al. using two 50-L working volume fermenters with calcium alginate immobilized biocatalysts (beads with alginate/ yeast ratio of 1.5 wt/dry wt, reintroduced into 1% alginate and supported by stainless steel pierced plates vertically disposed) with surface area to volume ratios 1.33 and 0.66, respectively. The immobilized-system kinetic parameters had similar values, and they were not influenced by the different surface area to volume ratios of the biocatalysts. Analytical data showed no detectable differences between the wines obtained by the immobilized system and wines produced traditionally. During the first phases of fermentation with immobilized cells, the must was kept perfectly clean in both bioreactors. Nevertheless, at the end of the fermentation process, a free-cell deposit, caused by yeast leakage from the beads, was found in the bioreactors.

Table 6.7 Chemical composition of sparkling wines obtained with Charmat methods by free and immobilized yeasts

	Free yeast	*Immobilized yeast*
Ethanol (v/v %)	10.60	10.62
Sugar (g/L)	4.5	4.7
pH	3.07	3.07
Total acidity (g/L)	5.70	5.78
Volatile acidity (g/L)	0.25	0.28
Tartaric acid (g/L)	2.37	2.28
Malic acid (g/L)	0.58	0.53
Lactic acid (g/L)	1.81	1.85
Succinic acid (g/L)	0.41	0.47
Glycerol (g/L)	4.79	5.08
Acetaldehyde (mg/L)	49	53
Totalnitrogen (mg/L)	205	199
Calcium (mg/L)	81	87
n-Propanol (mg/L)	30	30
Isobutyl alcohol (mg/L)	55	62
Amyl alcohols (mg/L)	125	125
Higher alcohols (mg/L)	210	217
OD_{660nm}	>1	0.003

Malolactic Fermentation

Malolactic fermentation is an important factor for many red wines or at least for those high acid wines produced from grapes grown in cool climate. This process is effected by a strain of lactic bacteria of the genera *Leuconostoc, Lactobacillus,* and *Pediococcus* and essentially consists of the degradation of malic acid into lactic acid and carbon dioxide, the consequence of which is a reduction in total acidity (deacidification). In addition, the transformation of malic acid is always accompanied by the fermentation of sugars, which although on a modest scale has the result of increasing aroma components such as acetaldehyde, acetic acid, acetoin, diacetyl, 2,3-butandiol, ethyl lactate, and high alcohols. As a whole, the wine gains in mellowness, roundness, and fullness and becomes more pleasing to the palate. The difficulties that hinder growth of lactic bacteria in wine have causes known only in part (pH, ethanol concentration, SO_2 concentration, fermentation temperature, competition between lactic bacteria and other microorganisms, lack of nutrients). Thus, spontaneous biological deacidification of wines is not always possible.

To help in the deacidification of wines, it is a common practice to make use of an inoculum of active indigenous bacteria effected through the addition of wines where malolactic fermentation is in process or has just finished. However, this system does not always give good results and is unsuitable for those wines that must maintain their peculiar characteristics unaltered. For quite some time, attempts have been made to induce malolactic fermentation by inoculating the wine with selected lactic acid bacteria. For this purpose, lactic bacteria have been isolated from musts and wines and have been selected for their tolerance of ethanol (13 to 14 vol.%) and sulphur dioxide, their capacity to develop in the pH of wine, their capacity to produce small quantities of acetic acid, and their capacity to not alter organoleptic characteristics. Dried or frozen cultures of selected bacteria suitable for malolactic fermentation have been on the market for some years. Research carried out thus far on the efficacy of these preparations has shown that malic acid is broken down completely if the strain is previously reactivated in suitable substrata.

Lonvaud-Funel observed that complete breakdown of malic acid is achieved by using an inoculum containing up to 10^7 cells/mL that has been allowed to develop previously for 24 h in either a synthetic substrate (40 g/L glucose, 3 g/L malic acid, 5 g/L of neopeptone, 50 mg/L of Mg^{2+}, 20 mg/L of Mn^{2+}, 50 g/L of yeast extract, pH 4.5) or in a grape juice base substrate. Some investigators have suggested the use of immobilized lactic bacteria or enzymes to convert malic acid into lactic acid. Spettoli et al. immobilized *Leuconostoc oenos* ML34 in 1.67% calcium alginate gels and tested their ability to convert L-malic acid into L-lactic in a red wine having 10.20% v/v alcohol; 7.90 g/L of titratable acidity; pH 3.15; 21 mg/L of total SO_2, and 1.50 g/L of L-malic acid. The cells, in batch reaction, maintained 56% conversion of L-malic acid after 16 h. Cuenat and Villettaz immobilized a strain of *Leuconostoc oenos* on alginate (1 vol of 15% dry weight cells in 5 to 12 vol of alginate, 5% (w/v)). A complete degradation of malic acid (from 6.1 to 0.07 g/L) was obtained in a continuous system where 25 L of wines (pH 3.3; 11.3% (v/v) ethanol, 139 mg/L total SO_2) was passed through the columns in 5.5 h. Naouri et al. compared the malic-acid-degrading ability of a strain of *Leuconostoc oenos* with that of a strain of *Lactobacillus* sp. immobilized in alginate in a high compacting multiphase reactor (HCMR) containing 170 g of alginate beads and 1.41 L of red wines.

After bioconversion in the reactor, malic acid was almost completely (82%) degraded by the immobilized *Leuconostoc* strain, whereas incomplete (45%) degradation was obtained with the *Lactobacillus* strain. Crapisi et al. reported on the efficiency of a continuous-flow bioreactor filled with *Lactobacillus brevis* immobilized on four types of alginate and κ-carrageenan for the control of malolactic fermentation in wine. Bioreactor efficiency seems to be dependent on gel properties ranging from 34 to 45%, depending on the immobilization matrix used. Attempts to deacidify wine using malolactic enzyme immobilized on gels have not been successful probably because of the inactivity of the enzyme at wine pH and because the required cofactor, NAD, is particularly unstable in wine.

FACTORS AFFECTING THE PRODUCTION

Factors limiting the application of standardized biotechnological processes to wine production include the following:

1. The seasonality and high variability in grape composition
2. The limited annual season of vintage
3. The length of the winemaking process for some wines, which delays experimental results
4. The variations in the amount of wine production (ranging from a few hundred hectolitres to more than 200.000)

In fact, under the generic term wine there is a diversity and hierarchy of quality that is quite unique among the products of the soil and that is connected to the cultivar, the area of cultivation, and the transformation technologies applied to the grapes. The latter includes the transformation of the must by yeast fermentation, which is certainly the oldest of biotechnologies. The principles of winemaking have been established empirically over the course of several centuries of rigorous and methodical observation. The production of quality wines came before all scientific knowledge. However, although its origins were empirical, the production of quality wine has, in the last quarter of a century, benefited most from research developments. In this area, biotechnological knowledge may well become more and more important, not only as regards the microorganisms involved in the fermentation process but also because of the use of enzymes in pressing grapes, prefermentation treatments, and refinement of the wines. Since the development of the use of pectolytic enzymes, the tendency has been to search for enzymic preparations that have precise effects and that allow us to reduce the use of chemical additives.

In this field, studies have made it possible to show that there are numerous limitations to the use of enzymes in the field of enology. For example, experiments with enzymes to improve the stabilization of white wines and to increase the aroma of some wines have demonstrated that the composition of the medium and the working conditions are not very favorable to the action of enzymes such as proteases and glycosidases. At present, the use of extracellular yeast enzymes, such as proteases, able to act on the constituents of the must are being studied. This is obviously of technological interest, because it would make it possible to take advantage of the enzyme activity of the yeast, not only for fermentation but also for stabilization of wines. In addition, the energy liberated during the fermentation process would give temperatures better suited to the activity of the enzymes. These days a selected yeast is required not only to ensure that undesirable characteristics do not arise but also in order to have a positive effect on the characteristics of the wine. Methods of genetic improvement offer a means of programming and constructing new strains of yeasts. However, certain problems remain when one attempts genetic improvement of the strains, not the last among which is the lack of objective methods of evaluation of the quality of the yeast.

The contribution of yeast to the quality of wine is complex because it is the result of interactions between grapes, yeast strains, and technology. In spite of this, the application of recombinant DNA technologies to the improvement of wine strains has undergone rapid development, and although practical applications have not yet been achieved, the basis for future advances has been established. In the field of bottled sparkling wine production, studies have led to the creation of technologies that permit the use of alginate-immobilized yeasts at a semi-industrial level. However, the technology of entrapped yeast for must fermentation requires further optimization of immobilized systems, in the areas of mass transfer, reactor design, and yeast physiology. The most logical strategy for providing the means of increasing productivity without changing the flavor characteristics of fermented products must be found. Reactors with immobilized lactic acid bacteria have also been shown to be potentially very interesting, but putting them into operation on an industrial scale still requires critical evaluation, not only of the efficiency of the breakdown of malic acid but also of the results on the organoleptic characteristics of the wines. Every biotechnological advance that is to be accepted in winemaking must allow one to maintain the quality of the product, must take into account the hygienic and legal aspects as well, and must not impair the flavor and bouquet of the final product.

7

MICROBIAL PRODUCTION OF CITRIC ACID

Citric acid ($C_6H_8O_7$) is a white or translucent solid with a molecular weight of 192.12. It occurs as a natural constituent in citron, lemon, lime, pineapple, pear, peach, and similar fruits. It is also found in animal tissues. The popular forms of citric acid are the anhydrous forms, the monohydrate and sodium salts of the acid. Citric acid was first isolated by Scheele in 1784 when he crystallized it from lemon juice. It was later synthesized (via symmetrical dicholoroacetone) from glycerol by Grimocex and Adam in 1880. It was first made in crystallized form from lemon juice. Calcium citrate was precipitated through the reaction of hot lemon juice with calcium carbonate, followed by decomposition of the product with sulphuric acid. Two hundred years later, this process is still being used. However, raw material is the principal limiting factor for production on a large scale. Thirty tons of lemon are needed to produce 1 ton of citric acid. The first commercial citric acid was prepared in 1860 in England from calcium citrate imported from Italy and Sicily. By 1880, France, Germany, and the United States had begun manufacturing citric acid using similar methods. In 1913, recovery of citric acid from calcium citrate began in Italy, and by 1922, Italy produced 90% of the world's supply.

COMMERCIAL PROCESS

In 1893, Wehmer recognized citric acid as a microbial metabolite. Although his attempt to develop a commercial fermentation process with a species of *Penicillium* was unsuccessful, he established the basis for the subsequent complete reorganization of the citric acid industry. It was later confirmed that the use of *Aspergillus niger* favors sucrose fermentation to citric acid. In 1923, a plant was started in New York to make citric acid by a fungal fermentation technique developed by Currie. This production effectively broke the power of the Italian cartel. In this process, *A. niger* was grown on the surface of a shallow pan of sugar medium. This surface fermentation process was subsequently used in England, The Netherlands, Belgium, Germany, Switzerland, Argentina, and the former Soviet Union. A new fungal fermentation process was introduced successfully in the United States in 1952. In this process, *A. niger* was allowed to grow in the entire volume of the culture solution in large deep tanks. This process was introduced in Mexico in 1959 and in Israel in 1961.

Table 7.1 Citric acid contents of various fruits and vegetables

Fruit/vegetable	*Citric acid (wt %)*
Lemon	4.0–8.0
Grape	1.2–2.1
Tangerine	0.9–1.2
Orange	0.6–1.0
Currant	
Black	1.5–3.0
Red	0.7–1.3
Raspberry	1.0–1.3
Strawberry	0.6–0.8
Apple	0.008
Tomato	0.25
Potato	0.35–0.5
Asparagus	0.08–0.2
Turnip	0.05–1.1
Pea	0.05
Corn	0.02
Lettuce	0.16
Eggplant	0.01

The earliest attempt to produce citric acid by the submerged growth technique was that by Amelung, who aerated the culture by bubbling air through the solution. Numerous studies, subsequent to the work of Amelung, resulted in the development of two approaches to the control of citric acid production in submerged cultures. In the first approach, Perquin produced citric acid by inducing the establishment of a deficiency of one of the major nutrients, such as phosphate. In another approach, Johnson et al. studied the effect of manganese and iron and related the concentration of these to the amount of inhibition of citric acid accumulation. Further development of methods for the control of iron and manganese levels in the culture solution led to the popularity of the submerged fermentation process, now used in the United States, Mexico, and Israel. Excess iron may be precipitated by ferrocyanide.

The removal of iron and the presence of specific enzyme inhibitors that control the destruction of citric acid form the basis for the successful submerged fermentation process. Other mineral nutrients required for the submerged fermentation are the same as those for surface fermentation.

The most interesting change in citric acid manufacture was done by Miles Laboratories. They changed the raw materials from molasses to glucose, doubling the production. The company patented a process in which starchy material is first converted to sugar by enzymatic action and then used as the raw material for citric acid production. They also claim that with the addition of various phenols and other compounds to check the growth of *A. niger,* it is possible to use less highly purified starting material. This is similar to the use of the lower alcohols, which has been known for many years and is being investigated now as an additive to blackstrap molasses.

Table 7.2 Chronology of events in citric acid production

Authors	*Investigation/remarks*
Wehmer 1893	Demonstrated citric acid to be a metabolic product of certain molds.
Currie 1917	First chose *Aspergillus niger* for deriving citric acid.
Doelger and Prescott 1934	Fermentation above 30°C decreased citric acid yield and increased accumulation of oxalic acid.
Cahn 1935	First to describe solid-state fermentation for production of citric acid.
Mazzadroli 1938	Potassium ferrocyanide could be added to eliminate excess Fe^{2+}.
Szucs 1944	Citric acid synthesis took place only after total assimilation of phosphates.
Moyer 1953	Suggested use of methanol as stimulant for increasing acid yield.
Usami and Takatomi 1958	Spores produced under submerged fermentation were poor, and spores from surface culture were better acid formers.
Usami et al. 1960	Increase in aeration resulted in higher yields and reduction in cycle time.
Kovats 1960	Demonstrated lower concentration of sugar leads to lower yields of citric acid and accumulation of oxalic acid.
Schweiger 1961	Iron concentration above 0.2 ppm affected yield of acid. However, addition of copper 0.1 to 500 ppm at the time of inoculation or during the first 50 h of fermentation countered the deleterious effect of iron.
Naguchi and Bando 1960	Concentration of ammonium nitrate greater than 0.25% lead to accumulation of oxalic acid.
Sanchez-Marroquin et al. 1963	Higher yields of citric acid occurred in a simple medium rather than a complex medium.
Clark 1962	Ferrocyanide addition on the acid yield 10–200 µg/mL was tolerated during growth. But less than 20 µg/mL helped acid production.
Millis et al. 1963	Addition of vegetable oils, fatty acids, etc., increased acid yield.
Czech. Acad. Sci. 1964	Mycelial digests from *A. terreus* or *A. niger* stimulated acid production by 60–70% when added to the medium.
Leopold 1965	Addition of pressed baker's yeasts to culture medium increased acid yield from 61.5 to 72.1%.

Authors	Investigation/remarks
Bruchmann 1966	Addition of mild oxidizing agents stimulated acid production.
Sussman and Halvorson 1966	Spore viability varies with age.
Wendel 1967	In the initial stages, citric acid diffused into the medium, which leads to a stratification on the mycelium. This stratification inhibited fungal metabolism, leading to decreased acid yields. This stratification can be eliminated by stirring.
Tabuchi et al. 1969	Production of citric acid from *n*-paraffin using yeast, (*Candida*) medium composition. Hydrocarbon, 40-60 gs; NH_4Cl_2, 2 gs; KH_2PO_4, 0.5 g; $MgSO_4$, 0.5 g; cornsteep liquor, 1 g; $CaCo_3$, 30 g/L.
Khan et al. 1970	Phosphates promoted more growth at less acid yields.
Fedoseev et al. 1970	Copper sulfate at 4.7 mg/100 g molasses resulted in better conversion of sugar to citric acid.
Kyowa Fermentation Industry 1970	Citric acid production from dodecane (or) C_{12}-C_{14}. Using arthrobacterium in aqueous medium, inoculum 5% air at 3 vvm at 28°C, yielded 28 mg/mL.
Sanchez-Marroquin et al. 1970	The use of ion-exchange resin for reduction of metal content was better than chemical treatment.
Fukuda et al. 1970	Use of *Corynebacterium* on *n*-paraffins; yield 41.4 mg/mL; culture period 64 h, 32°C.
Leopold 1971	Glycerol addition increased acid yield by 30%.
Dhankar et al. 1972	Acid yield improved by addition of peanut oil to molasses medium.
Kumamoto and Okamura 1972	Mn^{2+}, Ba^{2+}, Al^{3+} had an effect on fungal morphology.
Sardinas 1972	Patented a process for citric acid production involving *B. licheniformis*.
Chaudhary et al. 1972	A low pH in molasses medium was inhibitory to growth.
Ohmori and Ikeno 1973	Bacterial production of citric acid from media containing isocitric acid.
Dhankar et al. 1974	Sodium nitrate at a concentration of 0.4% was superior to ammonium nitrate.
Zhuravskii, 1974	Patented continuous multistage process for citric acid production.
Halama 1974 (34)	Chemicals such as pentachloralphenolate, semicarbozone, tetracycline, etc., used for control of bacterial growth.
Ohtsuka et al. 1975	Addition of malic hydrazide increased yield from 30 to 70%.
Wold and Suzuki 1973	The rate of acid production increased when AMP was added to the culture medium.
Wold and Suzuki 1976	Zn^{2+} regulated growth and production.
Brezhnoi et al. 1976	Continuous production of citric acid from molasses.
Choudhary 1978	A 3-day-old slant culture was as good as 7-to 8-day-old spore.
Krishnan and Natarajan 1987	Liquid–liquid extraction of citric acid.
Krishnan and Annadurai 1992	Precipitation of citric acid.
Krishnan et al. 1996	Reduction in cycle time for citric acid production.

ANHYDROUS FORM OF CITRIC ACID

The popular forms of citric acid are the anhydrous forms, the monohydrate form and sodium salts of the acid. Both citric acid and citric monohydrate are available in variety of sieve sizes. They are conventionally available as granular, fine granular, and powder forms. Crystalline anhydrous citric acid can be stored in dry form without difficulty, although highly humid conditions and elevated temperatures should be avoided to prevent caking. The product should be stored in tight containers to prevent exposure to moist air. Several granulations are commercially available with larger particle sizes having less tendency toward caking. Materials packed with desiccants are also available. Solutions of citric acid are corrosive to normal concrete, aluminium, carbon steel, copper alloys and should not be used with nylon, polycarbonates, polyamides, polyimides, or acrylics. Recommended materials of construction for pipes, tanks, and pumps handling citric acid solutions are 316 stainless steel, fiberglass- reinforced polyester, polyethylene, polypropylene, and polyvinyl chloride. Sodium and potassium citrates are not as corrosive as citric acid but they should be handled in the same type of equipment as citric acid.

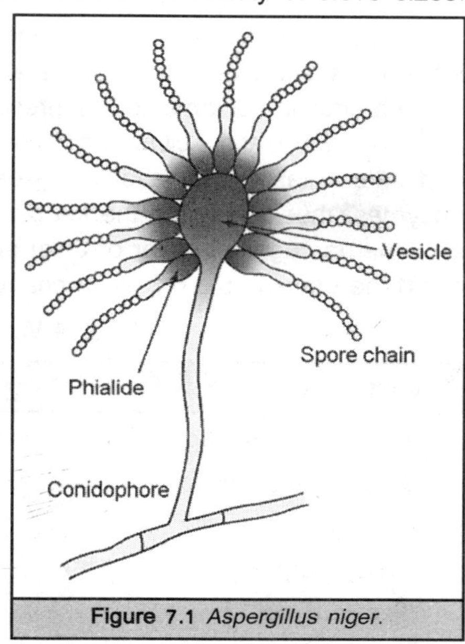

Figure 7.1 *Aspergillus niger.*

Table 7.3 Global production pattern of citric acid

Place	*Production (%)*
Western Europe	41
North America	28
South and Central America	11
Far East, Australia, and New Zealand	11
Other	9

PHARMACEUTICAL USES

The food and pharmaceutical industries use citric acid extensively because of its high solubility, pleasant sour taste, very low toxicity, and ready assimilability. Citric acid also finds application in some cosmetic preparations, metal cleaning, electropickling, copper plating, secondary oil recovery, and other industrial uses.

Application of Citric Acid

Citric acid is widely used in foods, beverages, and jams and jellies owing to its ability

1. To maintain desired pH levels

2. To impart refreshing and tingling taste
3. To act as a flavor-enhancing agent
4. To act as a color stabilizer

In candy manufacture, citric acid helps to enhance the flavor and taste of berries and other ingredients. As a very dependable agent for maintenance of pH, citric acid finds wide application in the production of jams, jellies, preserves, soft drinks, syrup, and soft drink tablets. For pH control and color stabilization, citric acid is used in fruit and vegetable juices. The pH prevents juice spoilage, and natural taste and flavor are greatly enhanced. Lowering of pH inactivates certain oxidative enzymes, thus improving the keeping qualities of frozen fruits, peaches, apricots, plums, pears, and cherries against flavor or color spoilage. Citric acid also combines with trace metals, helping avoid the undesirable oxidation changes.

Table 7.4 World consumption pattern of citric acid

Uses	*Consumption (%)*
Food and beverages	60
Pharmaceuticals	12
Detergents	12
Metal cleaning	6
Textile dyes	5
Cosmetics	3
Others	2

Enhancing Taste and Flavor

Citric acid's wide application in the manufacture of drugs and pharmaceuticals is for enhancing the taste and flavor. It is used as an anticoagulant for blood. Powders and tablets owe their effervescent property to citric acid. The free acid and its sodium and potassium salts are used as mild acidulants in astringent preparations. Several salts of citric acid are used in the manufacture of antianemic tonics, vitamins, liver tonics, antipyretic agents, cough syrups, antidysentery, and antidiarrheal agents. It is also used in inducing fertility.

Citric Acid in Cosmetics

Hair rinses and hair setting agents derive benefit from citric acid. It is also used in astringent lotions, bleaching lotions, and others.

Metallurgical Application

Citric acid finds extensive use in metallurgical application and as a sequestering agent for such metals as iron, copper, zinc, nickel, cobalt, chromium, and manganese. The acid and ammonium salts are added for scale removal in boilers; cleaning of reactors could be done very well using the acid. Citric acid is widely used in electropickling of copper and its alloys and in copper coating.

Any iron plugging in the oil-flow line could be removed by the acid. Tanning liquors, bottle washing compounds, and diazo printing paper all contain citric acid.

Industrial Uses

It is used in the manufacture of several esters, linoleum, inks, silvering compounds, and fabric dyes. Citric acid enjoys a superior status above phosphates in the manufacturing of detergents.

Other Applications

Some of the algicides, pesticides, fish preservation, chemicals, and insecticides are based on citric acid. Preservation latex from certain natural plants involves the use of citric acid.

MICROORGANISMS IN PRODUCTION

Citric acid production centers around one strain, *Aspergillus niger,* which is still extensively used. In the late 1970s, processes involving *Candida* yeast became commercial. *Candida guillermondi* has unique benefits over *A. niger* in that trace metal removal is not essential. It can also tolerate higher pHs (3.5 to 5) and sugar concentrations for acid production. The rate of acid production is also faster. As in the case of *A. niger,* nitrogen limitation triggers acid production. Certain bacteria also produce citric acid on a variety of substrates.

Citric acid can be manufactured commercially by three methods:

1. Surface culture technique
2. Submerged culture technique
3. Solid substrate technique

Commercially used substrates are citrus fruits (not adapted presently because the fruits are costly and seasonal), sucrose, glucose, molasses cane juice, and certain petroleum fractions. Shallow pan reactors, stirred tank fermenters, and airlift fermenters are commonly used depending on the process requirements. Fermenters work in a semibatch mode or a continuous mode. It will be of interest to state here that yields of product vary with the pretreatment to which the media are subjected. For example, yields for pretreatment with suitable exchange resins is 98% in the case of sucrose and 75% for molasses; ferrocyanide-treated molasses yields 68% as contrasted to the untreated molasses, for which the yield is 62%.

Citric Acid from Citrus Fruits and Pineapples

Lemon peels and the white layer are removed and separated for the recovery of oil and pectin, respectively. The remaining portion is pulped and filtered. The filtrate is likely to contain pectin and albumin, which are removed by spontaneous fermentation. In the case of pineapples, the waste and low quality fruits are crushed and filtered. The lemon juice and the pineapple juice, recovered separately, could contain 4% and 0.75% citric acid, respectively. Calcium carbonate or hydroxide is added to the individual juice to the required amount, and the temperature is maintained between 55 and 95°C. Any oxalate present is filtered preferentially before citric acid. The calcium citrate cake obtained is washed and reacted with a small amount of sulfuric acid. The resultant slurry is filtered to recover citric acid solution from the precipitated calcium sulfate. The clear filtrate is subjected to decolorization before recovering the citric acid crystals by evaporation. This process is on the decline, as stated earlier.

Citric Acid Production from Cane and Beet Molasses

Blackstrap molasses, about 40°Be with a sugar content of 52 to 57% sugar, or beet molasses, 41°Be containing 48 to 52% sugar, could be used for citric acid production. However, cane molasses should be purified before subjecting it to fermentation. Alternate methods for molasses purification are discussed below:

1. *Sulfuric acid treatment.* The pH of the molasses (10% total reducing sugar) is adjusted to 3.0 by adding 0.1 N sulphuric acid. This is allowed to stand for 1.5 h and then centrifuged at 3,000 rpm for 15 min. The supernatant is collected and used.
2. *Potassium ferrocyanide treatment.* The molasses (10% TRS) is heated to 85°C for 30 minutes and centrifuged at 3,000 rpm for 15 min; 100 mL of the supernatant is collected, and 0.5 mL of $K_4\,Fe(CN)_6$ (10% solution) is added. The pH is adjusted to 6.5.
3. *Tricalcium phosphate treatment.* The pH of the molasses (10% TRS) was adjusted to 7.0 by the addition of 0.1 N NaOH and treated with 2% (w/v) tricalcium phosphate followed by heating at 105°C for 5 minutes. The mixture was cooled and centrifuged at 3,000 rpm for 15 minutes. The supernatant was used.
4. *Tricalcium phosphate with hydrochloric acid treatment.* The TCP-treated liquor was adjusted to pH 2.0 by the addition of 0.1 N Hcl followed by vigorous shaking. The mixture was allowed to stand for 6 h.

   The supernatant was used after centrifugation at 3,000 rpm for 20 min.
5. *Bentonite treatment.* The pH of molasses (10% TRS) was adjusted to 7.0. Bentonite 2% (w/v) was added and kept in a boiling water bath for 30 min. The solution was then centrifuged at 3,000 rpm for 15 min. The supernatant was collected and used.

Table 7.5 Analyses of cane and sugar beet molasses

Component	*Beet (%)*	*Cane (%)*
Water	20.0	16.5
Sugar	62.0	53.0
Nonsugars	10.0	19.0
Ash	8.0	11.5

Beet Molasses

Beet molasses is the most widely used raw material in the United States and Europe. Latin American and Caribbean plants use sugar as raw material (3 tons of sucrose per ton of citric acid). Smaller plants in the Caribbean use citric wastes from citrus fruits. The production rates are Argentina, 2,000 tons per annum, Mexico, 1,000 tons per annum; and Uruguay, 500 tons per annum.

The beet molasses or pretreated cane molasses is taken in a mixing vessel where dilute sulphuric acid is added to register a pH of 5.5 of 6.5. Phosphorous, potassium, and nitrogen are added as nutrients to the required amount for growth and citric acid production. This mixture is sterilized with live steam. After this, a requisite amount of sterile water is added to give a sugar percentage of 15 to 20%.

The feed is admitted into shallow aluminium pans arranged in convenient stacks in a sterile room. Each tray is shallow (about 75-mm deep) but big enough to hold solutions up to 500 L. The temperature (28 to 32°C) and humidity (40 to 60%) are controlled for maximum yield. This sugar medium is inoculated with the spores of a selected strain of *A. niger*. In about 8 to 10 days, the fermentation process is over (pH being about 2). The acid might be contaminated with oxalic acid and gluconic acid. The tray contents are sent out for recovery of the acid, and the trays and the chamber are sterilized with steam. Dilute formic acid or sulfur dioxide could also be used for positive sterilization. The recovery consists of adding hydrated lime (1 part of lime to every 2 parts of liquor) to the fermenter broth kept at 95°C. The precipitated calcium citrate could be reacted with enough dilute sulfuric acid when citric acid remains in solution, leaving behind calcium sulfate. Based on the sugar content of the raw materials, the yield could be 35 to 65% by weight of sugar. In this process, pH maintenance is difficult.

Industrial Fermentation

The majority of industries have adopted this process. Conventionally, the growth phase and the production phase are two distinct regions. Two different media are used, one for growth and the other for production. After 3 to 4 days of lush growth, the mycelia are separated from the growth medium and added to the production fermenter under stirred conditions. In the production fermenter, the pH and temperature are maintained around 3 and 30°C, respectively. Air is sparged without interruption into the solution at about 1 to 1.2 volume air per volume medium per minute (vvm). In about 4 to 5 days, the production becomes nearly complete with about 65 to 70% of sugar being converted to the final product.

Table 7.6 Medium for citric acid

Component	*Sporulation (g/L)*	*Production (g/L)*
Sucrose	140	140
Bactoagar	20	0.0
Ammonium nitrate	2.5	2.5
Potassium hydrogen phosphate	1.0	2.5
Magnesium sulphate heptahydrate	0.25	0.25
Copper ion	0.0048	0.00006
Zinc ion	0.0038	0.00025
Ferrous ion	0.0022	0.0013
Manganous ion	0.001	0.001

The submerged process is admirably suited for a flexible operation. It is possible to introduce certain process accelerators or nutrients at desired and convenient time intervals; activated carbon or ascorbic acid could be used as the accelerator. The flexibility of such a fermenter is so good that sugar concentration and pH could be maintained at desired values. The acid recovery is through the calcium citrate process. For every ton of citric acid produced, one needs about 3,650 kg of molasses, 5 to 14 kgs of nutrient, 650 kg of sulfuric acid, and 450 kg of lime. Although one might use several organisms (*A. niger* is still the most sought after), the yield depends mainly on the nutrient used, presence of trace metals, and above all, the pH and air supply. In the new process by Miles Laboratories mentioned previously, sucrose has been replaced with glucose for acid production. Starch is converted to sugar through enzymatic reaction, and the resulting sugar is the raw material for acid production. Specific chemicals control the growth of the mold at the expense of citrate production. An interesting point is that their starting material is not of high purity. They use very large fermentation vats into which sucrose solution is pumped, and fungal spores are added to help the growth and production cycles. After the cycle is over, the vats are charged with fresh medium.

Role of *A. niger* in Fermentation

In the solid-substrate fermentation process, the mold *A. niger* ferments molasses or sucrose adsorbed on beet or cane pulp. Carriers such as beet or cane pulp offer large areas for microbial contact with the medium and ensure good aeration. Bagasse is preferred over beet pulp because of its larger surface area and lower cost. In addition, bagasse can be used several times over. Bagasse is broken down into fine pieces, 1 to 2 mm long, sterilized, and soaked in molasses or sucrose solution before being inoculated with the selected strain. The citric acid yields are about 45% of the sugar content of molasses and 55% of sucrose. This process is labor intensive and needs more floor space, plus there is the likelihood of contamination. It is difficult to maintain pH. The process requires good aeration with controlled humidities.

Interaction between Various Fermentation Process

The starting sugar cane concentration and the yield percentage are more or less the same for submerged and surface fermentation. However, the surface process is highly labor intensive and not easily amenable for pH or temperature control. Flexibility of operation is higher in the submerged process because addition or removal of chemicals is easier. The risk of contamination is less in the submerged process. The surface process needs a larger working area, and the chance of pollution (caused by spores) is also high. Oxalic acid formation is higher in surface fermentation. One advantage of the surface process is that it is less energy intensive. Although automation of the submerged process may be easier, in the event of contamination, the losses suffered will be enormous.

LABORATORY PRODUCTION OF CITRIC ACID

Successful production of citric acid depends on several factors. External factors can be controlled directly, and biochemical factors, the various biological reactions that go on inside the cell, can be controlled indirectly by controlling the external factors.

External Factors

1. The type of sugar used and sugar levels
2. The pH in the milieu
3. The temperature of production
4. The presence or absence of trace metals such as Fe, Cu, Zn, Mn
5. The presence or absence of phosphates and nitrates in the medium
6. The oxygen concentration in the environment

Biochemical Factors

1. Breakdown of sugars to pyruvic acid and acetyl CoA
2. Formation of oxalacetic acid from pyruvic acid and carbon dioxide
3. Promotion of the enzymatic reaction necessary for citric acid accumulation
4. Suppression of the enzymatic reaction that represses citric acid production
5. Oxygen concentration in the medium and cell
6. Favoring citric acid accumulation at the expense of cell growth

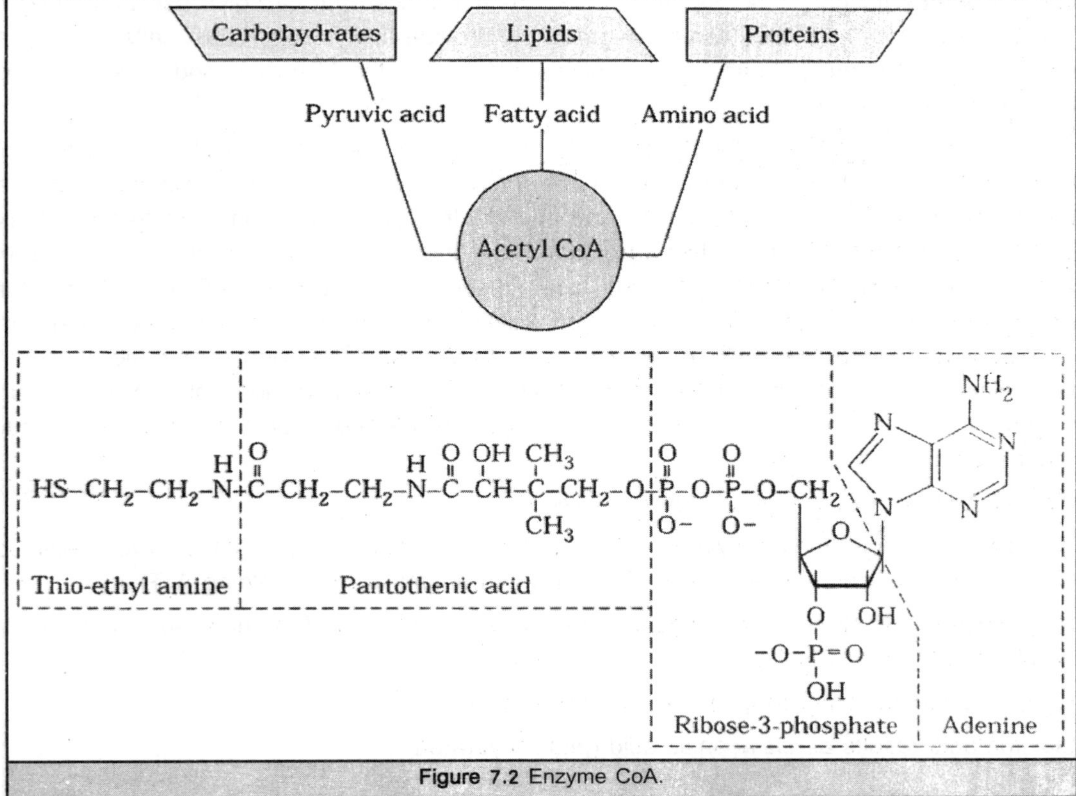

Figure 7.2 Enzyme CoA.

Based on the external and biochemical factors, one could draw the following conclusions:

1. Glycolytic and pyruvate enzymes should be activated for the citrate synthesis by maintenance of high sugar levels.
2. Repression of enzymes in the tricarboxylic acid (TCA) cycle would inhibit citrate production.
3. Manganese deficiency should be ensured for NH_4^{2+} generation, which would counter the inhibition of phosphofructokinase by citrate.
4. A copious oxygen supply will overcome possible cell starvation, even under emergent conditions.
5. Ideal pH maintenance encourages citrate production rather than oxalic or gluconic acid formation.

In most cells, catabolism of simple sugar is the major source of energy. It starts with the glycolytic pathway. Glucose is broken down into two molecules of three-carbon compounds called pyruvic acid. The cell gains two ATP molecules for each molecule of glucose entering the glycolytic pathway. The ATP molecules are storehouses of energy, releasing energy necessary for cell work. The catabolic reaction also gives rise to NADH, which is useful for several anabolic reactions. Depending on the type of cell and availability of oxygen, the pyruvic acid produced earlier can undergo two different types of reactions. When anaerobic conditions prevail, pyruvic acid is converted to lactic acid or ethyl alcohol. Under aerobic conditions, the pyruvic acid is converted to carbon dioxide and acetyl CoA. This acetyl CoA is further degraded in the TCA cycle or election transport system to produce energy for cell work. Several other substrates exist that can give rise to acetyl CoA. For example, proteins and lipids can be degraded to provide acetyl CoA. Thus, cells can degrade several chemical compounds to acetyl CoA and subject them to the TCA cycle and electron transport chain for obtaining energy.

The TCA cycle together with the electron transport system constitutes the cells' primary metabolic furnace. These two combine to burn acetyl CoA in oxygen to liberate carbon dioxide, water, and useful energy. This energy from the TCA cycle and electron transport chain is captured and taken up in ATP for storage of energy. During oxidation of one molecule of acetyl CoA (by the TCA cycle occurring in the mitochondrial matrix), one molecule of flavoprotein (FP or FAB) and three molecules of nicotinamide adenine dinucleotide (NAD) are reduced. These reduced coenzymes are oxidized by molecular oxygen by way of a system of enzymes and coenzymes called the respiratory chain or electron transport system, occurring in the inner mitochondrial membrane. During this oxidation process, enormous amounts of energy are released, some of which is utilized by the inner membrane subunits.

The TCA cycle occurs in the following ways:

1. The pyruvic acid is decarboxylated; the acetyl group combines with CoA to give acetyl CoA. The H and NAD combine to give NADH. Energy transfer occurs, yielding ATP.
2. Active acetyl combines with oxalacetic acid to form citric acid. The released SCoA can now combine with more CH_3CO.
3. Citric acid looses water to give rise to aconitic acid.
4. Isocitric acid results as *cis*-aconitic acid gets rehydrated.

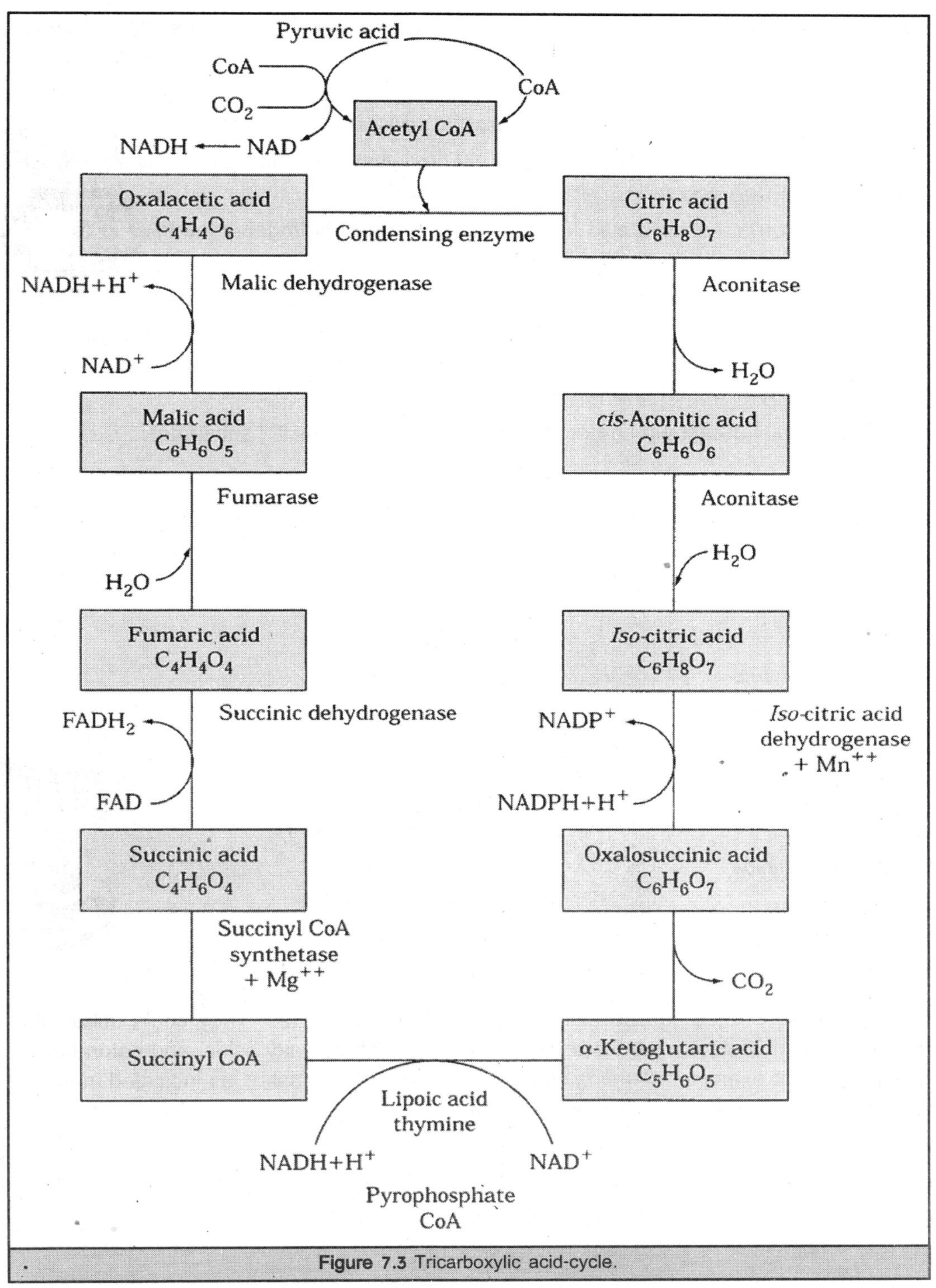

Figure 7.3 Tricarboxylic acid-cycle.

5. Isocitric acid loses hydrogen to yield oxalosuccinic acid. The H combines with NADP to form NADPH. Energy transport takes place, yielding ATP. Loss of CO_2 also occurs.
6. Decarboxylation of oxalosuccinic acid takes place to yield α-ketoglutaric acid. Hydrogen reacts with NAD to form $NADH_2$. Energy is released to convert ADP to ATP.
7. SCoA combines with the succinyl group of α-ketoglutaric acid and gets converted to succinic acid, yielding SCoA and ATP.
8. Dehydrogenation of succinic acid leads to fumaric acid. Hydrogen joins FAD to form $FADH_2$ and energy. This leads to formation of ATP.
9. Malic acid is formed as fumaric acid gets hydrated.
10. Malic acid is dehydrogenated to form oxalacetic acid. Oxalacetic acid combines with active acetyl to form citric acid and the cycle continues. The released hydrogen forms $NADH_2$ from NAD and energy released gives rise to ATP.

Several yeasts and bacteria can produce citric acid from specific substrates:

Yeasts in production

- *Candida lipolytica*
- *Candida tropicalis*
- *Candida zylenoides*
- *Candida fibrae*
- *Candida intermedia*
- *Candida parpsilosis*
- *Candida petrophylum*
- *Candida subtropicalis*
- *Candida oleophila*
- *Candida hitachinica*
- *Candida citra*
- *Candida guillermondi*
- *Candida sucrosa*

A variety of substrates can be utilized by the *Candida* species. They could utilize glucose, acetic acid, calcium acetate, hydrocarbons, molasses, alcohols, fatty acids, and natural oils (such as coconut oil). For example, *Candida* can utilize glucose or molasses as indicated in the list. *C. lipolytica* can use *n*-paraffins, *n*-alkanes, and alkenes.

From Glucose	From Molasses
Glucose, 10-18%	Sugar cane/beet molasses, 5-250 g/L
NH_4Cl	$(NH_4)_2SO_4$, NH_4Cl, NH_4NO_3
KH_2PO_4	0.3-1.5 VVM, 6 days

$MgSO_4 \cdot 7H_2O$	5.5-6.50 pH for growth
22-30°C, 3-6 days	2.8-4 pH for production
	3,154 kg sugar → 1,119 kg monocitrate

Bacteria in citric acid production

Certain bacteria are preferred for producing citric acid because of their low doubling time. Examples include:

Bacillus licheniformis
Bacillus subtilis
Brevibacterium flavum
Arthrobacter paraffinens can utilize dodecane (C_{12}–C_{14}) at 28°C over a period of 72 h; yield 28 mg/mL

They utilize glucose or isocitric acid or hydrocarbons
Bacillus licheniformis can utilize glucose medium: 30-37°C 36-120 h; pH 7; nitrogen salts act as nutrients; yield 42 g/L

Cyclic Behavior of Citric Acid

Two important routes are followed. One is the precipitation as calcium citrate. The second is the liquid–liquid extraction technique as discussed by Krishnan and Natarajan. Solvents such as esters, ketones, amines, and alcohols have been successfully used for the recovery, as outlined by Rieger and Kioustelidis.

CONCLUDING REMARKS

Citric acid enjoys a unique place in several important everyday applications. The capabilities of *A. niger* have been almost totally exploited by using several strains of its culture and their mutants. The application of *A. niger* for citric acid production has been stretched nearly to the limit. It is the wishful thinking of the author that some more work could be undertaken in producing such stains that could utilize unclarified raw materials in shorter periods. No effort should be spared in developing newer and cheaper chemicals (promoters) that could accelerate the production rate. Care has been taken to cover the entire world literature on citric acid production. However, some references might have been missed more because of specific constraints rather than because of oversight or willful omissions. We should like to end by saying that some more work on the application of bacterial culture would be rewarding. It may not be right to say that the last word has been spoken on citric acid production.

8

SIZE EXCLUSION CHROMATOGRAPHY

Size-exclusion chromatography (SEC) is a widely used means for separation of macromolecules of different sizes. Porath and Flodin in 1959 first demonstrated the use of SEC, or gel filtration chromatography, for separations by resolving glucose from two size fractions of dextrans. In addition, they applied the technique to desalting of serum proteins. Remarkably, these two applications—crude separations and desalting—remain the primary uses of SEC today. A wide range of materials, including sugars, proteins, and nucleic acids, have been analyzed or separated using this technique. In addition, SEC is used extensively in the characterization of polymers. However, in this article, the emphases are on analysis and on separations involving biological macromolecules.

ESSENTIAL PRINCIPLE

The essential principle of size exclusion chromatography is partitioning of species on the basis of preferential sieving into the pores of support particles. Sample is applied to the top of a bed packed with porous stationary phase particles. Elution results from flow of a fixed composition buffer through the bed (isocratic elution), in contrast with other modes of chromatography in which the buffer composition is varied (gradient elution). Molecules that are larger than the pores of the support access only the interstitial space between particles and therefore pass through the column quickly, whereas smaller particles access the volume within the pores and are retained for a longer time by the column. Therefore, a support must be selected with a pore size distribution so that there exists a differential partitioning of the molecules to be separated.

Retention Volume

In any chromatographic process, sample is injected into the column and eluted by continuous flow of a buffer through the column. The time between injection and the peak of any eluted solute is termed its retention time (t_R). An equivalent measure of retention is the retention volume (V_R), which, for the usual case of constant flow rate, is merely the product of the flow rate and retention time. The terms *elution time* and *elution volume* are used interchangeably with *retention time* and *retention volume*. For solute molecules larger than the size of the pores, the elution volume represents the volume of the column between the support particles plus that of extracolumn tubing between the injection loop and the column and from the column to the detector. This is known as the void volume (V_0), and it is frequently estimated using large, inert tracer solutes. The pore volume (V_p) is the volume within the particles occupied by fluid and, in principle, accessed entirely by very small particles, for which the retention volume is maximized:

$$V_{R,max} = V_0 + V_p \quad \ldots(1)$$

In general, the elution volume is intermediate between the void volume and the sum of the void and pore volumes and thus defines a distribution coefficient (K_d) through the relation

$$V_R = V_0 + K_d V_p \quad \ldots(2)$$

By definition, the distribution coefficient assumes values only between zero and unity. The other relevant volume is that of the support particles or chromatographic matrix (V_m). In principle, one can calculate the pore volume based on the amount and density of the support through the relation

$$V_p = \frac{W_r \rho_{gel}}{1 + W_r \rho_{H_2O}} (V_t - V_0) \quad \ldots(3)$$

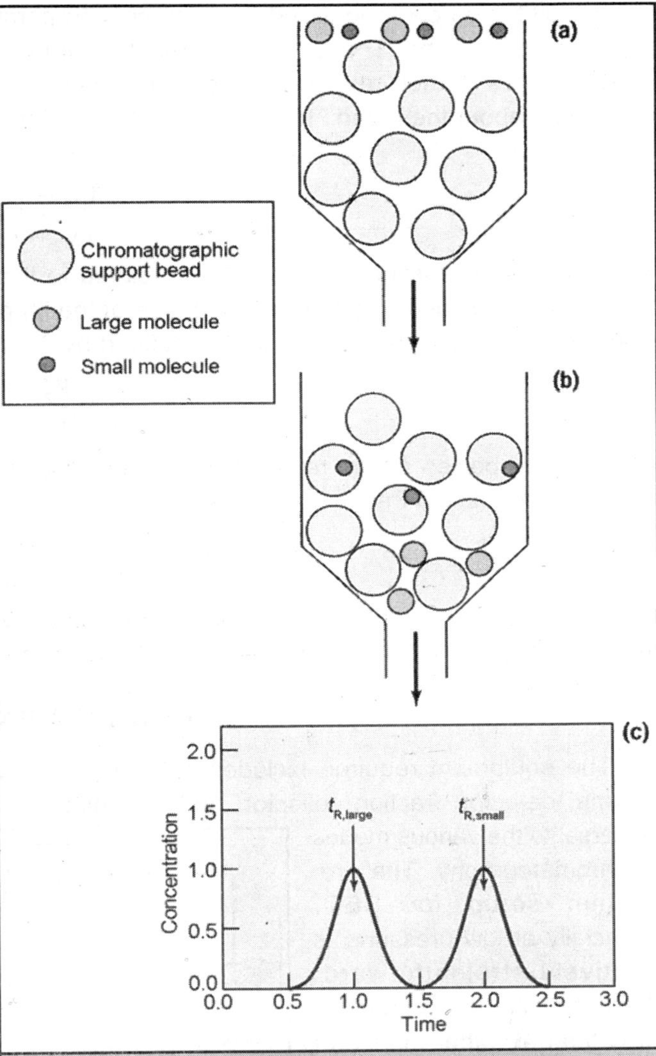

Figure 8.1 Principle of SEC. (a) A mixture of large and small particle is introduced together into the top of the column. (b) As the particles move down the column, the small particles are able to access the volume within the pores of the particles. (c) Schematic chromatogram showing the elution profiles of the large and small particles.

where V_t is the total volume (sum of void, pore, and matrix volumes, the first corrected for the volume of extracolumn tubing), ρ represents mass density, and W_r is the water regain of gel (grams of water per gram of dry gel divided by the density of water). Experimentally, V_p can be measured as the retention volume of a small molecule that does not interact with the support. With determination of V_0 and V_p, the distribution coefficient of any solute can be determined from its retention and equation 2; conversely, the retention time of a solute could be predicted from a model for its distribution coefficient and from the column parameters. The typical chromatographic peak is Gaussian in shape. The quality of separation of two (or more) species in SEC is characterized by differences in their retention times and by certain measures of the distribution of material in each peak about the mean. The resolution (R_s) is a measure of separation of two bands, defined by

$$R_s = \frac{t_{R2} - t_{R1}}{\frac{1}{2}(w_2 - w_1)} \quad \text{...(4)}$$

where w_i is the baseline width of peak i, related to the standard deviation of the Gaussian peak through $w_i = 4\sigma_i$, and t_{Ri} is the retention time of the ith peak. Another measure of separation is the selectivity, α, between two bands. It is defined by

$$\alpha = \frac{k_2'}{k_1'} \quad \text{...(5)}$$

where k' of species i is its retention factor (or capacity factor). The retention factor is a relative retention time defined as

$$k' = \frac{t_R - t_0}{t_0} = \frac{V_R - V_0}{V_0} \quad \text{...(6)}$$

For non-Gaussian bands, t_R and σ can be determined via statistical analysis of the chromatogram from the first and second moments, respectively, of a trace.

REQUIREMENTS

The equipment required includes solvent and reservoir, valves and tubing, pump, packed column, detector, fraction collector, and recording device. For the most part, this equipment is universal to the various modes of chromatography. The pre-column setup for SEC, especially at low pressure, is relatively straightforward. Solvent delivery is simpler in SEC than for other modes of chromatography because there is no need for gradient formation and the only pump requirement is for stable flow. Peristaltic pumps are adequate for low-pressure work, whereas piston or diaphragm pumps are used for high pressures. The most important aspect of the equipment is the column packing. The gel must be packed uniformly throughout the column to avoid channeling or other heterogeneous effects that act to blur separation of the solutes. Frequently, the gel is supplied as a powder or as a suspension in a storage solvent. In the latter case, this solvent is filtered off,

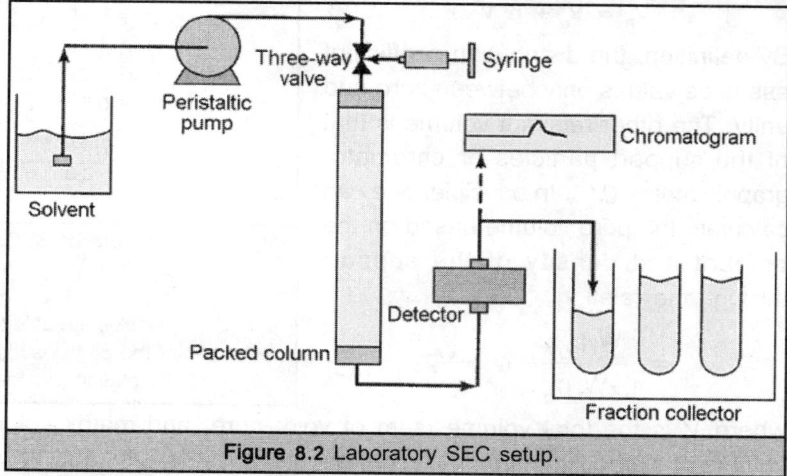

Figure 8.2 Laboratory SEC setup.

and in both cases the gel is then suspended in a packing solvent, which is usually an aqueous buffer with some NaCl added. A stable, uniform packing is attained by degassing the gel slurry before packing, pouring it at once, and then packing at a flow rate greater than used in normal operation. Detection of proteins and nucleic acids is most commonly accomplished with UV absorbance detectors, although more sensitive techniques such as light-scattering and mass spectrometry are also possible. The detection limit (C^*) is dictated by the absorbance sensitivity (A^*) of the instrument, the extinction coefficient (ε) of the solute, and the path length (L) of the detector, according to

$$C^* > \frac{A^*}{\varepsilon \ell} \qquad \ldots(7)$$

The sensitivity of absorbance detectors is usually about 5×10^{-4}, and their path lengths can be up to 1 cm. Typical proteins have extinction coefficients on the order of 1 cm^2/ mg; therefore, the detection limit is on the order of 1 μg/ mL. Nucleic acids exhibit strong absorbance, peaking around 260 nm, with extinction coefficients on the order of 25 to 50 cm^2/mg; consequently, lower concentrations down to 20 ng/mL may be detected. Carbohydrates, on the other hand, exhibit no significant absorbance in UV or visible light; refractive index detectors are often used for their quantitation.

Packing Materials

A wide variety of packing materials are available for use in SEC. The choice of gel for a particular application depends on a number of factors, including materials compatibility, operating conditions, and particularly the sizes of the molecules to be separated. Manufacturers typically provide a good deal of information about their packing materials, including fractionation range, maximum flow rate or pressure drop, and stable operating ranges of pH and temperature.

Key features

The key feature of the packing material with respect to a particular separation is the pore size distribution, because this determines the range of particles that may be fractionated. Very small pores are required for desalting applications, so that all macromolecules are excluded and low molecular weight species are retained. Materials possessing much larger pores are used for analysis and purification of biological macromolecules. Because the pore sizes are normally inferred from the sieving of test solutes, the useful molecular weight fractionation range, rather than the actual pore size, is generally reported. It is important to note that different types of molecules exhibit different fractionation ranges; therefore, a fractionation range reported using protein standards will not be identical to that of, for example, nucleic acids.

Some manufacturers report two fractionation ranges: one using protein standards and another using nucleic acids or dextrans. Usually, these two values will be sufficient to at least select an appropriate gel for the application of interest. Once a fractionation range is determined, a material appropriate for the solutes of interest must be identified. Although the gel materials used in SEC have been developed to exhibit minimal adsorption, biological macromolecules inevitably interact to some extent with these materials. Proteins in particular will adsorb to a variety of surface functionalities. A general rule is that the gel material should be hydrophilic and uncharged. However, most hydrophilic materials exhibit at least a weak charge. Undesirable electrostatic interactions caused by this charge can be minimized by eluting with a buffer of moderate ionic strength, 0.3 to 0.5 M. In addition, the column must be operated under conditions at which the material is stable.

Packing materials can deteriorate when subjected to extremes of pH, ionic strength, flow rate, and temperature. For some materials, enzymatic degradation is also possible in the event of bacterial contamination. The size of the particles and their degree of compressibility determine the pressure drop across a bed and limit the flow rate that may be used. Particles ranging from 5 to 600 μm are available. Some SEC media, particularly cross-linked materials, are rigid over their normal operating range. For these rigid materials, the pressure drop (Δp) across a gel of length (L) is inversely proportional to the square of the particle diameter (d_p) and proportional to velocity (v_0) (flow rate divided by cross- sectional column area), in accordance with Darcy's law:

$$\Delta p = \frac{\mu v_0 L}{k d_p^2} \qquad \ldots(8)$$

where k is a constant and μ is the fluid viscosity. However, as the particles compress, the pressure drop increases dramatically, effectively setting an upper limit on the maximum attainable velocity. From the standpoint of materials selection, manufacturers supply information regarding particle size, fraction range, and maximum allowable flow.

Variety of materials

Materials comprised of dextran, agarose, acrylamide, or combinations have traditionally been the workhorses of SEC. However, novel polymeric materials and polymer-coated silica media have been developed. Silica possesses excellent mechanical strength, but it is prone to adsorption of biomolecules. Dextrans are used in a wide variety of protein and nucleic acid separations. They are polysaccharides that are made into a porous gel by cross-linking with epichlorohydrin. These materials are quite stable for pH > 2 and for temperatures up to at least 120°C and can be used with either aqueous or organic solvents. Dextran gels with small pore sizes exhibit low compressibility, but versions with larger pore sizes are quite compressible. As a result, some dextran gels are limited to flow rates that can be less than 1 mL/min. Because dextrans are polysaccharides, they are susceptible to bacterial degradation; however, because of their thermal stability, they can be autoclaved.

A variation of the traditional dextran gel is provided by cross-linking dextran with *N,N'*-methylenebisacrylamide. This provides a material that is substantially more rigid but has a smaller stable pH range and is more susceptible to adsorption of proteins. The latter obstacle can often be overcome by eluting with a higher ionic strength solvent. Acrylamide gels are formed by cross-linking acrylamide monomer with *N,N'*-methylenebisacrylamide. The resulting matrix is quite stable over the pH range from 2 to 10 and to high temperatures and is not prone to bacterial degradation. It is less adsorptive than the corresponding dextran material. Consequently, polyacrylamide gels are commonly used for protein separations. Their main disadvantage is that they are incompatible with organic solvents. A copolymer of D-galactose and 3,6-anhydro-L-galactose is used to form an agarose gel. The composition and degree of cross-linking have a profound effect on the properties of agarose-based media. With less cross-linking, the gels are typically stable only in the pH range of 4 to 9, are compressible at flow rates as low as 1 mL/min, and must be sterilized chemically, but they exhibit little adsorption of macromolecules and broad fractionation ranges that make them much more suitable than acrylamide or dextran for the separation of large proteins and nucleic acids. With a higher degree of cross-linking, the material is stable over a wider range of pH and flow rate, and it can be autoclaved; however hydrophobic interactions may occur between proteins and the support, and the available fractionation range is reduced. A number of variations on traditional materials have been established. Generally, these are produced to combine the desirable sieving properties of one material with the mechanical or chemical stability of another.

The three most commonly used materials—dextran, acrylamide, and agarose—have been combined in all possible pairs. The dextran–bisacrylamide material (Sephacryl) mentioned earlier is one such example. Another is a composite of dextran and cross-linked agarose, covalently bound together to form a material that has the fractionation range of a dextran support with the rigidity of agarose. The resulting material is autoclavable and stable in the range $3 < pH < 12$ and to a variety of strong solvents. Acrylamide has also been entrapped in an agarose gel to form the commercial Ultrogel matrix. Other supports, based on materials such as silica and hydrophilic polymers, have been produced for SEC. Although porous silica is inherently rigid and can be manufactured in a variety of pore sizes, it exhibits marked adsorption of proteins and is unstable at basic pH. Other polymer gels exist commercially, such as Toyopearl and Spheron, which are based on polyethers and poly(2- hydroxyethylmethacrylate), respectively. The latter is a material approved by the Food and Drug Administration and can thus be used to produce biocompatible supports.

COMMERCIAL VALUES

A number of applications exist for SEC in the research laboratory and in industrial processes. These can be divided into the two categories of sample analysis and purification. The former is most often performed on small columns with high resolution, and the latter is a bulk operation in which the separations are usually fairly crude.

Method of Analysis

At the analytical level, SEC is used as a tool in characterizing the size, shape, and interactions of biological macromolecules. For example, SEC has been used to study the kinetics and equilibrium dimerization of proteins. Likewise, it can be used to study the hybridization of nucleic acids. Similar applications have been made to reversible interactions between macromolecules and low molecular weight species. It has also been used to identify and characterize molten globule states in studies of protein unfolding. In addition, it is an excellent tool for identification and separation of small molecules that are conjugated to biomolecules, such as probes attached to antibodies or drugs conjugated to targeting molecules. SEC is often used for molecular weight determination, although this is an approximate method.

Molecules that are similar to the solute of interest but with known molecular weight are used to create a calibration curve of retention time versus molecular weight. The unknown molecular weight of the solute is then estimated from its retention time. The degree of success of this approach is highly dependent on how similar the unknown solute is to the molecular weight standards. That is, a calibration curve prepared with globular proteins will give an inaccurate molecular weight for an oligonucleotide or even for a glycoprotein. Nonetheless, for similar molecules, molecular weights can be successfully estimated.

Table 8.1 Examples of SEC applications

Analysis
Molecular weight estimation
Identification and analysis of protein dimers
Analysis of reversible protein–small molecule interactions
Studies of protein unfolding
Identification of products in molecular conjugates
Evaluation of the stability of macromolecular complexes
Separations
Desalting
Crude protein separations
Purification of molecular conjugates
Removal of endotoxins and viruses from pharmaceutical formulations

The retention time of α_1-acid glycoprotein is 8.70, from which an apparent molecular weight of 57 kDa is calculated, significantly greater than its actual molecular weight of 40 kDa. For molecules of similar chemistry, such as globular proteins, the deviation of the observed elution volume from the calibration curve can be an indication of molecular shape. Deviation from spherical to ellipsoidal shape results in increased molecular friction, and thus the molecule sieves as if its apparent molecular weight is greater than its actual value. Although exact molecular weight determination is complicated by the shape and chemical nature of different protein molecules, changes in a particular structure can be discerned with high resolution. For instance, the extent of modification of a protein by a dye or other ligand can be monitored by the relatively small decrease in retention time if the same column, buffer, and operating conditions are used.

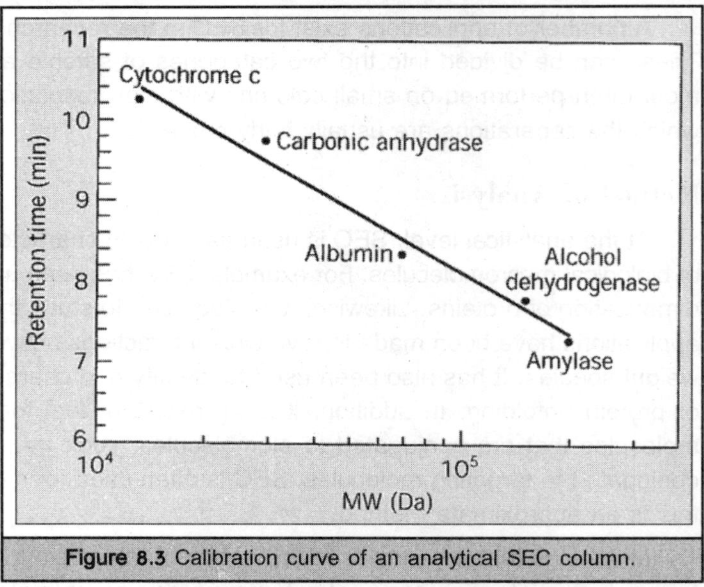

Figure 8.3 Calibration curve of an analytical SEC column.

Isolating Technique

By design, SEC separates molecules based on their size, so it is not possible to achieve well-resolved separations of biomolecules from a mixture of similar species using this technique. Although fair discrimination of molecules according to molecular weight was seen in the previous section, small differences in retention time are insufficient for resolution in purification. Nonetheless, crude fractionation of a particular size range is possible with SEC. Thus, it is best suited either for purification of one molecule of interest from a small set of dissimilar contaminants in the laboratory or as an early or late step in conjunction with other steps in an industrial process. SEC finds many laboratory applications in manipulations of biological molecules. For example, unreacted species in molecular conjugates are often removed via SEC. Cleanup of materials involved in the synthesis of peptides and oligonucleotides is also performed frequently via SEC. It is also used for removal of aggregates resulting from a variety of biochemical preparations. For example, an intermediate or final step in purification of a monoclonal antibody is often removal of dimers and aggregates that have been generated as a result of prior processing. In the production of recombinant proteins, a certain fraction may be clipped by proteolysis; these smaller molecules can be separated by SEC. The most common application of SEC at both laboratory and process scales is desalting or buffer exchange. Many processing steps (e.g., extraction, crystallization, other modes of chromatography) involve the use of a solvent that is undesired for future processing or formulation steps. Very small pore SEC columns are used to replace the undesired buffer by the desired one. Usually, a wide disparity in molecular sizes exists between the biological solute and the buffer salts, so short columns can be used, limiting the dilution to two-fold or less.

ANALYZING THE SAMPLES

Selection of a support material for an application of interest is only the first step in performing an analysis or separation. The amount of sample to be analyzed or purified must be considered. The column length and total amount of material to be injected into the column must be determined. Finally, a flow rate must be found that will provide adequate throughput and resolution. In the process of designing a configuration, convective dispersion, transport of the solutes to and within the particles, and partitioning of the solutes between the mobile and gel phases must all be considered. For these reasons, a substantial body of literature has been devoted to the engineering issues associated with SEC. The most general and applicable models are discussed in this section.

Plate Theory

An operational description of chromatography can be obtained from plate theory, which is a phenomenological description of zone spreading in chromatography, first proposed by Martin and Synge in 1941. The essential concept is that as the solute traverses the column, it undergoes repeated equilibration with the stationary phase, so that each equilibration defines one stage, or plate. The column length over which equilibration occurs is known as the height of an equivalent theoretical plate (HETP), or simply plate height. In each stage, fluid in the mobile phase is mixed, and partitioning occurs between the stationary and mobile phases. A mass balance on a particular stage takes the form

$$Q(C_{j-1} - C_j) = v_m \frac{dC_j}{dt} + v_p \frac{dC_{pj}}{dt}, \quad ...(9)$$

where Q is the flow rate, C_j is the concentration of solute in the jth stage, v_m is the mobile phase volume per stage, v_p is the pore volume per stage, and $C_{p,j}$ is the average pore concentration in the jth stage. For linear chromatography, the average pore concentration is proportional to the bulk concentration, and equation 9 becomes directly analogous to the mass balance on stirred tanks in series.

The input to a chromatography column is essentially a pulse; consequently, the output from a series of N stages is analogous to the residence time distribution (E) from N tanks in series. The output from a single stage is an exponential distribution, but, as the number of plates increases, the residence time distribution takes on a Gaussian shape. It can be shown that the variance σ^2 of the output peak is dependent only on the retention time and number of stages through

$$\sigma^2 = t_R^2/N \quad ...(10)$$

Therefore, the number of plates (N) in chromatography is defined as

$$N = t_R^2/\sigma^2 \quad ...(11)$$

and the plate height (H) as the length of the column (L) divided by the number of plates

$$H = L/N = L\sigma^2/t_R^2 \quad ...(12)$$

Size to Solutes

Because the purpose of SEC is to separate molecules on the basis of their size, interest developed from the time of its introduction to understand quantitatively the relationship between the size of a solute molecule and its retention volume. This is of importance in the use of SEC for molecular weight determination, in its interpretation for applications involving monitoring changes in molecular structure, and in the selection of SEC media and operating conditions to perform a separation.

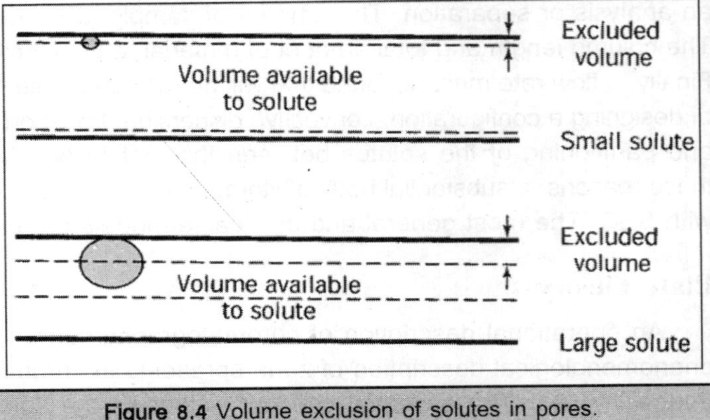

Figure 8.4 Volume exclusion of solutes in pores.

Volume of particles

The primary mechanism of sieving is based on the excluded volume of particles in the pores. At equilibrium, the solute molecules will attain a concentration within the pore based on the volume accessible to the center of the molecule. When the pore size is of the same order as the solute molecular size, the accessible volume becomes significantly less than the pore volume. As a result of the excluded volume effect, even a monodisperse pore size distribution will produce a continuum of distribution coefficients with increasing molecular size. To develop this concept into a quantitative model, a geometry is usually assumed. An early result was that of Laurent and Killander, who used the expression of Ogston for the available volume in a network of rigid rods to develop an expression for the distribution coefficient:

$$K_d = \exp[-\pi L(r_s + r_r)^2] \qquad \dots(13)$$

where L is the total length of rods per volume, r_s is the radius of the (assumed) spherical particles, and r_r is the radius of the rods. The parameters L and r_r are generally not known and must be fit to experimental data. Other expressions were derived for cylindrical, conical, and slit- like pores. The problem of ascribing a particular pore geometry was circumvented by Ackers, who represented a pore by the maximum characteristic radius particle that could pass through it and postulated that the distribution of pore sizes is distributed normally. For this model, the distribution coefficient takes the form

$$K_d = \operatorname{erfc}\left[\frac{a - a_0}{b_0}\right] \qquad \dots(14)$$

where a is the radius of the solute particle, a_0 is effectively an average pore radius, and b_0 is a measure of the variance of the distribution. These parameters must still be determined for each gel from sieving experiments on at least two (preferably more) solutes.

Diffusion through pores

The models based on volume exclusion (see previous section) are entirely thermodynamic; that is, they do not include any transport limitations and do not predict a dependence of distribution coefficient on flow rate. In practice, a slight dependence is observed, which led to the development of theories that account for transport of solute molecules into the pores. An approach to accounting for diffusion is to model the distribution of molecules between the mobile and gel phases using the one-dimensional (1-D) unsteady-state diffusion equation, which has the solution

$$C_g = C_m \operatorname{erfc}\left[\frac{x}{2(Dt_d}\,1/2\right] \qquad \dots(15)$$

where C_g and C_m are concentrations in the gel and mobile phases, respectively, x is distance into the 1-D gel phase, t_d is the diffusion time, and D is the diffusion coefficient of the solute. With t_d inversely proportional to the mobile phase velocity (v_0) and b representing the power to which diffusion coefficient is inversely proportional to molecular weight (M), the distribution coefficient can be determined to be

$$K_d = \frac{k}{(\pi v_0 M^b)^{1/2}}(1 - e^{-v_0 M^b/k^2}) + \operatorname{erfc}\left[\frac{(v_0 M^b)^{1/2}}{k}\right] \qquad \dots(16)$$

where k is a grouping of constants fit to experimental data.

Correlative studies

The equations initially used to correlate SEC data were essentially empirical. The most commonly used forms are

$$V_R = A - B\log(M) \qquad \text{...(17)}$$

and

$$\log(V_R) = A' - B'\log(M) \qquad \text{...(18)}$$

where M is the molecular weight of the solute and A, B, A', and B' are empirical constants for the particular column configuration. As discussed in "Analysis", very good correlations of data for similar species are possible, but a priori prediction or extrapolation beyond the measured range of molecular weights or to molecules of different chemistry is problematic. Note that the more sophisticated theoretical models discussed earlier also make use of essentially empirical constants that require calibration of the column using proteins (or other molecules) of known molecular weight. Calibration of the column must be repeated if the pH, ionic strength, or temperature is changed significantly or if the column is repacked. Furthermore, the calibration standards must be well characterized and of a similar nature to the solutes of interest. Nonetheless, the use of theoretical models attaches a physical significance to the parameter's fit from experimental data and provides some insight into the relative contributions of the volume-exclusion versus pore-diffusion mechanisms.

Dispersion

The previous section dealt with the retention time, which mostly results from the equilibrium partitioning of species into the gel support. For separations, it is necessary to consider that chromatography is conducted in a packed bed, under flow, sometimes at high pressures. The configurational aspects of the chromatography column lead to band broadening, or spreading of the eluted solute band. The main contributions to band broadening are eddy diffusion, axial dispersion, and mass transfer resistance at the surface of the particles and within the pores. Eddy diffusion results from the tortuosity of extra particle paths that solute particles take from the top to the bottom of the column.

Axial dispersion is essentially molecular diffusion along the length of the column. The mass transfer resistance at the surface arises from the boundary layer, or film, through which solute must pass to reach the gel particle, and the mass transfer resistance of the pores is due to diffusion of solute into them and is enhanced by their nonuniformity. A chromatography column is a packed bed of gel particles and is subject to the same types of scaling analysis as for packed beds in other chemical processes. Most of the resistance to mass transfer is at the level of the particle; consequently, the particle diameter is the characteristic length for scaling. Hence, a reduced velocity or Peclet number, (Pe) can be defined in terms of the particle diameter as

$$\mathrm{Pe} = \frac{d_p u}{D_0} \qquad \text{...(19)}$$

where u is the interstitial velocity of the fluid and D_0 is the bulk solute diffusion coefficient. Likewise, the reduced plate height (h) is

$$h = \frac{H}{d_p} \qquad \text{...(20)}$$

where H is the plate height defined in equation 12. The relation of interest is that between plate height and flow rate (or velocity) or, equivalently, between reduced plate height and Peclet number.

This is obtained from solution to the governing differential mass balance equations in lesser or greater complexity.

Operating Variables

Analysis of mass transport in chromatography columns has made it possible to make quantitative predictions regarding plate heights as a function of flow rate in SEC. The original analysis of mass transport was conducted by van Deemter. His equation has served as the basis for all subsequent analyses:

$$h = 2\lambda + \frac{2\gamma}{Pe} + CPe \qquad \text{...(21)}$$

where λ, γ, and C are constants accounting for eddy dispersion, axial diffusion, and mass transfer resistance at the particle surface, respectively. This equation predicts a minimum plate height with respect to increasing fluid velocity. More detailed accounting of the mass transfer contributions, including that caused by intraparticle diffusion, have produced alternative expressions for the plate height, for example, the expanded form of Horvath and Lin:

$$h = \frac{2\gamma}{Pe} + \frac{2\lambda}{1+\omega Pe^{-1/3}} + \frac{\kappa k_0^2}{(1+k_0)} Pe^{2/3} + \frac{\theta k_0}{30(1+k_0)^2} Pe \qquad \text{...(22)}$$

where θ is a geometric factor for a given particle, γ, λ, ω, and κ are parameters dependent on the details of the particular column packing, and k_0 is the ratio of intraparticle to interstitial volumes available to the solute of interest. The third and fourth terms represent mass transfer resistances at the particle surface and within the pores, respectively.

For liquid chromatography involving macromolecules, the contributions from axial dispersion and resistance at the particle surface are small, and the largest contribution is from pore diffusion. Consequently, the minimum observed by van Deemter for GLC is typically not observed in SEC of biomolecules, and the plate height is roughly proportional to velocity. Plate height also depends on particle size:

$$H \sim \frac{\mu d_p^2}{D_0} \qquad \text{...(23)}$$

From this approximate equation, two practical points can be inferred. First, because plate height is a measure of zone spreading and should generally be minimized, smaller particles are preferred because of the strong dependence of plate height on diameter. However, pressure drop is inversely proportional to the square of particle diameter, so the mechanical properties of the material limit the extent to which particle size can be reduced. Second, lower velocities, that is, slower flow rates, will result in less zone spreading. This advantage may be offset where the speed of a separation is important. Furthermore, the flow rate cannot be reduced too drastically or the effects of axial diffusion will become limiting.

An important measure of the separation is the resolution between two or more solutes to be separated. The dependence of the resolution on column variables can be determined from the equations already developed. Combining equations 4, 6, and 12 gives an expression for resolution in terms of capacity factors and plate heights:

$$R_s = \frac{L^{1/2}}{2} \frac{k_2' - k_1'}{H_2^{1/2}(1+k_2) + H_2^{1/2}(1+k_1')} \quad \text{...(24)}$$

Employing the approximation that the plate heights of the two solutes are equal, an expression for resolution as a function of retention, selectivity, and column efficiency can be derived by combining the definition of selectivity (equation 5) with equation:

$$R_s = \frac{1}{2}\left(\frac{\alpha - 1}{\alpha + 1}\right)\frac{\bar{k}'}{1+\bar{k}'}\frac{L^{1/2}}{H^{1/2}} \quad \text{...(25)}$$

where $\bar{k}' = (k_1' + k_2')/2$ is the average retention time of the solutes.

Equation 25 expresses resolution as a function of three terms representing selectivity, retention, and zone spreading, respectively. Selectivity is the most important determinant of resolution, up to a point. An increase in α from 1.1 to 1.2 results in an 83% increase in resolution, all other factors being equal; however, an increase from 2.1 to 2.2 results in only a 6% further increase in resolution. Selection of a column best suited to fractionate the species of interest is thus critical to obtaining high resolution. Because the capacity factor is greater than or equal to unity, the middle term takes values only between 0.5 and 1; nonetheless, some improvement in resolution is seen to be possible by increasing the retention time. Because of the weak dependence of resolution on length, significant increases in column length are often required to effect a modest improvement in resolution.

The inverse relationship between resolution and plate height suggests that minimizing zone spreading by the methods discussed earlier—reducing particle diameter or flow rate—will concomitantly improve resolution. This is true, subject to the same limitations on flow rate and pressure drop. Furthermore, if speed of separation is important, a trade-off exists. In addition to the particle size and composition, column length, and operating flow rate, the SEC practitioner must choose column diameter, sample volume, and buffer composition. Column diameter does not appear explicitly in the equations describing resolution, but they are derived with the assumption of a uniform packing, which is increasingly difficult to achieve as the diameter-to-length ratio increases. Therefore, columns used for fractionation are prepared with a length-to-diameter ratio of about 10:1. But in industrial scale desalting applications, where the selectivity is great and resolution not an issue, much wider columns can be used.

Sample volumes are usually chosen to be less than 5% of the total column volume, because values greater than a couple of percent tend to produce significant increases in zone spreading. In principle, the solute molecules should not interact with the stationary phase in SEC, but in practice biological molecules exhibit some interaction with a variety of materials, including gels used for SEC. Variation of buffer composition can be used to modulate adsorptive interactions. For example, adsorption is often mediated by charge interactions between solute and sorbent and can be reduced by increasing the ionic strength of the buffer, which provides double-layer screening of the electrostatic interactions.

COMMERCIAL ISSUES

SEC is used in large-scale industrial processes for the production of proteins for agricultural and pharmaceutical use. At this scale, economic considerations become imperative, and in particular it is important for the chromatographic operation to achieve a high level of productivity and throughput while maintaining high resolution.

Strategies

For large-scale operation, the throughput required necessitates reuse of chromatography columns in a repeated- batch operation. Usually, the primary goal is to maximize productivity, which is defined as the amount of purified material per column area per time. This can be achieved by maximizing flow rate and minimizing the amount of time during which the column is not actively performing a separation. From a productivity standpoint, the time between elution of the valuable solute and all others is wasted time. One route to improving productivity is to reduce the resolution of the solutes to the minimum necessary. Likewise, the time after elution of all solutes before the next injection can be used more efficiently by applying the next sample before the column volume of the present run is finished. These general guidelines govern both desalting and fractionation; further optimization depends on the particular application. Another strategy to maximize throughput is to maximize the sample size. However, dispersion effects are enhanced for large sample volumes. Consequently, for fractionation applications, it is advisable to use sample sizes less than 3% of the total column volume.

Although it is often desirable to have highly concentrated samples, higher concentrations result in higher viscosities, which result in slower diffusion into and equilibration with the gel. The requirement of sample equilibration with the gel also limits the flow rate that may be applied. For desalting, on the other hand, larger sample volumes can be applied, because the sample elutes in the void volume and the salts are retained in the pore. Because of this large discrepancy in elution volumes, the increased dispersion resulting from large sample sizes is not much of an issue, and sample sizes as great as 25 to 30% of the column volume may be used. Furthermore, small molecules (e.g., ions) equilibrate quickly with the pores because of their small excluded volume and high diffusion coefficients, and macromolecules are completely excluded, so high flow rates may be used without compromising the separation.

Desirable Results

The principle of scale-up is to determine the values of new process variables that will allow a desired change in one process variable while keeping others constant. Generally, it is desired to increase the throughput while maintaining the same resolution, but sometimes it is necessary to change resolution while keeping throughput constant. In any case, the key is to have a description of the relationship among process variables. It is here that the analysis of the preceding section is most useful.

Scaling rules for isocratic elution chromatography, which includes both SEC and adsorption chromatography, have been described by Wankat and Koo. The starting point is the definition of scaling parameters describing the ratios of process variables in the new and old designs, that is,

$$\delta = \frac{d_{p,new}}{d_{p,old}}$$

$$\lambda = \frac{L_{new}}{L_{old}}$$

$$\rho = \frac{D_{new}}{D_{old}}$$

$$\pi = \frac{\Delta p_{new}}{\Delta p_{old}}$$

$$\theta = \frac{Q_{new}}{Q_{old}} \quad \ldots(26)$$

where D is the column diameter and Q is the buffer flow rate. The capacity factor in SEC is a thermodynamic quantity representing the partitioning of solutes in the pores of the support particles and therefore should not vary with column configuration or process variables. Therefore, the description of column resolution in equation 24 can be used for both the new and old designs to define a ratio of design resolutions, ψ:

$$\Psi \equiv \frac{R_{s,new}}{R_{s,old}} = \left(\frac{L_{new}}{L_{old}}\right)^{1/2} \frac{[H_2^{1/2}(1+k_2') + H_1^{1/2}(1+k_1')]_{old}}{[H_1^{1/2}(1+k_2') + H_1^{1/2}(1+k_1')]_{new}} \quad \ldots(27)$$

The difference between plate heights of the different solutes is usually ignored and equation 27 is greatly simplified, that is,

$$\Psi = \lambda^2 \left(\frac{H_{old}}{H_{new}}\right)^{1/2} \quad \ldots(28)$$

As discussed in "Dependence of Separation Parameters on Operating Variables", transport resistances at the particle surface and within the pores usually dominate, with the result that the plate height can be represented by a scaling of the form equation 21. Substituting in equation 28 gives a simple design equation:

$$\Psi = \left(\frac{\lambda}{\theta}\right)^{1/2} \frac{\rho}{\delta} \quad \ldots(29)$$

Note that more complete plate height expressions (derived from either theoretical or empirical correlations) may be substituted into equation 28 to develop more sophisticated design equations. Equations such as equation 29 can be used to derive new sizes and operating conditions for a given change in any of the others. As a simple example, suppose that a design involving a reduction in particle size is desired, while maintaining constant column dimensions and throughput. In this case, λ, ρ, and θ are equal to unity, and the resolution is inversely proportional to the particle size

$$\Psi = \frac{1}{\delta} \quad \ldots(30)$$

Alternatively, the objective of reducing particle size may be to allow a greater flow rate while maintaining constant resolution. In this case, Ψ, ρ, and λ are equal to unity, and the flow rate scales as the inverse square of particle diameter

$$\theta = \frac{1}{\delta^2} \quad \ldots(31)$$

However, this is only valid in the regime where the particles are rigid, and it may not be feasible because of the large increase in pressure drop that would result.

The application of scaling rules to the interrelationship of process variables in equation 24 is an idealization that is subject to a number of constraints. Arbitrary column diameters and lengths are not possible, and certain length-to-diameter ratios (on the order of 10:1) are desirable to achieve a uniform packing. Also, only certain particle sizes are available, and the range in which they are manufactured is dependent to a large degree on their resistance to flow (i.e., pressure drop). As discussed in "Packing Materials", many packings are rigid only within a certain range of pressure drops. If a new column design takes the pressure drop out of this range, the performance may be drastically altered. For rigid particles, the pressure drop is described by equation 8; therefore, its scaling is

$$\pi = \frac{\theta\lambda}{\rho^2\delta^2} \qquad \text{...(32)}$$

For compressible particles, the same procedure may be used with a correlation that describes the relationship between pressure drop and flow rate.

CONCLUDING REMARK

SEC has been applied extensively to biological macromolecules in the laboratory and has found utility as a step in the production of proteins at large scale as well. Further growth of the biotechnology industry and the ease and applicability of the technique ensure that SEC will be used widely in the future as well. Many packing materials have been developed to enable the chromatographer to perform an analysis or separation quickly, efficiently, and unambiguously through improvements in the inertness and mechanical stability of the support. Engineering analysis has enabled quantitative descriptions to be made regarding the influence of mass transfer resistances to the efficiency (plate height) of the column. These can be incorporated into simple design equations for the development and optimization of column configurations and operating conditions for a number of industrially important separations.

9

ELECTROPHORETIC METHOD

This article is intended as a guide for scientists and engineers who are considering using electrophoresis to process proteins or other biologicals for the biotechnology industry at bench or larger scales. An overview of several types of equipment, a description of what each piece of equipment is designed for, and the problems the end user should anticipate before operating that equipment are discussed. The discussion of each class of apparatus focuses on design principles rather than experimental protocols, emphasizing the engineering aspects of each chamber rather than the details of operation. This type of treatment provides insight into the general operating characteristics of each device and is intended to aid the reader in (i) choosing an appropriate instrument for a given separation and/or (ii) setting operating conditions (e.g., column length, sample volume, or power levels) that avoid excessive dispersion (e.g., band smearing).

Table 9.1 Physical constants used in illustrations

Thermal conductivity	k_{gel} (W/cm °C)	0.00578
Thermal conductivity	k_{glass} (W/cm °C)	0.011
Thermal conductivity coefficient	λ [equation 4] (°C^{-1})	0.02874
Thermal viscosity coefficient	α [equation 30] (°C^{-1})	−0.02874
Thermal expansion coefficient	β [equation 40] (°C^{-1})	0.000088
Electrophoretic mobility in gel	μ (cm^2/V s) (gel)	0.00002
Electrophoretic mobility in free solution	μ (cm^2/V s) (buffer)	0.0001
Diffusion coefficient in gel	D_{gel} (cm^2/s) (gel)	1×10^{-7}
Diffusion coefficient in free solution	D_m (cm^2/s) (buffer)	5×10^{-7}

For the purposes of illustration, it is assumed that the buffer has the kinematic and electrical properties of pure water at 10°C, with the exception of its electrical conductivity, σ_e, which is determined from electrolyte composition. Most of the values used are sufficiently accurate for purposes of illustration. However, the mobilities and diffusion coefficients found in practice may be quite different from those listed, since they depend primarily on the solutes processed. The free-solution values of the mobility and diffusion coefficient correspond roughly to those of serum albumin and tend to be high compared with other proteins. The values of the mobility and diffusion coefficient given for gels are reduced from the free-solution values by a factor of 5 in order to exaggerate the retardation effects found in gels. In practice, the retardation factor depends on the composition and degree of cross-linking in the gel, and can be substantially larger or smaller than 5. The ratio of the electrophoretic mobility to the diffusion coefficient in the gel and in free solution, a parameter that frequently appears in theoretical models of electrophoresis, has been kept constant in these calculations.

DISPERSIVE MECHANISM

Diffusion

Before describing the various apparatuses, it is worthwhile to briefly review those phenomena that can lead to deterioration of a separation, that is, band broadening or dispersion (which is a general term for these types of effects). The most ubiquitous dispersive phenomenon is diffusion; one should always take precautions to minimize the effects of diffusion, especially when other band-broadening effects are small. An important characteristic of diffusion-based dispersion is that it increases band width in proportion to the square root of the residence time according to the formula

$$\chi^2_{\text{diffusion}} = 2D_m t \text{ or } \sigma^2_{\text{diffusion}} = \frac{2D_m Z}{\langle u \rangle^3} \quad \ldots(1)$$

where the spatial variance χ^2 represents band spreading in space, as would be observed by scanning bands on a gel that had been run for time t. The temporal variance σ^2 represents band spreading in time, as would be observed as bands migrate at a mean electrophoretic velocity $\langle u \rangle$ past a fixed detector located at mean position $z = Z$ relative to a starting point $z = 0$. The width of a solute band in space can be approximated as 4σ and its duration in time as 4χ. When several different mechanisms contribute to dispersion, their net effect is usually taken into account by summing the appropriate variances, in either space or time. This assumes that the individual effects are independent and is often a reasonable approximation.

Effect of Temperature

The next major source of dispersion comes from the effect of spatial temperature variations on the electrophoretic mobility. All forms of electrophoresis require that a current be passed through a resistive medium, so electrical energy is necessarily converted to thermal energy in the medium. To efficiently remove the joule heat generated by the current, one spatial dimension of the chamber is generally kept thin, and heat is removed through the bounding surfaces along this axis. The mobilities of charged solutes and aqueous electrolytes usually vary inversely with the buffer viscosity, which is itself quite similar to water in its temperature dependence. The viscosity of water decreases by about 25% with a 10°C rise in temperature, so the warmer parts of a solute band migrate faster than the cooler parts and often take on parabolic or crescent shapes. This deformation results in a broadening of the fractionated sample, which usually overwhelms diffusive spreading, especially at the field strengths used in high-performance equipment (e.g., >200 V/cm).

Process of Electroosmosis

Electroosmosis can be a severe problem in free-flow systems (i.e., those with no anticonvective medium). It arises when mobile counter ions in the double layer adjacent to a charged wall are moved parallel to the wall surface by an electric field. These ions drag solvent with them, setting up a bulk fluid motion that can smear solute bands. Although it is difficult to eliminate electroosmosis entirely, for example, by using surface coatings or additives, steps may be taken to control it or even to take advantage of it. In homogeneous gels, electroosmosis plays a lesser role in band distortion but may act to displace purified bands, thereby giving inaccurate estimates of solute mobilities. In capillary zone electrophoresis at alkaline pH, electroosmosis typically provides the convective flow used to transport solutes through the capillary and does not contribute significantly to dispersion. Other problems associated with Kohlrausch effects and conductivity gradients can have a major impact on band broadening.

The former gives rise to stable band broadening or narrowing due to differences in mobilities between the target proteins and the electrolyte ions, but without inducing bulk flows. The latter may generate stable or unstable flows that can deform solute bands in a number of ways. The instability associated with conductivity gradients is particularly deleterious. As a rule of thumb, to avoid these problems with native zone electrophoresis, the electrolyte composition and conductivity should closely match those of the sample and should provide adequate buffering; the sample proteins should not be too heavily concentrated and the electric power applied should not exceed the effective cooling capacity of the column.

Continuous Flow

Although problems with thermally induced natural convection have long hampered scale-up of the thin-film continuous-flow electrophoresis device, these problems are largely mitigated in recycling electrophoresis equipment and the membrane-based chambers developed by Faupel et al. and Horvath et al. This is not meant to imply that natural convection is a trivial problem but rather that dispersion due to natural convection can be ameliorated by innovative and careful design.

Dielectric Flows

Although flows induced by gradients in the conductivity and dielectric constant are still largely a curiosity in electrophoretic separations, they will become major barriers to device development as electric fields are increased to boost chamber performance. The cusp-like distortion of conductive dye streaks containing 100 mM NaCl in a clear solution containing 10 mM NaCl with the flow stopped and the streamlines exposed to a nominal electric field of about 26 V/cm. When the salt is replaced with different concentrations of ethanol, the induced flows have essentially the same morphology, suggesting a common mechanism for both flows. One important feature of this instability is that conductive dispersion can be explosive, smearing finely resolved protein bands over a distance of several centimeters along the electric field lines in a few seconds. A stable form of these flows has been described and analyzed by Roberts and coworkers and the unstable form recently analyzed by Baygents and Baldessari. These papers support the empirical rule of thumb that bands in zone electrophoresis or isoelectric focusing should have conductivities as close to those of the surrounding buffer as possible to avoid conductivity gradients. These flows can be stabilized by applied shear stresses, but at fields above 400 V/cm even shear rates in excess of about 100 s^{-1} are only marginally effective at damping these flows.

Measurements

Chamber design

In order to facilitate comparison of chamber designs and separation techniques, it is useful to have a yardstick or metric that can be universally applied not only to electrophoresis but also to other techniques. The simplest and most commonly used performance metric is the plate number *N*:

$$N_{spatial} = \frac{Z^2}{\chi^2} \text{ or } N_{temporal} = \frac{t^2}{\sigma^2} \qquad \text{...(2)}$$

for spatial or temporal analysis, respectively, where the various mean displacements and variances are taken from a single peak. Strictly speaking, the plate efficiency is only applicable to rate-based separations, for example, zone electrophoresis or isocratic chromatography, as opposed to (pseudo)equilibrium-based separations, for example, isoelectric focusing or gradient-elution chromatography. Typical values of the plate number in zone electrophoresis range from below 1,000 for the high-capacity BioStream apparatus to above 10^7 for PAGE in gel-filled capillaries. A closely related concept to plate number is plate height, HETP = $Z/N_{spatial}$ or = $(\langle u\rangle t)/N_{temporal}$. Although neither plate height nor plate number can be considered a truly universal metric for comparing different separation columns, they are both useful ways of checking performance within an individual column, for example, after a packing has been changed in an electrochromatography column.

Resolution technique

A more universal metric for separation columns that can be applied to both rate-based and equilibrium-based techniques is resolution. An approximate form for the resolution that is sufficiently accurate for most applications is

$$R_{spatial} = \frac{|Z_1 - Z_2\rangle}{\frac{1}{2}(W_1 + W_2)} \approx \frac{|Z_1 - Z_2|}{2(\chi_1 + \chi_2)} \text{ or } R_{temporal} = \frac{|t_1 - t_2|}{\frac{1}{2}(W_1 - W_2)} \approx \frac{|t_1 - t_2|}{2(\sigma_1 + \sigma_2)} \qquad \text{...(3)}$$

for spatial or temporal analysis, respectively, where the relationship between the peak width at its base and the variance is $W = 4\chi$ (spatial) or 4σ (temporal). Two key components with an *R* equal to or slightly greater than 1 can be considered to be baseline separated, so *R* can be used either to compare data from different separation devices or to modify the column so that complete separation can be achieved.

REQUIREMENTS

Capillary Electrophoresis

Capillary electrophoresis (CE) offers important advantages over conventional gel electrophoresis (GE) in the characterization of proteins and nucleic acids. CE is much faster than GE, and can be fully automated and instrumented to include sample injection, detection, and collection. In addition, CE can separate solutes at a resolution significantly greater than that of HPLC. Very small sample loads (e.g., <10 ngs/run) make CE suitable primarily for analytical work. Automated equipment for CE is commercially available from several vendors, including units that allow collection of fractionated samples or direct coupling to a secondary separation device (e.g., HPLC or mass spectrometer).

Principle

In principle CE is quite simple. It can be carried out at very high power densities because the large surface area to volume ratio of the capillary lumen allows it to shed joule heat efficiently. However, in practice it has not yet reached the level of maturity of automated chromatography (i.e., where stand-alone columns can be operated by individual staff on an as-needed basis). CE is best operated by a trained technician who retains responsibility for developing protocols. The basic apparatus consists of a fused-silica capillary with a lumen diameter typically ranging from 25 to 150 μm, an outer diameter of several hundred microns (e.g., 400 μm) and a thin coat of polyimide that makes the otherwise brittle capillary quite flexible. A small window is burned in the polyimide near one end of the capillary, and this window is then placed in the beam of a spectrophotometer.

To avoid damaging the sensitive electronics in the detector, this end of the capillary is dipped into a vial filled with running buffer and the electrode placed in this vial is set to electric ground. Before loading a sample into a bare capillary the lumen is usually precleaned with 0.1–0.5 M sodium hydroxide and then rinsed sequentially with pure water and running buffer. The capillary is loaded by dipping the high-voltage end into a vial containing the sample and using a pressure head or electric field to load 10–100 nL of sample, which is roughly 0.0 1–0.1 μg at a concentration of 1 mg/mL. The high-voltage end of the capillary is then moved to a vial containing running buffer, the hot electrode is dipped into this vial, and the voltage is increased until the electric field falls to a range of 200–400 V/cm. These large fields drive solutes through the capillary at electrophoretic velocities on the order of 1–4 cm/min, yielding run times of a few minutes with plate efficiencies approaching 10^6 plates per meter of capillary in free solution.

Basis of electrophoresis

Two important points that must be considered when discussing CE performance are the temperature profile in the capillary and dispersion due to temperature variations. The temperature profile can be calculated using the steady-state form of the thermal energy equation in cylindrical coordinates

$$\frac{k_{buffer}}{r}\frac{d}{dr}r\frac{dT}{dr} + \sigma_{buffer} T_{ref} E^2[1+\lambda(T - T_{ref})] = 0 \quad \text{where } \lambda \equiv \left[\frac{d \log_e(\sigma_{buffer})}{dT}\right]_{T = T_{ref}} \quad ...(4)$$

where the first term accounts for radial conduction and the second for generation of joule heat. The second term in brackets accounts for the autothermal effect, that is, the tendency of warm buffer to conduct more current than cold buffer. The thermal energy equation, equation 4, must be solved together with the boundary conditions

at $r = 0$ $\quad T$ is finite

$$\text{at } r = a \quad -k_{buffer}\frac{dT}{dr} = \frac{1}{2}h_{CE}(T - T_{ref})$$

$$= \frac{1}{2a}\frac{T - T_{ref}}{\left[\frac{1}{k_g}\log(a_g/a) + \frac{1}{k_p}\log(a_p/a_g) + \frac{1}{a_p h_{fluid}}\right]} \qquad \ldots(5)$$

The boundary condition at $r = 0$ states that the temperature must remain finite at the centerline, and the wall condition at $r = a$ indicates that heat transfer from the capillary buffer to the environment outside the capillary is controlled by the serial resistances of the glass capillary, the polyimide coat, and the boundary layer in the external cooling fluid surrounding the capillary.

These three resistances correspond respectively to the three terms in the denominator of the boundary condition at $r = a$. The overall heat transfer coefficient between the lumen OD and the coolant is h_{CE}. Equation 4 with the boundary conditions in equation 5 has the solution

$$\Theta_{buffer}\eta = \frac{1}{\kappa^2}\left(\frac{BiJ_0(\kappa\eta)}{BiJ_0(\kappa) - 2\kappa J_1(\kappa)} - 1\right) \qquad \ldots(6)$$

with dimensionless parameters defined as

$$\Theta_{buffer} = \frac{k_{buffer}(T - T_{ref})}{\sigma_{buffer}E^2a^2}\eta = \frac{r}{a}Bi = \frac{2h_{CE}a}{k_{buffer}}\kappa^2 = \lambda\frac{\sigma_{buffer}E^2a^2}{k_{buffer}}$$

where Bi is the Biot number. J_0 and J_1 are Bessel functions of the first kind of order 0 and 1, respectively. The temperature profile in the capillary lumen is roughly parabolic with its maximum at the centerline. The temperature in the lumen can vary markedly with applied voltage, as well as with the overall heat transfer coefficient h_{CE}.

However, the variation across the radius of the lumen is usually less than 1°C, and even this small variation can have a marked impact on dispersion in CE. This occurs because the electrophoretic mobility varies inversily with viscosity. Viscosity, in turn, decreases with increasing temperature, so solutes migrate faster in the hot center of the lumen than near the lumen wall,

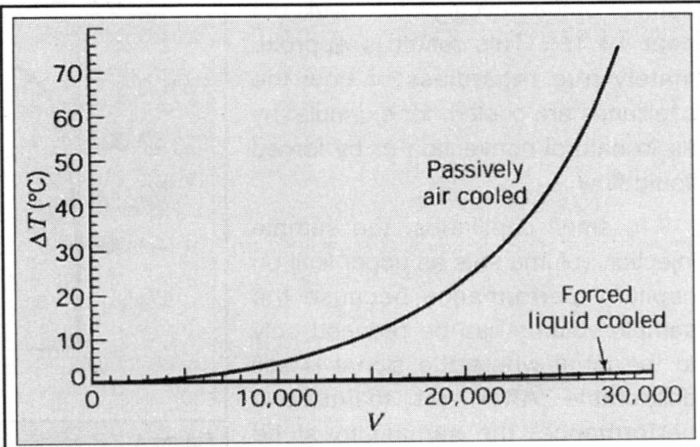

Figure 9.1 Temperature rise in CZE lumen as a function of applied voltage V.

distorting the solute band into a parabolic shape. When this is taken into account, the effective dispersion coefficient in the capillary becomes

$$D_{CE} = D_m + \frac{a^2(\mu(T_{ref})E^2)}{3072D_m} \frac{\kappa^4 Bi}{(BiJ_0(\kappa) - 2\kappa J_1(\kappa))} \quad ...(7)$$

The first term on the right accounts for pure axial diffusion, and if only this term is included it predicts a temporal variance equal to that given in equation 1. Under normal operating conditions the second term is roughly the same order of magnitude as the first, indicating that thermal band broadening is generally significant in CE.

Performance benchmarks

Modifying the expression for the number of theoretical plates to include sample volume, wall electroosmosis, and the extended dispersion coefficient, yields the formula

$$\frac{1}{N_{CE}} = \frac{2D_{CE}}{|(\mu + \mu_{eo})\Delta\Phi|} + \frac{1}{12}\frac{(q/\pi a^2)^2}{Z^2} \quad ...(8)$$

where μ_{eo} is the electroosmotic mobility and q is the volume of sample loaded into the capillary. The first term on the right gives the contribution from distortion due to thermal effects and the second term gives the contribution from the variance of the initial sample injection volume q. When hydrostatic loading is used, the injection volume can be calculated as

$$q = \frac{\rho g H}{8\nu Z}\pi a^4 t_{load} \quad ...(9)$$

where ρ and v are the density and viscosity of the buffer, respectively, and H is the difference in height between the liquid levels in the vials into which the two ends of the capillary are immersed, assuming both vials are at atmospheric pressure. Alternatively, if a pressure differential is used to load the capillary, then ρgH can be replaced with ΔP. As illustrated in Figure 9.2, for capillaries whose lumen ID is less than 100 μm and for a loading of about 1% of the column volume, the number of theoretical plates levels out near 1×10^5. This result is approximately true regardless of how the capillaries are cooled, for example, by air in natural convection or by forced liquid flow.

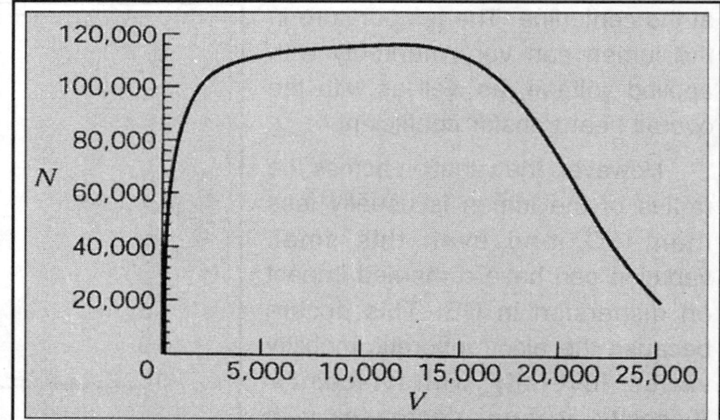

Figure 9.2 Predicted plate number as a function of applied voltage for a 75-µm ID x 350-µm OD x 100-cm capillary in air using an injection volume, q = 50 nL.

In small capillaries, the sample injection volume sets an upper limit on capillary performance because the sample volume can be reduced only to the point where the signal is still detectable. After that, to improve performance, the sample must be preconcentrated or stacked or the sensitivity of the detection system somehow improved. Clearly CE will benefit from the development of protocols for stacking or focusing injected solutes in free solution before starting a run. The main advantages of forced cooling are that it (i) dramatically reduces the temperature rise in the capillary lumen by reducing the heat transfer resistance in the external fluid and (ii) reduces the thermal fluctuations associated with natural convection. Using conventional sample-loading protocols, there is little benefit in plate efficiency gained by forced liquid cooling. However, as stacking procedures become more common, liquid-cooled CE systems should eventually outperform both passively and actively air-cooled units.

Gel Electrophoresis

For several decades now, polymer gel electrophoresis has been a standard method used for laboratory analysis of proteins, nucleic acids, and other biological materials. The primary advantages of GE over other applicable techniques are (i) moderately high resolution, that is, native PAGE routinely provides more than 10,000 theoretical plates and (ii) the protocols are well established and widely accepted in the field. All engineers and scientists who enter the biotechnology industry will, at some point in their careers, find themselves running or reading results from electrophoresis gels.

The primary disadvantages of conventional GE are that it is (i) labor intensive; (ii) slow, often requiring 24 h between initial buffer preparation and final destain; (iii) difficult to recover purified solutes for subsequent steps; and (iv) difficult to scale beyond a few milligrams of sample. My experience with PAGE is that it takes several weeks of practice to be able to produce publishable gels, and even then users may experience problems with gel-to-gel reproducibility, especially between departments within a company. Given these limitations, it is not surprising that an effort would be made to develop instrumentation that retains the advantages of GE while minimizing its disadvantages. One can reduce the size and thickness of the gel, actively remove joule heat from one or both sides of the gel, and automate the stain and destain cycle. This greatly reduces labor requirements, turnaround time, and many of the problems associated with gel reproducibility. Also, a variety of prefabricated electrophoresis gels are now commercially available from a number of vendors for about $10 each.

Detailed protocol

Detailed protocols for preparing, running, and analyzing gels are available in any number of references and are not discussed further here. In all likelihood, the end user will run some combination of SDS or native PAGE with the sample in reduced or nonreduced form, in Tris or tricine buffer, with or without a gradient in cross-linking agent, using one of several stacking procedures and Coomassie blue or silver stain to visualize the fractionated bands. Within a biotech company, it is also likely that the gel protocols, including stain and destain procedures, will be specified by a quality assurance or analytical chemistry department or their equivalent. These protocols will almost certainly include a ladder of molecular weight markers in one lane and additional lanes containing controls appropriate to the separation. IEF-PAGE gels or capillary IEF should be run in conjunction with the molecular weight gels to obtain pIs for the various protein isoforms and glycoforms that inevitably accompany production of recombinant therapeutics. Not only does this give a clearer picture of both contamination and heterogeneity of the purified protein, but it may also help in preparing a rational design for downstream processing.

Principles

Although some progress has been made in understanding how proteins migrate through tight gels from a practical standpoint it is not necessary to understand the underlying theory to produce publishable gels. Once an appropriate gel and running buffer have been selected for the molecular weight range of interest, it is best to run the gel as quickly as possible to minimize diffusive band spreading. As a rule of thumb, an electrical power density of 0.1 W/cm^3 per gel should be reasonably low for use with air-cooled, glass-walled, vertical slab PAGE. (*Note*: To calculate watts, multiply amps times voltage. If the gel is set up correctly, nearly all of this power will be dissipated in the working portion of the gel slab.) However, if the apparatus comes with a run protocol, these directions should be followed closely, at least during the initial stacking phase of the run. Once the proteins have entered the separating gel, the user has a bit more flexibility with the applied voltage, depending on how well the gel box is cooled, but the voltage should probably not exceed 0.2 W/cm^3 per gel in an air-cooled box.

Marking the results

Once the stacked proteins have entered the separating gel, the conventional wisdom is that band spreading in near-isothermal gels is controlled by diffusion. Because noneluting gels are analyzed in space, the number of theoretical plates is given by equation 2 as

$$N_{\text{SDS-PAGE}} = \frac{Z^2}{\chi^2} = \frac{Z^2}{2D_{\text{eff}}t} = \frac{(\mu_{\text{eff}}Et)Z}{2D_{\text{eff}}t}$$

$$= \frac{(\mu_{\text{eff}}\Delta\Phi)Z}{2D_{\text{eff}}L_{\text{gel}}} = \frac{\mu_{\text{eff}}\Delta\Phi}{2D_{\text{eff}}}\frac{Z}{L_{\text{gel}}} \approx 10{,}000 \quad ...(10)$$

where the approximation is for a 5-cm displacement. The penultimate term, which is a classical result for electrophoresis, shows how applied voltage, gel length, and so on, affect performance.

This formula also implies that high performance can be maintained as the gel is scaled down to about the size of a glass slide if the voltage drop is fixed during miniaturization. The higher-power densities generated can be accommodated by replacing the passive air-cooled, glass-encased gel by an actively cooled (e.g., Peltier system), ceramic-encased gel and by making the gel as thin as is reasonable for handling and detection. Actively cooling both sides of the gel would allow the user to apply about twice the voltage and would therefore be expected to double performance, that is, double the run speed and increase resolution by 41%.

METHODOLOGY

Laboratory Preparation

Continuous elution preparative gel electrophoresis (pGE) is a relatively new entry in the field of preparative bioprocessing and is intended to address the limitations of slab GE. The essence of this technique is that a pulse of multicomponent sample can be loaded onto a column and fractionated using well-known protocols on a standard electrophoresis gel and then continuously recovered in free solution as it migrates off the end of the gel. This technique may be viewed as a melding of conventional gel electrophoresis with electroelution. In combining these two steps, it is possible to relieve several of the disadvantages of GE mentioned earlier, especially by automating and accelerating the otherwise labor-intensive sample-recovery step.

Principle

A generic pGE apparatus consists of three basic parts: (i) a tube- or annular-shaped electrophoresis gel, (ii) an elution chamber at the base of the gel, and (iii) a product off take line leading to a detector and/or fraction collector. In practice, a sample is loaded into the liquid layer immediately adjacent to the top of a standard gel and then electrophoresed, at low power to avoid mixing due to natural convection, into the top of the gel.

Once inside the gel, electrophoresis of the solutes is carried out at power densities at or below 1 W/cm^3. The solutes migrate through the gel at different velocities, and as each solute band reaches the end of the gel it elutes into free solution in the elution chamber and is carried to a fraction collector by the elution buffer. Once the effluent has been aliquoted by the fraction collector, the recovered solutes, which are suspended in the elution buffer, are ready for biochemical analysis or for further processing. Although only a single batch sample can be processed in each run, the gel can be reused several times before it must be replaced. A typical column load ranges from less than 100 μg to more than 10 mg per run depending on the cross-sectional area of the gel; typical run times are somewhat less than 30 min per centimeter of column.

Gel electrophoresis

In the electrophoresis gel, the central issue is resolution. This is the heart of the apparatus, and once the purified solutes have eluted from the gel, whatever resolution was achieved there can only be degraded in subsequent processing steps. In this part of the system the major contributors to peak variance are sample volume, molecular diffusion, and thermally induced dispersion.

Sample bandwidth is directly related to the volume of the sample band as it begins electromigration through the gel. However, if a stacking gel is used, sample volume is usually not an important contributor to peak variance. Assuming isothermal conditions and ignoring the contribution to the variance due to sample injection volume, molecular diffusion dominates band spreading by contributing a temporal variance

$$\sigma^2_{\text{diffusior.}} = \frac{2D_{\text{gel}}Z}{\langle u_{\text{gel}}\rangle^3} \qquad \ldots(11)$$

as measured by a detector located at a fixed position Z. Here D_{gel} is the diffusion coefficient of the solute in the gel and $\langle u_{gel}\rangle = \langle \mu_{gel}\rangle E$ is the average electrophoretic velocity of the solute in the gel. Noting that the average holdup time for a gel of length Z is $Z/\langle u_{gel}\rangle$, the number of theoretical plates is

$$N_{diffusion} = \frac{t^2}{\sigma^2_{diffusion}} = \frac{u_{gel}Z}{2D_{gel}} \quad \text{...(12)}$$

For native PAGE, the number of theoretical plates routinely falls in the range $10^3 < N < 10^4$.

Elution steam chamber

As solutes leave the bottom of the gel tube, they are rapidly swept away from the gel, out of the electric field, and into an elution stream. In a properly designed elution chamber, dispersion is not an important issue, but dilution is. If the elution stream is too slow, some or all of each solute will be entrained by the electric field and permanently removed from the elution stream. If the elution stream flow rate is too high, the solute will be diluted, making it difficult to detect or use in subsequent processing steps. Many different strategies could be used to reduce dilution in the elution chamber, and this part of the system offers the greatest potential for innovation. Here only two generic approaches to elution, cross-flow and counter flow, are considered to illustrate how an elution chamber is designed. In cross-flow elution a hydrodynamic flow is directed perpendicular to the electric field at the base of the gel tube and flows out of the elution chamber through an opening on the right.

The minimum elution flow rate is determined by finding the conditions under which a molecule exiting the gel tube and entering the elution stream at the farthest point from the elution channel just avoids being captured by the electric field. In counter flow elution, the hydrodynamic flow is directed opposite to the electrophoretic migration in the elution chamber. The minimum elution flow is found when a molecule's electrophoretic velocity in the elution buffer exactly balances the counter flow. The lower trajectory is captured on the electrode membrane and the upper trajectory is cleared into the elution stream. It turns out that in both of these cases the minimum elution flow rate that discourages capture of the solute by the electrode is

$$Q_{min} = \langle u_{buffer} \rangle A_c \quad \text{...(13)}$$

where A_c is the cross-sectional area of the elution chamber through which the current flows and $\langle u_{buffer} \rangle$ is the electrophoretic velocity in free solution. This is the free solution electrophoresis velocity in the elution buffer and may be substantially different from the velocity in the gel. In practice, the elution flow rate is often several times higher than this minimum value, leading to unnecessary dilution of the sample.

Fraction collector

After being flushed from the elution chamber, the solute is carried by the elution buffer through a length of tubing to a detector and/or fraction collector. In this stage the main consideration is loss of resolution, since the purified solute bands are now subject to Taylor (diffusive–convective) dispersion as they travel through the tubing. This occurs because the flow in the system is laminar, parabolic with a maximum velocity at the tube centerline. Solutes near the wall move slowly compared with those near the centerline so the solute band is elongated by the flow as it travels through the tube. Aris–Taylor theory, which is strictly valid only for long tubes, predicts a dispersion coefficient that goes as

$$D_{Taylor}D_m + \frac{\langle v\rangle^2 d_t^2}{192D_m} = D_m + \frac{Q^2}{12\pi^2 d_t^2 D_m} \qquad \ldots(14)$$

where $\langle v\rangle$ is the average hydrodynamic velocity of the buffer in the tube. Because of Taylor dispersion the off take line can generate significant band spreading if it is improperly designed. The variance predicted by Taylor's theory for dispersion in a tube of length L is

$$\sigma^2_{Taylor} = \frac{D_m L}{32}\frac{\pi^2 d_t^6}{Q^3} + C\frac{\pi d_t^4 L}{384QD_m} \qquad \ldots(15)$$

which indicates that Taylor dispersion may be reduced by using short, thin tubes. For tubes of length $L < Q/\pi D_m$, Aris–Taylor theory overpredicts dispersion and must be adjusted before it can be applied to the lengths of elution channel typically used with these instruments. To put this in perspective, for Aris–Taylor theory to apply in commercial pGE equipment, one would need $L > 1$ m when $Q = 10$ μL/min and $L > 100$ m for $Q = 1$ mL/min. The constant C, which corrects for short tubes, where $\zeta = D_m L/\langle v\rangle a_t^2$.

For example, a 1-m-long tube with a flow rate of 1 mL/min has an adjusted variance of 8,214 s^2 at a dimensionless distance of $\zeta = 0.009425$ ($C = 0.185803$). A pulse of solute placed at the entrance of the tube will take about 6.1 min to flush through the tube entirely. The implication here is that solute bands that come off the gel less than 6.1 min apart will tend to overlap as they come out of the elution stream tube. If this becomes a problem, it is best addressed by decreasing the length of the tube. The effect of Taylor dispersion in short tubes is summarized in Figure 9.3 for the two cases, $Q \approx 1$ mL/min, which corresponds roughly to the annular device, and $Q \approx 10$ μL/min, which corresponds roughly to the tubular device. The temporal bandwidth, that is, the time it takes for a pulse of solute to completely elute from the tube, is approximately equal to 4σ, so this has been plotted against the tube length for each of these flow rates.

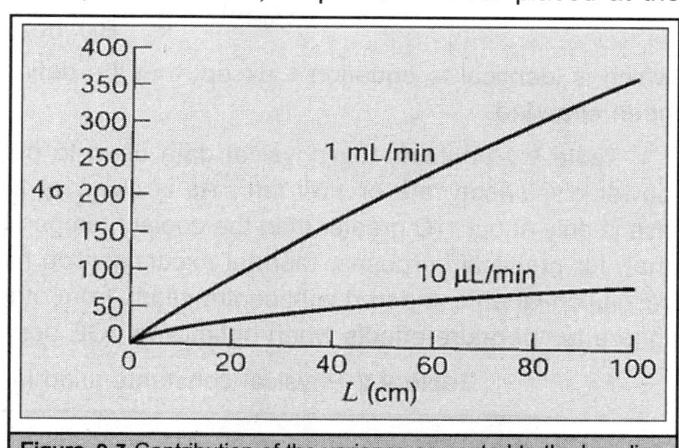

Figure 9.3 Contribution of the variance computed to the baseline width, in seconds, for flow in the elution stream tube.

Note that the temporal dispersion is independent of tube diameter (a counter intuitive result) for a fixed volumetric flow and that dispersion can be decreased by reducing Q and L. Although existing continuous elution systems do not meet the long-tube condition required for Taylor's analysis to apply rigorously, the variance can still be significantly decreased by reducing the inner diameter and length of the elution stream tube.

Basic principles

As with CE, temperature profiles in pGE can be calculated using equation 4, except that the electrical parameters used are those that pertain to the gel. The steady-state form of the thermal energy equation in cylindrical coordinates is

$$\frac{k_{gel}}{r}\frac{d}{dr}r\frac{dT}{dr}+\sigma_{gel}(T_{ref})E^{2}[1+\lambda_{gel}(T-T_{ref})]=0 \qquad ...(16)$$

For a tubular gel, this equation is put in dimensionless form using the transformations

$$\Theta_{gel}(\eta)=\frac{k_{gel}(T-T_{ref})}{\sigma_{ref}E^{2}a_{gel}^{2}}\ Bi=\frac{2h_{pGE}a^{gel}}{k_{gel}}\approx\frac{2k_{glass}}{k_{gel}\log\left[\frac{a_{gel}}{a_{glass}}\right]}\kappa^{2}=\frac{\sigma_{gel}E^{2}a_{glass}^{2}}{k_{gel}}\left[\frac{d\log(\sigma)}{dT}\right]_{T_{ref}}$$

where a_{gel} is the radius of the gel and $\eta = r/a_{gel}$. The heat transfer coefficient h_{pGE} for the micropreparative apparatus includes resistances due to the glass wall of the tube and the external cooling fluid. If cooling is by forced air and the glass wall is thick, the heat transfer resistance is typically controlled by the glass. This is reflected in the approximate term on the far right in the relation for *Bi*.

These transformations yield a second-order ordinary differential equation that, when solved with the boundary conditions from equation 5, has a solution of the form

$$\Theta_{gel}(\eta)=\frac{1}{\kappa_{2}}\left(\frac{BiJ_{0}(\kappa\eta)}{BiJ_{0}(\kappa)-2\kappa J_{1}(\kappa)}-1\right) \qquad ...(17)$$

which is identical to equation 6 except that the definitions of the dimensionless parameters have been changed.

Table 9.2 contains the physical data used to predict temperature profiles in pGE at a base power dissipation rate of 1 W/cm³. As is clear, at this power setting the centerline temperature rise is only about 1°C greater than the coolant temperature. The 1°C variation in temperature implies that, for practical purposes, thermal excursions do not have a significant impact on dispersion or resolution when compared with contributions from other sources. Consequently, it is acceptable to ignore temperature effects when optimizing pGE performance at power densities below 1 W/cm³.

Table 9.2 Physical constants used in the tubular pGE illustrations

	Gel tube
$\sigma(T_{ref})E^2$ (W/cm³)	1.0
Z (cm)	5.0
$2a_{gel}$ (cm)	0.25
d_i (cm)	0.8
Run time (h)	3–5

Basic tube gels

In cases where larger amounts of material must be processed in a single run, the basic tube gel used can be significantly scaled by switching to an annular gel. The gel in the annulus should be kept thin to avoid excessive temperature excursions, but the annulus perimeter can be enlarged to increase capacity. For example, an annulus with a 2-cm inner diameter and a 3-cm outer diameter would have a typical column load of roughly 20 mg/run.

An annulus with a 4-cm outer diameter and a 2-cm inner diameter could process about 50 mg/run. For a thin annulus with a large diameter and an annular gap equal to the diameter of a tubular column, the relative increase in scale of the annulus over the tube is approximately $4a_o/a_{gel}$, where a_o is the outer diameter of the annulus. In the more general case where the annulus is not thin, the relative increase in scale is given by the ratio S of the cross-sectional areas of the two columns

$$S = \frac{a_o^2 - a_i^2}{a_{gel}^2} \qquad \text{...(18)}$$

The capacity can be increased further by increasing the thickness of the annulus and by lengthening the column, but the former admits some degradation of performance due to thermal dispersion and the latter requires longer run times.

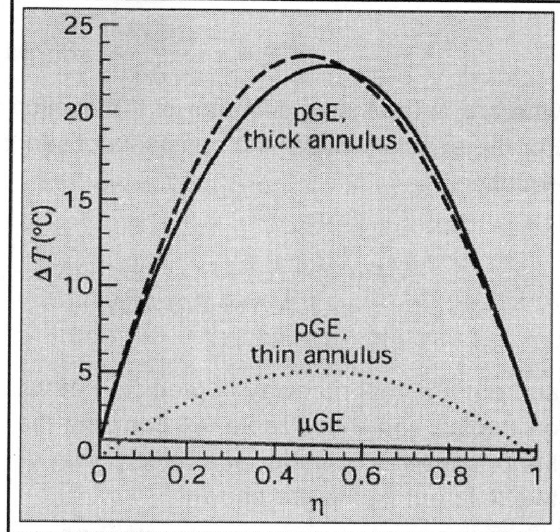

Figure 9.4 Temperature rise (°C) in continuous elution gel electrophoresis at a base power density.

For an annulus, equation 16 is put in dimensionless form by using different transformations than in the previous cases. These are

$$\Theta(\eta) = \frac{k_{gel}(T - T_{ref})}{\sigma(T_{ref})E^2(a_o - a_i)^2} \qquad \eta = \frac{r - a_i}{a_o - a_i}$$

$$\kappa^2 = \frac{\sigma(T_{ref})E^2(a_o - a_i)^2}{k_{gel}}\left[\frac{d\log(\sigma)}{dT}\right]_{T=T_{ref}} \qquad \Lambda = \frac{a_i}{a_o - a_i}$$

where a_i is the radius of the inner surface of the gel and a_o is the radius of the outer surface. Note that (i) the dimensionless radial coordinate η is defined so that the domain extends over $0 < \eta < 1$ and (ii) a new geometric parameter, Λ, appears in the governing equations.

These transformations yield a second-order ordinary differential equation

$$\frac{d^2\Theta}{d\eta^2} + \frac{1}{\eta + \Lambda}\frac{d\Theta}{d\eta} + \kappa^2\Theta = -1 \qquad \text{...(19)}$$

that has solutions of the form

$$\Theta(\eta) = AJ_0[\kappa(\eta + \Lambda)] + BY_0[\kappa(\eta + \Lambda)] - \frac{1}{\kappa_2} \qquad \text{...(20)}$$

where A and B are constants that must be fit using appropriate boundary conditions. J_0 is a Bessel function of the first kind of order 0, and Y_0 is a Bessel function of the second kind of order 0.

Appropriate boundary conditions in this case are

$$\text{at } \eta = 0 + \frac{\partial\Theta(\eta)}{\partial\eta} = Bi_i\Theta(\eta) \text{ where } Bi_i = \frac{k_i}{k_{gel}}\frac{2}{\log(2a_i/d_i)}$$

$$\text{at } \eta = 1 - \frac{\partial\Theta(\eta)}{\partial\eta} = Bi_0\Theta(\eta) \text{ where } Bi_0 = \frac{k_0}{k_{gel}} \frac{2}{\log(d_0/2a_0)} \quad \ldots(21)$$

where d_i is the inside diameter of the cooling finger and d_o is the outer diameter of the glass tube. For the annular device, the constants of integration, *A* and *B*, are obtained by solution of the matrix equation

$$\begin{bmatrix} -\kappa J_1(\kappa(1+\Lambda)) + Bi_0 J_0(\kappa(1+\Lambda)) & -\kappa Y_1(\kappa(1+\Lambda)) + Bi_0 Y_0(\kappa(1+\Lambda)) \\ +\kappa J_1(k\Lambda) + Bi_i J_0(\kappa\Lambda) & +\kappa Y_1(\kappa\Lambda) + Bi_i Y_0(\kappa\Lambda) \end{bmatrix} \begin{bmatrix} A \\ B \end{bmatrix} = \begin{bmatrix} \frac{Bi_0}{\kappa^2} \\ \frac{Bi_i}{\kappa^2} \end{bmatrix}$$

and autothermal runaway is predicted at values of κ for which the determinate of the 2 x 2 matrix on the left vanishes. Table 9.3 contains the physical data used to predict temperature profiles in the preparative chamber at a base power dissipation rate of 1 W/cm³. For purposes of illustration, two different cases are shown.

The first is an instrument with a thin annular gap and the second has a thick annular gap. Since the temperature rise in each of these devices is roughly proportional to $s(T_{ref})E^2(a_o - a_i)^2/k_{gel}$, these results can be readily extrapolated to other operating conditions and gel geometries. To maintain performance levels while scaling up, the product of the electric field times the gap thickness, $E(a_o - a_i)$, should be held constant in order to limit temperature variations across the transverse axis, thereby discouraging excessive thermal dispersion.

Table 9.3 Physical constants used in the preparative GE illustrations

	Thin annulus	*Thick annulus*
L (cm)	10.0	10.0
2 a_i (cm)	1.905	1.905
2 a_o (cm)	2.8	3.7
d_o (cm)	1.565	1.565
d_i (cm)	3.14	4.04
k_{gel} (W/cm °C)	0.00578	0.00578
k_{glass} (W/cm °C)	0.011	0.011
$k_{ceramic}$ (W/cm °C)	0.0578	0.0578
Run time (h)	5–15	5–15

This implies that, while the scale may be doubled by doubling the annular gap thickness, the field must be decreased by half so the run takes twice as long. Consequently, overall productivity in preparative GE remains roughly independent of gap thickness, with the major savings coming from the reductions in the time and effort associated with reloading and purging the gel. As stated earlier, the gel apparatus is most effectively scaled by increasing the perimeter of the annulus while keeping the annular gap thickness constant.

Concentration profiles of solutes

In the absence of thermal dispersion, the concentration profiles of solutes exiting the gel can be calculated by solving the diffusion–convection equation

$$\frac{\partial C}{\partial t} + u_{gel}\frac{\partial C}{\partial z} = D_{gel}\frac{\partial^2 C}{\partial z^2} \quad \text{...(22)}$$

subject to the boundary and initial conditions

$$\begin{array}{lll} \text{at } t = 0 & \text{all } z & C = 0 \\ \text{at } z = 0 & \text{all } t & C = H(0, s)C_o \\ \text{as } z \rightarrow \infty & & \text{all } t \ C \text{ is bounded} \end{array} \quad \text{...(23)}$$

where the distribution function $H(0, t) = 1$ if $0 < t < \tau$ and $H(0, t) = 0$ if $t > \tau$. This problem is most conveniently solved using Laplace transforms to obtain

$$C = \frac{C_0}{2}\left[e^{Pe_{gel}Z/a_{gel}\zeta}\left(\text{Erfc}\left[\sqrt{\frac{Pe_{gel}Z}{4a_{gel}}}\left(\frac{\zeta}{\sqrt{\theta}} + \sqrt{\theta}\right)\right]\right.\right.$$
$$\left.-\text{Erfc}\left[\sqrt{\frac{Pe_{gel}Z}{4a_{gel}}}\left(\frac{\zeta}{\sqrt{\theta - \Theta}} + \sqrt{\theta - \Theta}\right)\right]\right)$$
$$+\text{Erfc}\left[\sqrt{\frac{Pe_{gel}Z}{4a_{gel}}}\left(\frac{\zeta}{\sqrt{\theta}} - \sqrt{\theta}\right)\right]$$
$$\left.-\text{Erfc}\left[\sqrt{\frac{Pe_{gel}Z}{4a_{gel}}}\left(\frac{\zeta}{\sqrt{\theta - \Theta}} - \sqrt{\theta - \Theta}\right)\right]\right] \quad \text{...(24)}$$

where the dimensionless parameters in equation 24 are defined as

$$\theta = \frac{\langle u_{gel}\rangle t}{Z} \quad \zeta = \frac{z}{Z} \quad \Theta = \frac{q}{\pi a_{gel}^2}\frac{\langle u_{gel}\rangle}{u_{buffer}} \quad Pe_{gel} = \frac{\langle u_{gel}\rangle a_{gel}}{D_{gel}}$$

Here θ is the dimensionless time, ζ is the dimensionless distance, Pe_{gel} is the gel Peclet number, and Θ is a dimensionless loading factor. Note that the leading fraction in the definition of Θ is the ratio of the load volume to the gel volume and is normally kept below 5%. The trailing fraction is the ratio of the electrophoretic velocities in the gel and in the buffer. If a stacking gel is used, the estimated volume of the sample as it leaves the stacking gel should be used in calculating Θ. At the lower injection volumes (leading peak) the peak concentration drops below the value in the feed, its breadth decreases, and the center of the peak approaches a symmetric distribution about the mean elution time, $\theta = 1$, that is, $t = Z/\langle u_{gel}\rangle$.

At higher loadings the peak width expands to accommodate the added mass of solute, thus making it more difficult to resolve closely spaced peaks. To minimize the influence of the injection volume, q should be adjusted so that $\Theta; < 1/300$. This is most conveniently done by using a stacking procedure.

Radial temperature variation

To estimate the effect of radial temperature variations on axial dispersion, it is necessary to solve the extended form of the convective–diffusion equation

$$\frac{\partial C}{\partial t} + u(r)\frac{\partial C}{\partial z} = \frac{D_{gel}}{r}\frac{\partial}{\partial r}r\frac{\partial C}{\partial r} + D_{gel}\frac{\partial^2 C}{\partial z^2} \quad \text{...(25)}$$

where $u(r)$ is the axial electrophoretic velocity distribution in the gel. We can estimate the contribution of axial velocity distribution to the variance by dropping the diffusion terms, assuming that the injection volume is infinitesimally small and then using moment analysis to compute the variance. On eliminating the diffusive terms, the mass transport equation reduces to

$$\frac{\partial C}{\partial t} + u(r)\frac{\partial C}{\partial z} = 0 \quad \text{...(26)}$$

A very thin volume of concentrated solute injected onto the column can be represented by the Dirac distribution (impulse function) $C_0\delta(z)$ as an initial condition and the resulting system readily solved (e.g., using Laplace transforms) to obtain

$$C(t,r,z) = \begin{bmatrix} C_0\delta\left(t - \dfrac{z}{u(r)}\right) & \text{when}\ \dfrac{z}{u_{max}} < t > \dfrac{z}{u_{min}} \\ 0 & \text{when}\ \dfrac{z}{u_{max}} > t > \dfrac{z}{u_{min}} \end{bmatrix} \quad \text{...(27)}$$

These solutions are then used to calculate the radially averaged flux J and the temporal moments M_n at the outlet of the gel. The formulas for these are

Radially averaged flux

$$J(t,z) = \int_{a_t}^{a_o} u(r)C(t,r,z) \Big/ \int_{a_t}^{a_o} r\,dr \quad \text{...(28)}$$

$$\text{Temporal moments } M_n(z) = \int_0^{\infty} J(t,z)t^n dt \quad \text{...(29)}$$

The electrophoretic velocity distribution $u(r)$ varies with the buffer viscosity, which is a function of the temperature. For relatively small temperature variations (i.e., less than 10°C), the velocity may be calculated by using a Taylor expansion in the temperature rise to obtain the formula

$$\begin{aligned} u(r) &= \frac{\nu(T_{ref})}{\nu(T)}\mu(T_{ref})E \\ &= \mu(T_{ref})E\left[1-(T-T_{ref})\left(\frac{d\log(\nu)}{dT}\right)_{T_{ref}}\right] \\ &= \mu(T_{ref})E\,[1-\alpha(T-T_{ref})] \end{aligned} \quad \text{...(30)}$$

where μ is the electrophoretic mobility and v is the viscosity of the buffer.

Strictly speaking, the temperature profiles are given by equation 20 for annular gels. However, to generate a relatively simple analytical form for the fluxes, Galerkin's method has been used to find an approximate solution of equation 19 in the form of a parabolic polynomial in η

$$\Theta(\eta) \approx A_0[Bi_b(\eta - \eta^2) + 1] \qquad \text{...(31)}$$

where

$$A_0 = -7[10(\Lambda^2 + \Lambda)(Bi_b + 6) + 3Bi_b + 20]$$
$$\div \{14(\Lambda^2 + \Lambda)[Bi_b^2(\kappa^2 - 10) + 10Bi_b(\kappa^2 - 6) + 30\kappa^2]$$
$$+ Bi_b^2(4\kappa^2 - 49) + 14Bi_b(3\kappa^2 - 25) + 140\kappa^2\}$$

To simplify the analysis somewhat, only the symmetric form of the solution for the temperature profile is considered here. Although this introduces a small error in the variance computed for thick-gap devices, it greatly simplifies the analysis, allows closed-form solution of the resulting equations, and gives the correct qualitative tendency as the power density is increased.

For an electrophoretic velocity profile that is symmetric about the centerline of the annulus and that has the parabolic form $u(\eta) = A + B\eta + C\eta^2$, the radially averaged concentration and flux measured at position Z are

$$C(t,Z) = \begin{cases} \dfrac{\left(B\sqrt{t} + \sqrt{B^2t - 4C(At - Z)}\right)Z}{Ct^2\sqrt{B^2t - 4C(At - Z)}(a_o^2 - a_i^2)} C_c & \text{when } t_{min} = \dfrac{Z}{u_{max}} < t < \dfrac{Z}{u_{min}} = t_{max} \\ 0 & \text{when } t_{min} = \dfrac{Z}{u_{max}} > t > \dfrac{Z}{u_{min}} = t_{max} \end{cases} \qquad \text{...(32)}$$

$$J(t,Z) = \begin{cases} \dfrac{\left(B\sqrt{t} + \sqrt{B^2t - 4C(At - Z)}\right)Z^2}{Ct^3\sqrt{B^2t - 4C(At - Z)}(a_o^2 - a_i^2}C_o & \text{when } t_{min} = \dfrac{Z}{u_{max}} < t < \dfrac{Z}{u_{min}} = t_{max} \\ 0 & \text{when } t_{min} = \dfrac{Z}{u_{max}} > t > \dfrac{Z}{u_{min}} = t_{max} \end{cases} \qquad \text{...(33)}$$

where the parameters, A, B, and C, calculated for the annular columns.

Equation 28 for the radially averaged flux is used to compute the first three temporal moments, and these are then used to compute the mean elution time and the temporal variance. The results shows the reduction in plate efficiency due to thermal dispersion in pGE as a function of the power density. The most interesting aspect of the thermal dispersion is its dependence on column length. The variance associated with thermal dispersion varies with the square of the column length rather than the first power, as is common with convective–diffusive dispersive processes. As a result, thermal dispersion dominates other sources of dispersion in long columns and at high power densities. As a rule of thumb, the optimum power density in existing commercial equipment for pGE should occur at a characteristic temperature rise, $\sigma E^2(a_o - a_i)^2/k_{gel} \approx 8°C$. Although the formulas given are approximate to the extent that molecular diffusion is ignored and only symmetric parabolic temperature distributions are admitted, they provide an excellent qualitative description of the available pGE columns and a rational quantitative basis for designing new units or upgrading existing columns. If higher resolution is needed, the user has a choice of reloading the thin-annulus device several times or building a new instrument with a larger perimeter.

Dilution of solutes

As solute leaves the gel phase, it enters the elution chamber and is then flushed into the elution stream. There is some loss of resolution at this point, but the main consideration here is dilution of the fractionated samples as they come off the gel. The dilution factor $F_{dilution}$ is calculated by taking the ratio of the solute velocity J/C as it comes off the gel to the elution stream velocity Q/A_c. This is

$$F_{dilution} = \frac{J(Z,t)}{C(Z,t)}\frac{A_c}{Q} = \frac{Z}{t_{elution}}\frac{A_c}{Q} \lessapprox \frac{\langle u_{gel}\rangle}{\langle u_{buffer}\rangle} \qquad ...(34)$$

from which it is seen that the dilution factor is a function of elution time.

The approximate inequality on the far right has been included to give perspective to the dilution factor. This approximation is based on the assumption that one would like to operate under conditions in which no solute is captured by the electrodes in the elution chamber. Roughly speaking, the largest attainable dilution factor (i.e., that giving least dilution) is equal to the ratio of the electrophoretic mobility in the gel near the outlet to the electrophoretic mobility in the elution buffer. The utility of this expression is that it may be rearranged to compute the lowest admissible value of the elution stream flow rate Q

$$Q \gtrapprox Q_{min} = \frac{ZA_c}{t_{elution}}\frac{\langle u_{buffer}\rangle}{\langle u_{gel}\rangle} = A_c\langle u_{buffer}\rangle \qquad ...(35)$$

Then, if the electrophoretic velocity of a solute in free solution is known, the elution buffer flow rate can be set very near this minimum value. If a collection of solutes is being processed at a constant elution flow rate, then Q should be calculated using the mobility of the fastest solute (i.e., the first one off the gel) to avoid the loss of fast solutes to the electrode.

Note that the dilution factor depends not only on the tightness (i.e., the reduction of the solute's diffusion coefficient) of the gel being run but also on the individual electrophoretic mobilities of the solutes being processed. This implies that, to keep the dilution factor constant during a run, the elution buffer flow rate would need to be decreased in inverse proportion to time. This latter type of correction could be important in applications where peak width or height must be accurately measured, for example, so that a direct comparison of bands may be made with those on a standard gel. It may also be important in minimizing dilution of the slower solutes.

Important steps

The continuous elution pGE annular apparatus is capable of processing several hundred times more solute in a single pass than the tubular apparatus described earlier. To accomplish this, some resolution is sacrificed for capacity, and because the optimal run conditions in the annulus are found at a fraction of the power density used in the tube the runs generally take longer to complete. It is up to the user to decide for a given separation whether this trade-off between time and resolution is worthwhile.

CONTINUOUS-FLOW ELECTROPHORESIS

Thin-Film Electrophoresis

When still greater amounts of material must be processed, the apparatus that has the next largest capacity is a thin-film continuous electrophoresis device, such as the VaP 22, which was until recently available through BenderHobein, FRG. This type of apparatus, which dates back to the early 1960s, consists of a thin film of buffer that flows down between two closely spaced (e.g., 0.5-mm transverse) vertical glass plates about 20 cm across (lateral) and 50 cm long (axial), as illustrated in Figure 9.5. The CFE is usually run in free-flow mode, that is, without anticonvectant media. Multicomponent sample is injected into the carrier buffer at the top of the column through a thin capillary.

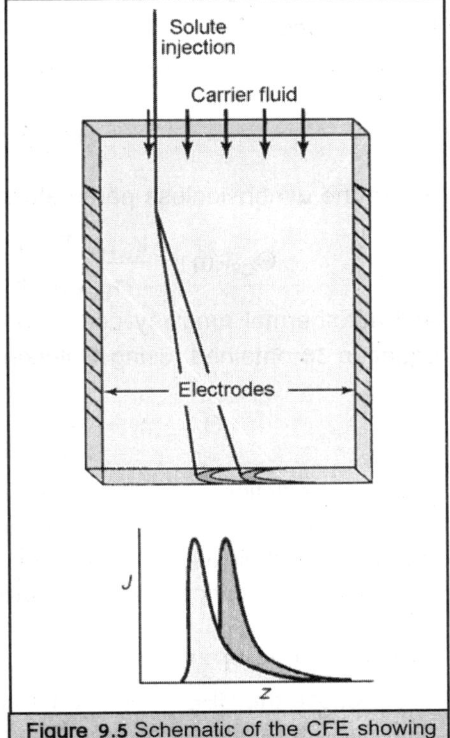

Figure 9.5 Schematic of the CFE showing electrode placement, multicomponent sample injection, carrier fluid flow, and product distribution.

The solutes are convected down with the buffer, flow between electrodes that are located at the lateral edges of the plates, and are deflected by the electric field according to their electrophoretic mobilities. As the fractionated solutes exit the electric field, they are swept into discrete effluent ports by the carrier buffer and collected in separate vials.

A major advantage of the CFE over equipment described earlier is that it operates in true continuous mode, whereas the others operate in semibatch mode (i.e., samples are loaded batchwise, but product fractions are eluted continuously). Processing rates can approach 50 mg/h, and theoretical plate efficiencies can in theory exceed $N = 10{,}000$, although in practice they are generally somewhat lower.

Basic principles

The first step in analyzing the performance of the CFE is to calculate the temperature profile along its thin (transverse) axis. The governing differential equation is

$$k_{buffer} \frac{d}{dy}\left(\frac{dT}{dy}\right) + \sigma_{buffer}(T_{ref})E^2[1+\lambda(T - T_{ref})] = 0 \qquad \ldots(36)$$

with the boundary conditions

$$\text{at } y = \pm b \quad -k_{buffer}\frac{dT}{dy} = \pm h_{CFE}(T - T_{ref}) \approx \pm\frac{k_{glass}}{d_{glass}}(T - T_{ref}) \qquad \ldots(37)$$

Here it is assumed that both transverse surfaces are cooled, that the major resistance to heat transfer is the thickness of the transverse surfaces, and that the heat transfer coefficients are equal on both sides of the unit. Note that asymmetric conditions could significantly impair CFE performance, so every effort should be made to match the electrokinetic and thermal properties of these surfaces.

Equation 36 with symmetric boundary conditions as in equation 37 has the solution

$$\Theta_{CFE}(\eta) = \frac{1}{\kappa^2}\left[\frac{Bi\cos(\kappa\eta)}{Bi\cos(\kappa) - \kappa\sin(\kappa)} - 1\right] \qquad \ldots(38)$$

where the dimensionless parameters used are

$$\Theta_{CFE}(\eta) = \frac{k_{buffer}(T - T_{ref})}{\sigma_{buffer}E^2b^2} \quad \eta = \frac{y}{b} \quad Bi = \frac{h_{CFE}b}{k_{buffer}} \quad \kappa^2 = \lambda\frac{\sigma_{buffer}E^2b^2}{k_{buffer}}$$

and autothermal runaway occurs when $Bi\cos(\kappa) - \kappa\sin(\kappa) = 0$. The parabolic approximation to equation 38 obtained, using Galerkin's method is

$$\Theta_{CFE}(\eta) \approx \frac{5(Bi + 2)(Bi + 3)}{2[5Bi(Bi + 3) - \kappa^2(2Bi^2 + 10Bi + 15)]}\cdot\left(1 - \frac{Bi\,\eta^2}{Bi + 2}\right) \qquad \ldots(39)$$

and the predictions of both of these formulas are typical operating conditions. Because of the symmetry inherent in this problem, the exact and approximate solutions differ by only a few percent over most of their useful range. The agreement between equations 38 and 39 breaks down as K approaches the critical value for autothermal runaway.

Axial velocity profile

Because the CFE does not normally use an anticonvective packing, the velocity profiles in the axial and lateral directions have a significant impact on performance. Along the *x* axis, the velocity profile between the front and back plates is, to a first approximation, laminar and parabolic. However, because of the small temperature difference between the walls and the centerline, there is a slight distortion of the flow due to stable natural convection, which can either improve or degrade performance. Assuming that the base flow is parabolic, the perturbation to the flow is found by solving the Boussinesq approximation to the equations of motion

$$\nu_{buffer}\frac{d}{dy}\left(\frac{dv}{dy}\right) + \rho_{buffer}g\beta(T - T_{ref}) = \frac{\Delta P'}{L} \qquad \ldots(40)$$

where

$$\beta = \left[-\frac{d\log(\rho_{buffer})}{dT}\right]_{T = T_{ref}}$$

with no-slip conditions on the bounding surfaces

$$\text{at } y = \pm b \quad v(y) = 0 \qquad \ldots(41)$$

and subject to the integral constraint that there is no net increase in the average velocity associated with natural convection, that is,

$$\int_{-b}^{b} \nu(y)dy = 0 \qquad \ldots(42)$$

Note that P' is a perturbation pressure that determined by the integral constraint of equation 42. When the exact form for the temperature profile is used, this system of equations yields the solution

$$\nu(y) = \frac{3}{2}\langle\nu\rangle\left((1-\eta^2) + \frac{Gr}{Re}\cdot\frac{Bi\{3(\eta^2-1)[\sin(\kappa)-\kappa\cos(\kappa)]+2\kappa[\cos(\kappa\eta)-\cos(\kappa)]\}}{3\kappa^5[Bi\cos(\kappa)-\kappa\sin(\kappa)]}\right) \qquad \ldots(43)$$

while the polynomial approximation yields

$$\nu(y) = \frac{3}{2}\langle\nu\rangle\left\{(1-\eta^2) + \frac{Gr}{Re}\cdot\frac{Bi(Bi+3)[5(1+\eta^2)-6](1-\eta^2)}{36[\kappa^2(2Bi^2+10Bi+15)-5Bi(Bi+3)]}\right\} \qquad \ldots(44)$$

where

$$Re = \frac{\rho_{buffer}\langle\nu\rangle b}{\nu_{buffer}} \text{ and } Gr = \frac{\rho^2_{buffer} g\beta}{\nu^2_{buffer}}\frac{\sigma_{buffer}E^2 b^5}{k_{buffer}}$$

Under normal CFE operating conditions the ratio of the Grashof number to the Reynolds number in equations 43 and 44 is somewhat less than 1, and the functional portion of the perturbation velocity never exceeds ±0.01, so the contribution of natural convection in commercial CFEs, $b < 1$ mm, can be neglected when $Gr/Re < 1$. Note, however, that $Gr/Re \propto b^4$, so this term rapidly becomes important as the slit thickness is increased.

Lateral directions

Because the transverse surfaces of the CFE carry a fixed charge, the electric field induces wall electroosmosis, which drives a bulk recirculating flow along the lateral axis. The shape and magnitude of this flow can be calculated from the equation of motion

$$\nu_{buffer}\frac{d}{dy}\left(\frac{dw}{dy}\right) = \frac{\Delta P'}{L} \qquad \ldots(45)$$

together with the integral constraint

$$\int_{-b}^{b} w(y)dy = 0 \qquad \ldots(46)$$

Assuming that the induced electroosmosis is identical on both transverse surfaces, perhaps due to application of a surface coating, a suitable set of first-kind boundary conditions is

$$\text{at } y = \pm b \quad w(y) = w_{eo} \qquad \ldots(47)$$

Solving the differential equation 45 subject to boundary conditions as in equation 47 and constraint from equation 46 yields the polynomial

$$w(\eta) = -\frac{3}{2}w_{eo}(1-\eta^2) + (w_{eo} + \langle u\rangle) \qquad \ldots(48)$$

where the electrophoretic velocity of the solute $\langle u \rangle$ has been tacked on to account for displacement of the solute relative to the electroosmotic flow. Note that when $w_{eo} = -u$, the term on the far right vanishes and the lateral flow becomes parabolic.

Dispersion of solutes

In practice, a multicomponent feed is injected into the CFE carrier flow through a capillary that extends over one-third to one-half of the transverse thickness of the chamber. In most instances it is possible to neglect the effects of molecular diffusion on dispersion because convective distortion of the sample band dominates band spreading. The criterion for neglecting molecular diffusion is that the time required for a molecule to diffuse from the centerline to the transverse will be much greater than the solutes' average holdup time in the chamber. This ratio is characterized by the dimensionless Graetz number

$$Gz = \frac{t_{diffusion}}{t_{convection}} = \frac{b^2/D_m}{L/\langle v \rangle} = \frac{\langle v \rangle b}{D_m} \cdot \frac{b}{L} \geq 10$$

which for proteins typically has values well in excess of 10.

Because this apparatus operates continuously and solutes are separated in space along the lateral axis rather than in time, it is appropriate to use spatial moments rather than temporal moments to characterize performance. If diffusive dispersion is ignored, the spatial moments can be found using the formula

$$M_n(L) = \left(\frac{L}{b}\right)^n C_o \int_{-\eta^*}^{+\eta^*} \frac{w(\eta)^n}{v(\eta)^{n-1}} d\eta \qquad \ldots(49)$$

where η^* is the fraction of the transverse thickness over which sample injection extends. In this calculation the feed band is assumed to be infinitesimally thin along the lateral axis. Neglecting the effects of natural convection in equation 44, we can readily generate analytical expressions for the mean lateral displacement and spatial variance. Analytical formulas for the mean lateral displacement, the spatial variance, the theoretical plate number, and the resolution in the CFE, taking into account only pure convective dispersion. Several points are brought to light in the moment analysis that are crucial to CFE operation.

First, as is clear from the premultiplicative terms in the variance and the plate number, performance is optimized when $\langle u \rangle = -w_{eo}$, a condition that can be approximated by coating the transverse surfaces with a material that has a zeta potential similar to that of the key solutes being processed.

Second, the plate number N_{CFE} is a function neither of column length nor of the carrier fluid velocity, and is therefore not a function of residence time. This occurs because, like thermal dispersion, convective band spreading due to electroosmosis is faster than a diffusion-like process; that is, CFE variance increases with the first power of the residence time rather than the square root.

Third, as $\eta^* \to 1$, the variance increases rapidly, and, from a purely theoretical standpoint, if diffusive effects are ignored it becomes infinite in this limit. Performance degrades rapidly as η^*

approaches 1, so that there is a stark trade-off between capacity and resolution. Although plate efficiencies in excess of $N = 20,000$ can be obtained in continuous operation, this is only possible under rather stringent conditions. However, for $\eta^* < 3/4$ and $Gz > 100$, the analysis given earlier should be sufficiently accurate to allow the user to optimize a given separation.

The contribution of the breadth of the sample injection stream to the variance is

$$\chi^2_{\text{injection}} = \frac{W^2}{12} \qquad ...(50)$$

where W is the lateral breadth of the injected stream. This expression is readily incorporated into the calculation for the plate efficiency N by adding this contribution to the spatial variance.

Figure is a three-dimensional plot of N versus $\langle u \rangle$ and η^* for $w_{eo} = -0.01$, which shows how CFE performance varies with operating conditions. Here we see that N is greatest along the line $w_{eo} = -\langle u \rangle$ and for small values of η^*, but for reasonable injection capacities ($\eta^* > 1/3$) rarely exceeds 20,000 plates. In practice one should choose conditions that give at least baseline resolution, $R_{CFE} > 1$, of key components and differences in the mean lateral displacements z, which are at least as broad as the effluent ports.

Dilution factor

For the CFE, the dilution factor that corresponds to that discussed with the GE systems is determined from the size of the sample injection stream as $F_{\text{dilution}} = \eta^{*-1}$, and typically ranges from 2 to 4 depending on the injection flow rate. However dilution by band smearing along the lateral axis in the CFE is often an important consideration and is relatively easy to estimate by using the formula

$$F_{\text{gross}} = \frac{1}{\eta^*} \frac{\sqrt{\chi^2_{\text{injection}} + \chi^2}}{\chi_{\text{injection}}} \qquad ...(51)$$

where $\chi_{\text{injection}}$ is given by equation 50. The numerator in equation 51 is proportional to the breadth of the product in the effluent stream, and the denominator is proportional to the breadth of the sample injection stream. The premultiplicative factor $1/\eta^*$ is simply the dilution factor given earlier. Minimum dilution occurs when $\langle u \rangle = -w_{eo}$, with the broadest operating region centered near $\eta^* = 1/3$. In practice, plate efficiency and dilution must be weighed against capacity.

Recycle Continuous-Flow Electrophoresis (RCFE)

Effluent recycle was introduced for isoelectric focusing by Bier and Egen in 1979. At that time it was believed that recycle could not be adapted to zone electrophoresis because of fundamental differences in the transport processes governing these two techniques. Specifically, isoelectric focusing is an equilibration process in which solutes placed in a pH gradient migrate to the point in the gradient where they have zero net charge. Separation depends primarily on the steepness of the pH gradient and the difference in isoelectric points (pIs) of the solutes. Zone electrophoresis, on the other hand, is a rate process in which solutes migrate at fixed velocities, so separation depends primarily on the length of the run, the electric field strength, and the difference in electrophoretic mobilities.

Recycle was first applied to zone electrophoresis in 1983. The major obstacle that had to be overcome to accomplish this was to find a way to get two solutes with like charges but different electrophoretic mobilities to move in opposite directions. Our solution to this problem works in the following way: Two proteins with similar electrophoretic mobilities can be made to migrate in opposite directions by opposing the electrophoretic motion with a counter flow velocity set close to the average of the electrophoretic velocities of two key components. Solutes migrate at their effective velocities (i.e., electrophoretic plus counter flow) until they reach the "regenerators" on either side of the recycle section.

In the regenerators the counter flow of solvent is altered by means of the recycle lines, to keep solutes from migrating past the outlet port. At the regenerator on the left, the cross-flow rate is decreased; it is increased at the regenerator on the right and adjusted so that solutes cannot escape the off take ports. With proper design, complete separation can be achieved with virtually no loss of biological activity. The RCFE was set up so that on each cycle through the unit the solutes would separate slightly. However, after recycling through the chamber 100 or more times, the solutes are completely separated and can be continuously drawn off at either end of the slit with very low losses. The ultimate resolution of solutes in the RCFE depends largely on the number of recycle stages designed into the chamber. Because the design engineer has complete control of this variable, any degree of product purity can, in principle, be attained. The primary advantage of this configuration is that it can process feeds in excess of 1 g/h and operates in true continuous mode. Its major drawback is that it will split a multicomponent feed into just two products—one that contains components that are faster than the cross-flow and the other one having components that are slower than the cross-flow.

Consequently, complex feeds will generally require a minimum of two passes to isolate the desired product: The first pass could be used to remove faster contaminants and the second pass to remove slower contaminants. Separation is carried out in native buffer with background conductivities near 0.01 mho/cm (50 mM buffer), so low protein solubility is not generally a problem. Power requirements are kept low by using small electric fields (less than 40 V/cm), so the temperature rise in the chamber on a single pass is typically less than 5°C. A multichannel heat exchanger removes joule heat from the recycle streams and sets the base temperature of the separation. In addition, the RCFE is hydrodynamically stable because the chamber is operated adiabatically with upflow. The RCFE is not commercially available. However, commercial recycle IEF chambers, such as Rainin's MinipHor, can be readily modified to operate in RCFE mode.

Important steps

When the RCFE was first conceived, it was believed that zone electrophoresis, which can provide fine resolution of proteins in analytical equipment, would find its niche in the final polishing stages of industrial purification processes. In practice, this apparatus appears to be more useful in the early stages of downstream processing, where complex mixtures must be handled and emphasis is placed on removing solutes that are difficult to handle by other means, for example due to fouling or precipitation, or removing bulk proteins (such as serum albumin from blood plasma) inexpensively and without organic solvents or large amounts of precipitating salts.

CONCLUDING REMARKS

Since 1980, the field of applied electrophoresis has undergone revolutionary changes, spurred in large part by the development and automation of capillary electrophoresis. Instrumentation is available for application of zone electrophoresis at scales up to pilot, so it is now possible to decide, on a rational basis, whether to incorporate electrophoresis as part of a purification train. One could imagine, for instance, applying CE (along with analytical HPLC) early in process development to determine which contaminants could be removed from a product stream by electrophoresis and the optimum pH and buffer conditions at which this might be carried out. The results of this inexpensive and rapid preliminary screening step would allow the user to decide whether zone electrophoresis should be considered at all.

As a rule of thumb, if the mobility difference between key components, typically the desired product and a major contaminant, measured by CE is greater than 25%, then zone electrophoresis may be considered further. Following CE, one could scale up to native pGE in order to recover enough purified protein to check for product heterogeneity, activity, and toxicity, as well as to determine the amino acid sequence. When carried out in PAGE, the resolution achieved is boosted by the filtering power of the gel; however, separations conducted in native agarose gels can give insight to the presence (or absence) of problems with zone electrophoresis at larger scales. As before, one should look for a 25% difference in mobilities of key components before considering continuous zone electrophoresis as a possible process step.

Mobility differences greater than 25%, as measured in native agarose gels, should yield a straightforward and clean separation by zone electrophoresis at pilot scale. This preparative step is somewhat more expensive and time-consuming than CE but is a better predictor of success or failure at larger scales. If an automated GE unit is not available, the user can always resort to vertical or horizontal slab gels. Assuming the preparative step is successful, further scale-up requires a return to free-solution zone electrophoresis in either the CFE or the RCFE. (*Note*: In this step one can also use an RF3 or MiniPhor that has been outfitted for isotachophoresis. However, both the equipment itself and the protocols used must be modified for zone electrophoresis. The advantage of using this equipment is that it is commercially available and relatively inexpensive.) This phase of the study is again more expensive and time- consuming than previous ones, and should be carried out at processing rates between 10 and 100 mg/h. In free solution, the user is forced to confront potential problems that might arise as a result of heat generation or electroosmotic dispersion, hopefully for the last time during scaleup.

As the conductivity of the buffer is reduced, the electric power and chamber cooling requirements decrease, and throughput generally increases because electrophoretic mobilities increase. However, as the conductivity is reduced, wall electroosmosis eventually increases to the point where the separation is ruined; an optimum conductivity should thus be sought. In our experience, it is best to begin this search with a buffer that has a conductivity near 500 μmho/cm, which corresponds roughly to 10 mM Tris-acetate at pH 8.0, and raise the electrolyte concentration if it is necessary to reduce electroosmotic dispersion or to discourage protein aggregation. If the separation is successful at 500 μmho/cm, then the conductivity should be halved and the experiment repeated until a suitable electrolyte concentration is found. Keep in mind that whether Tris or more exotic electrolytes or additives are used, these materials are likely to constitute the major electrophoretic processing costs in zone recycle electrophoresis rather than, say, electric power or cooling.

Specific remarks

Over the past two decades there have been a number of important developments in the area of analytical electrophoresis, the most important and most active of these centering on the commercialization of capillary electrophoresis. In addition, gel electrophoresis has metamorphosed from a labor-intensive laboratory technique to one where the instrumentation is automated and the gels are mass-produced, changes that significantly reduce operator training and improve reproducibility. During this same period, preparative and larger-scale zone electrophoresis has remained largely stagnant, with few innovations surviving to commercialization. There is no agreement over the precise reasons for the lack of interest in electrokinetic processing, although it is clear that the biotech industry is convinced that the tools it already has, specifically the chromatographies, are superior to those currently available via electrophoresis.

It will be nearly impossible to move this technology forward without the cooperation and support of the industry it will serve. What will it take to change this mindset? In my opinion it will take at least one case study where electrophoresis shows itself to be clearly superior to existing separations technologies and offers a straightforward route to scale-up. A number of different electrophoretic devices have not been discussed in this article. For example, neither Hoefer's Isoprime nor BioRad's Rotofor are discussed because they are designed strictly for isoelectric focusing, a topic treated elsewhere in this book. Margolis' Gradiflo is not treated both because it is not yet commercially available in this country and because I am not sufficiently familiar with that device to give it due justice. CACE and various other forms of electrochromatography have been omitted here because they could not be adequately treated in the space allotted to this article. Other recent innovations by Shiue and Pearlstein, Lochmuller and Ronsick, and Shea et al. were omitted from this article because they are still in the early stages of development.

10

BULK PURIFICATION

There are three principle areas of modern biotechnology where crystallization plays important roles. These are (i) The crystallization of proteins as well as viruses, nucleic acids, and macromolecular complexes for X-ray diffraction analysis; (ii) the crystallization of specific proteins as a part of their formulation in a therapeutic compound, agricultural chemical, or drug; and (iii) the crystallization of enzymes, and in some cases other proteins, as a means of purification from bulk processes. Although the objectives in each of these situations may be quite different, with the exception of scale, the underlying principles and approaches are fundamentally the same. The X-ray analysis is a singular event or study, confined to the research laboratory, and the product is basic scientific knowledge.

The crystals themselves have no material or medicinal value, but serve as intermediaries in the crystallography process. They provide X-ray diffraction patterns that in turn serve as raw data allowing the visualization of the macromolecules composing the crystals. Thus, they serve as an essential component in the elucidation of molecular structure. Macromolecular structure is of formidable value in biotechnology because it is the essential knowledge required to apply the technique of rational drug design in the creation and discovery of new drugs and pharmaceutical products. It is the basis of powerful approaches now being applied in small, emerging biotechnology companies as well as major pharmaceutical companies to identify lead compounds to treat a host of human ailments, veterinary problems, and crop diseases in agriculture.

The underlying hypothesis is that if the structure of the active site of a salient enzyme in a metabolic or regulatory pathway is known, then chemical compounds, such as drugs, can be rationally designed to inhibit or otherwise affect the behavior of that enzyme. A second approach, of equal importance to biotechnology, that also requires X-ray quality crystals is the genetic engineering of proteins. Although recombinant DNA techniques provide the essential synthetic role that permits modification of the proteins, crystallography provides the analytical function. It serves as the structural guide for the introduction of intelligent and purposeful changes, in the place of random and chance amino acid substitutions. Crystallography allows direct visualization of the structural changes introduced and offers new directions for chemical or physical enhancements.

Bulk crystallization, along with the detailed consideration of crystallization processes and products, has long been a staple in the pharmaceutical and chemical industry, where it has been applied to conventional small molecules and even more complex biological molecules such as glycosides, peptides, and antibiotics. A few examples from the literature where the crystallization procedures are described in some detail are for L-alanine, fructosylxyloside from *Scopulariopsis brevicaulis,* cephalosporin, and a flavanone, hesperidin from orange peels. Here, it has long been recognized that manipulation of crystalline formulations could dramatically affect solubility, shelflife, impurity content, thermal stability, release, efficacy, or a host of other commercially significant properties. Much research has, therefore, been devoted to the crystallization of commercially important conventional, small molecules during the past hundred years. Only more recently has it become evident that crystalline formulations hold substantial promise for the administration and delivery of biological macromolecules as well.

The first major example was that of insulin, a small protein of striking physiological importance to nearly all animals and crucial to the treatment of diabetes. In this instance, it was found that injection of crystalline protein resulted in a more prolonged, slower release of the protein drug and a significant improvement in its efficacy. This timed-delivery feature, first realized for insulin, now appears possible for other protein drugs. In the coming years, a majority of biotechnology products will likely be proteins, particularly genetically modified proteins designed for introduction into living organisms, including humans. Their delivery and release will be a major problem, and their formulation as crystals will serve a major role in solving those problems. Here, unlike the growth of protein crystals for X-ray diffraction analysis, the crystals are in fact the product, and their production must be reproducible.

The physical properties of the crystals, such as size, morphology, solubility, and impurity levels will be rigorously specified. They will be produced, under strict regulatory control, as is any drug. Thus, the objectives are quite different than for crystals produced for diffraction analysis. In terms of pharmaceutically important proteins, natural insulin from animal tissues was certainly the first to be produced in bulk quantities by crystallization, and recombinant insulin followed. Others, however, are now known, including a large-scale facility for the production of crystalline Fab fragments from monoclonal antibodies. The third area for crystal growth, the downstream processing of proteins produced in large fermentations for the purpose of purification, has associated with it yet another set of objectives and problems. Crystallization is one of the oldest chemical purification technologies and one that has been seriously underutilized to this point in biotechnology. Here, the goal is to promote the crystallization of macromolecules from extremely large volumes, commonly reaching several thousands of liters, of rather impure solutions or culture broths. The ideal is to achieve a very high degree of purification in a single, economical step, and at the same time generate a concentrated, stable, and active product that appeals in its form to a broad customer base.

Although certain crystal sizes, morphologies, size distribution, and other features of the crystals may be preferred, these are not the most important parameters. The major considerations are yield, reproducibility, degree of purification, and the ability to crystallize the material in spite of very high and variable contaminant levels. Major difficulties accompany the crystallization of proteins in downstream processing and purification that are not seen in X-ray diffraction or formulation crystal growth. For example, processes such as vapor diffusion, which is applicable to microliter volumes, are completely inapplicable to thousand liter volumes. Exotic precipitants may not be economical, sensitive procedures such as column chromatography may be out of the question, and speed is often of the essence. Waste product disposal can be a significant consideration. The process must be carefully conceived and developed to be as economical and efficient as possible and still yield the desired crystalline product at the required purity. There are many examples in the literature of crystallization used for the purification of proteins from bulk fermentations, and these provide useful guides to the design of new procedures for other macromolecules.

In one of the earliest reports, Fukumoto et al. reported the purification of lipase from a strain of *Aspergillus niger* by fractionation with ammonium sulfate and its ultimate crystallization from acetone. Tsujisaka et al. in 1973 purified the same enzyme from *Geotrichum candidum* using salt fractionation, DEAE Sephadex, and gel chromotography. The protein was crystallized from its aqueous solution. Buckel et al. in 1981 and Mack et al. describe a scheme based on salt fractionations and chromatography for the production of glutaconate CoA-transferase from *Acidaminococcus fermentans*. In the 1990s, the appearance of reports of bulk crystallizations accelerated, including those on procedures for protein iminopeptidase from *Bacillus coagulans*, chymosin mutants from *Trichoderma reesei*, and pyroglutamyl peptidase from *Bacillus amyloliquefaciens*. Although only abstracts are currently available, some additional enzymes have also been crystallized for commercial production, these include cellulase, subtilisin protease and glucose isomerase, and alcohol oxidase.

CHARACTERISTICS OF CRYSTALS

Macromolecules, being unique in their properties, both in terms of size and complexity, give rise to crystals that are unique as well. We cannot, therefore, expect that macromolecular crystals will necessarily develop according to precisely the same mechanisms and principles as do conventional crystals. Indeed, recent investigations using advanced physical techniques such as inelastic light scattering, interferometry, and atomic force microscopy have yielded evidence of new mechanisms and properties not before encountered with conventional crystals. Macromolecular crystals are relatively small in comparison with conventional crystals, rarely exceeding 1 mm on an edge and generally smaller. Because only one stereoisomer of a biological macromolecule naturally exists, they do not form crystals possessing inversion symmetry and, therefore, generally exhibit simple shapes that lack the polyhedral character of small molecule crystals. They are extremely fragile, often crushing at the touch; degrade outside a narrow temperature, ionic strength, or pH range; generally exhibit weak optical properties; and diffract X-rays to resolutions far short of the theoretical limit.

The reason for most of these character deficiencies is that macromolecular crystals incorporate large amounts of solvent in their lattices, ranging from about 30% at the lower limit to 90% or more in the most extreme cases. Proteins also have, as individual molecules, an array of water molecules surrounding them that are relatively tightly bound both in solution and in the crystal. There are two other crucial differences between macromolecular and conventional crystal growth that have important practical consequences. The first is that macromolecular crystals are usually nucleated at extremely high levels of supersaturation, often several hundred to a thousand percent. Small molecule crystals, on the other hand, usually nucleate at only a few percent supersaturation. Virtually every quantitative aspect of crystal growth is a direct function of supersaturation. Although high super saturation may be essential to promote nucleation, it is far from ideal for growth, and the many problems observed for macromolecular crystals attest to this. Furthermore, supersaturated macromolecular solutions, in addition to crystal nuclei, produce alternate solid states that we refer to collectively as amorphous precipitates.

Unlike conventional systems, competition exists at both the nucleation and growth stages between crystals and precipitate. This is particularly acute because competition is promoted by high levels of supersaturation. Because amorphous precipitates are kinetically favored, though of less-favored energy states, they tend to dominate the equilibration process and often inhibit or preclude crystal formation. Given the complexities that beset macromolecules, can we reasonably expect their crystallization to resemble that of conventional molecules? Evidence to this point suggests that the answer is in principle yes, but in practice no. It appears that the fundamental mechanisms and pathways of macromolecular crystal growth are the same as for conventional crystals but that the magnitudes of the underlying kinetic and thermodynamic parameters that govern the process differ dramatically.

TECHNIQUE FOR MACROMOLECULAR CRYSTALS

The growth of macromolecular crystals has been studied for more than 150 years, initially as a purification tool, and more recently for other purposes. It has been the topic of several texts and reviews and, with the increasing importance of protein crystallography, has become a major factor in the biotechnology industry. Because this body of literature exists, no attempt will be made here to exhaustively review all methods and aspects of macromolecular crystal growth. An effort will be made, however, to describe the underlying principles, present the most commonly employed laboratory approaches, and suggest ways in which the field can be addressed by novice crystal growers. The strategy used to bring about crystallization is to direct a macromolecular solution slowly toward a state of supersaturation by modifying the properties of the solvent or the solute. This is generally achieved by altering the concentration of a precipitating agent, such as salt, or by altering some physical property, such as pH.

In concentrated solutions, macromolecules, unlike most conventional molecules, may aggregate as an amorphous precipitate, a result to be avoided. It is usually indicative that supersaturation has proceeded too far or been reached too swiftly. One would like to slowly and marginally exceed the equilibrium point of saturation and thereby allow the macromolecules reasonable opportunity to assemble into a crystalline lattice. The traditional procedure, for inducing proteins to separate from solution as a solid phase is to gradually increase the level of saturation of a salt. Classically, the salt was ammonium sulfate, but others are also in common use. Although the protein may separate as an amorphous precipitate, with appropriate care, manipulation of salt concentration is an effective means for growing protein crystals. This approach has, in fact, probably yielded more varieties of protein crystals than any other. The accepted explanation for this salting-out phenomenon is that the salt ions and macromolecules compete for chemical interactions, principally hydrogen bonds, with the solvent molecules, that is, water. This occurs because both salt ions and protein molecules depend on hydration layers to maintain solubility.

Table 10.1 Methods for attaining a solubility minimum

Methods
Bulk crystallization
Batch method in vials
Evaporation
Bulk dialysis
Concentration dialysis
Microdialysis
Liquid bridge
Free-interface diffusion
Vapor diffusion on plates or slides
Vapor diffusion in hanging drops
Sequential extraction
pH-Induced crystallization
Temperature-induced crystallization
Crystallization by effector addition

When competition between ions and proteins becomes sufficiently intense, the protein molecules begin to self-associate in order to satisfy, by intermolecular interactions, their electrostatic requirements. Thus, dehydration, or the elimination and perturbation of solvent layers around protein molecules, reduces solubility. For a specific protein, the solubility minima are usually quite dependent on the pH, temperature, chemical composition of the precipitant, and properties of both the protein and the solvent. In the low ionic strength range, a phenomenon known as salting-in occurs in which the solubility of the protein increases as the ionic strength rises from zero. Protein solubility at low ionic strength is diminished by the removal of ions essential for satisfying the electrostatic requirements of the molecules. As ions are removed, and in this region of low ionic strength cations are most important, the protein molecules again balance their electrostatic requirements by forming interactions among themselves.

The salting-in effect can be applied in the opposite sense, that is, reducing ionic strength can serve as a crystallization tool. In practice, a protein is optimally solubilized at moderate ionic strength and then extensively dialyzed against distilled water. Many proteins, such as catalase, concanavalin B, immunoglobulins, and plant seed proteins, have been crystallized by this means. Proteins may be forced from solution, at constant pH and temperature, by increases in ionic strength, or they can be crystallized or precipitated at constant ionic strength by changes in pH or temperature. Reduction in solubility occurs because the charge distribution on a macromolecule, its surface features, or its conformation may change as a function of these variables. By virtue of its ability to inhabit a range of states, a protein may exhibit a number of environmentally dependent solubility minima, and each of these minima may afford the opportunity for crystal growth. Similar effects may be achieved as well by the slow addition to the mother liquor of some relatively mild alcohols or other organic solvents such as ethanol, acetone, or methylpentanediol. The only essential requirement for the precipitant is that at the specific temperature and pH of the experiment, the additive not adversely affect the structure and integrity of the protein. This is a stringent requirement and deserves some consideration.

An organic solvent will compete, to some extent like salt, for water molecules. It also reduces the dielectric shielding capacity of intervening solvent. Reduction of bulk dielectric increases the effective strength of the electrostatic forces responsible for one protein molecule associating with another. Polymers such as polyethylene glycol also serve to dehydrate macromolecules in aqueous solution, as do salts, and they furthermore alter the dielectric properties in a manner similar to organic solvents. They produce, however, an additional important effect. Polyethylene glycol perturbs the natural structure of the solvent and creates a more complex network having both water and itself as structural elements. A consequence of this restructuring of solvent is that macromolecules, particularly proteins, tend to be excluded, and phase separation is promoted. Combinations of approaches using both precipitating agents, and variation of physical parameters are frequently effective in promoting the crystallization of macromolecules.

Increasing the concentration of a precipitating agent to a point just below supersaturation and then adjusting the pH or temperature to further reduce the solubility of the protein is a common approach. Modification of pH can be brought about quite well using the vapor diffusion technique with volatile acids and bases such as acetic acid and ammonium hydroxide. Because of the complexity of interactions between solute and solvent and the shifting character of the protein, the methods of crystallization must usually be applied over a broad set of conditions with the objective of discovering the particular minimum (or minima) that yield crystals. In practice, one determines the precipitation points of the protein at sequential pH values with a given precipitant, repeats the procedure at different temperatures, and then examines the effects of different precipitating agents and their concentrations.

MECHANISM OF SUPERSATURATION

The oldest technique used for the crystallization of proteins and nucleic acids, with a history of more than 150 years, is what is commonly called batch crystallization. It is the simplest, most thoroughly tested, and when optimized, the most reliable of all crystallization methods. Lysozyme, for example, is most generally crystallized by the batch method. It relies on the mixing of a protein solution at undersaturated concentration with precipitant components. The latter alter the electrolyte properties of the solution to produce a mother liquor supersaturated with respect to protein. From such a solution the solid state may be expected to eventually form, and if the conditions and physical parameters have been chosen correctly, this solid state will be crystalline rather than amorphous. Alternatively, the physical state of an undersaturated protein solution may be altered by, for example, lowering or raising the temperature. The supersaturated solution produced as a result may yield crystals. The batch method is attractive because of its inherent simplicity and reproducibility. It requires nothing more than the combination of two or more solutions into one and a period of time until nucleation commences and growth follows.

The concept is appealing because it can be carried out with an absolute minimum of human intervention. Because the apparatus is nothing more than a container for a small quantity of fluid, which can be varied as circumstances require from a few microliters to hundreds of milliliters, it is fully amenable to interface with both simple and complex observational systems designed to monitor and record the process of protein crystal growth. Thus, the technique is very useful, for example, for inelastic light scattering studies of nucleation, or time-lapse photography investigations of protein crystal growth. It is particularly attractive as well for macromolecular crystallization experiments in space. Batch crystallization requires no complex device to produce equilibration, either through the liquid or the vapor phase. It relies strictly on the energetics of the system—the activation barriers to nucleation—to regulate the process. Although it has not, in recent years, been extensively used to screen for protein crystallization conditions, this reflects more a prejudice of investigators and the popularity of other methods rather than any deficiency in the batch method itself. Because it was commonly used in the past with fairly large volumes of protein solution, investigators have mistakenly come to believe that it is inappropriate for small samples, which is untrue.

It can even be carried out using microliter amounts in glass capillaries or under oil. It can be a useful screening technique as well as a good method for the preparation of large quantities of protein crystals once optimal values for parameters are established. There are a number of ingenuous devices, procedures, and methods for dynamically altering the degree of saturation of a protein solution. Generally, these rely on the slow increase in concentration of some precipitant such as salt or polyethylene glycol. These same approaches can of course also be used for salting-in, modification of pH, and the introduction of materials that alter protein solubility. These techniques have been extensively reviewed elsewhere, and only a selection will be dealt with here. Dialysis is familiar to nearly all biochemists as a means of modifying the components of a protein solution. The solution containing the protein is enclosed in a membrane casing or in a container having a semipermeable membrane partition. The membrane selectively allows the passage of small molecules and ions, but pore size prevents passage of the much larger protein molecules. The vessel or dialysis tube containing the protein is submerged in a larger volume of liquid that has the desired properties of pH, ionic strength, ligands, and so forth. With successive changes of the exterior solution and concomitant diffusion of small molecules and ions across the semipermeable membrane, the protein solution gradually acquires defined properties.

The dialysis approach, in some manifestation, has been used to crystallize many proteins. It was usually applied on a large scale, and only when substantial amounts of protein were available. It has the advantage that liquid–liquid diffusion through a semipermeable membrane exposes the protein to a continuum of potential crystal-producing conditions without physically perturbing the mother liquor. Diffusion through the membrane is slow and controlled. Because the rate of exchange of substituents in the mother liquor is a function of the concentration gradient across the membrane, the nearer the system to equilibrium, the more slowly it changes. This method has been adapted to much smaller amounts of protein by crystallographers who use almost exclusively microtechniques. These typically involve no more than 5 to 50 μl of protein solution in each trial.

First described by Zeppenzauer et al. and Zeppenzauer and subsequently modified and refined by numerous others, the method confines a protein solution within a glass capillary or the microcavity of a small plexiglass button. The cavity of the button or the ends of the microcapillary tube are closed off by a dialysis membrane. The whole arrangement, charged with protein solution, is then submerged in a much larger volume of an exterior liquid that is held within a closed vessel such as a test tube or vial. If the exterior solution is at an ionic strength or pH that induces the mother liquor to become supersaturated, crystals may grow. If not, the exterior solution may be exchanged for another and the experiment repeated.

The dialysis buttons, are particularly ingenious. Not only are they compact and easy to examine with a dissecting microscope, but they have a shallow groove about their waist. A section of moistened dialysis membrane is placed over the mother-liquor-filled cavity, and it is held securely in place by simply slipping a rubber O-ring over the top of the button and seating it in the groove. These buttons are in wide use and have proven quite successful. The technique of liquid–liquid diffusion, also known as free-interface diffusion, for the crystallization of macromolecules is in principle and practice simple and straightforward. It relies on the direct interdiffusion of a precipitant solution into a protein solution across their common liquid–liquid interface. When the two solutions are juxtaposed, precipitant gradients form in the region of the interface and produce local supersaturation. This promotes the formation of crystal nuclei and subsequent growth, often to sizes useful for analysis. The fundamental idea of liquid–liquid diffusion of a solvent or solute into a macromolecular solution is similar to dialysis.

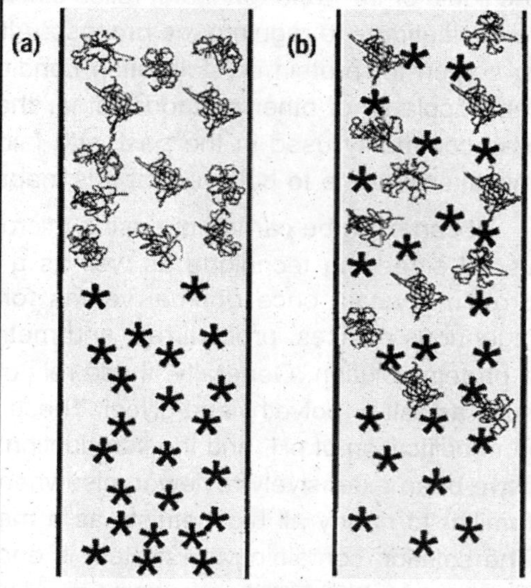

Figure 10.1 In free-interface diffusion the protein solution is (left) initially layered atop a precipitant solution, such as a highly concentrated salt. Diffusion of the small molecules into the protein solution (right) induces crystallization to occur.

The method is particularly applicable in microgravity because the necessity of a membrane is eliminated and the two liquids can be maintained relatively stable with respect to one another for long periods of time. Because the diffusion rate of macromolecules is so much less than that of the small molecules of most precipitants, there is relatively little penetration of protein into the precipitant during the course of most experiments. The technique of liquid–liquid diffusion has been applied in laboratories on earth to grow macromolecular crystals, and with some success. A search of the National Institute of Standards and Technology (NIST) database shows 35 entries using this technique with another 193 ascribed to various kinds of dialysis. Direct liquid diffusion has not, however, been as extensively used in conventional laboratories as have some other techniques, such as batch or vapor diffusion, because of some inherent disadvantages. Significantly, most of these methodological deficiencies are relieved or eliminated in the absence of gravity.

On earth, density-driven convective currents produce immediate and continuous turbulence at the liquid–liquid interface. Transient mixing, if layering of solutions is used, or the necessity of a membrane, as in dialysis, lessens the methods' effectiveness. Sedimentation produces a particularly deleterious effect with the free-interface method because nuclei and small growing crystals, under the influence of gravity, fall from the liquid–liquid interface. Thus, as a crystal forms, gravity increasingly pulls it downward from the region of the solution where growth would most favorably continue. Gravity essentially removes growing crystals from an optimal growth environment and thereby imposes a self-limiting character on the process, in turn restricting crystal size. Currently, the most widely used method for bringing about supersaturation in microdrops of mother liquid is vapor diffusion. This approach has provided a diversity of applications and may be divided into those procedures that use a "sitting drop" and those that use a "hanging drop."

In either form, the method relies on the transport of water or some volatile agent between a microdrop of mother liquor, generally 2 to 25 μL volume, and a much larger reservoir solution of 0.75 to 25 mL in volume. Through the vapor phase, the droplet and reservoir come to equilibrium. Because the reservoir is of much larger volume, the final conditions at equilibrium are essentially those of the initial reservoir state. In the vapor phase, water can be removed slowly from droplets of mother liquor, pH may be changed, or volatile components such as ethanol may be gradually introduced. As with those methods already described, the procedure may be carried out to further advantage at a number of different temperatures. According to one popular procedure, droplets of 10 to 20 μL are placed in the nine wells of siliconized glass depression plates. The samples are then sealed in transparent containers, such as Pyrex dishes or plastic boxes that also hold reservoirs of 20 to 50 mL of the precipitating solution.

The plates bearing the protein or nucleic acid samples are supported above the reservoirs by the inverted halves of disposable Petri dishes. Through the vapor phase, the concentration of salt or organic solvent in the reservoir equilibrates with that in the sample. In the case of salt precipitation, the droplet of mother liquor must initially contain a level of precipitant lower than the reservoir, and equilibration proceeds by diffusion of water from the droplet into the reservoir. This holds true as well for nonvolatile organic solvents, such as methylpentanediol and for polyethylene glycol. In the case of volatile precipitants, none need be added initially to the microdroplet, because distillation and equilibration proceed in the opposite direction. The method described here has the advantage that it requires only microliter amounts of material and is well suited for screening large numbers of conditions. A disadvantage is that all the samples in a single container must be equilibrated against the identical reservoir solution. It

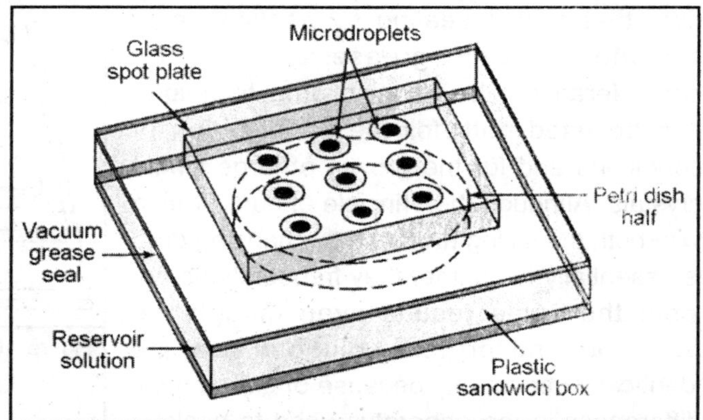

Figure 10.2 Schematic diagram of a nine-well depression spot plate enclosed in a plastic sandwich box used for the crystallization of macromolecules by the vapor diffusion method.

does, however, permit some flexibility in varying conditions once the samples have been dispensed, by modification of the concentration or pH of precipitants in the reservoir. When clear plastic boxes are used, large numbers of samples can be quickly inspected under a dissecting microscope and conveniently stored. The disadvantage of identical reservoir conditions throughout a single container was overcome by the introduction of plastic plates specifically designed for protein crystallization. One of these, sponsored by the American Crystallographic Association, is a plastic plate having accommodation for 15 protein samples.

Each chamber has a separate reservoir compartment and the mother liquor microdroplet may be either suspended from the underside of a glass cover slip, as in the hanging drop method described later, or sandwiched between two glass cover slips. Sealing of the chambers from air requires silicon grease or oil between cover slips and the plastic rims of the chambers. With these plates, the optical properties are very good but equilibration tends to be slow, and the setup is quite tedious. A second crystallization plate, is supplied by Charles Supper. With these plates, the drop sits atop a clear support post that protrudes upward from a circular moat containing the reservoir solution. The chambers can be rapidly and conveniently sealed from air by transparent plastic tape pressed over the upper surface of the plate after the reservoirs have been filled and the drops of mother liquor dispensed. Equilibration, as with the other plate, is through the vapor phase. Although the optical properties are less favorable with these devices, they are convenient and compact and can be used for rapid screens of crystallization conditions.

The hanging drop procedure also uses vapor phase equilibration, but with this approach a microdroplet of mother liquor (as small as 2 μL) is suspended from the underside of a microscope cover slip, which is then placed over a small well containing a milliliter or less of the precipitating solution. The wells are most conveniently supplied by disposable plastic tissue culture plates that have 24 wells with rims that permit sealing by application of silicone vacuum grease around the circumference. The hanging drop technique can be used both for the optimization of conditions and for the growth of large single crystals. Although the principle of equilibration with both the sitting drop and the hanging drop is essentially the same, they frequently do not yield the same results even though the reservoir and protein solutions may be identical. Presumably because of geometrical differences in the apparatus used to achieve equilibration, the path to supersaturation is different, even though the end point may be the same. In some cases there are striking differences in the degree of reproducibility, final crystal size, morphology, or time required for growth. These observations emphasize the point that in crystal growth the route leading to supersaturation and the kinetics of the process may be as important

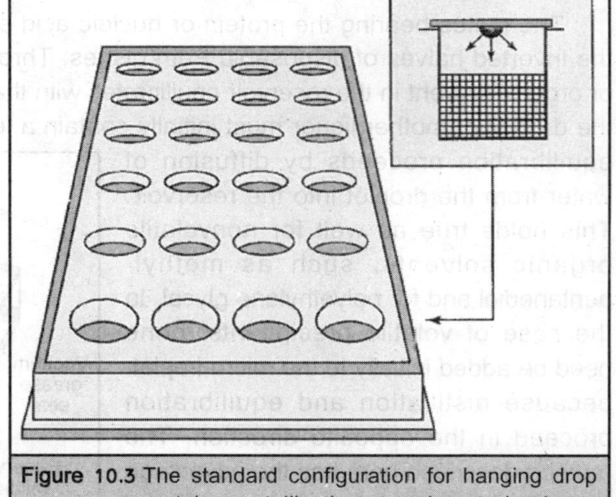

Figure 10.3 The standard configuration for hanging drop protein crystallization experiments is shown here.

as the end point achieved. There are nearly limitless variations on these themes for liquid–liquid or vapor diffusion, ranging from the very simple to the unnecessarily complex. You need only use a bit of imagination to devise a novel approach for yourself. With some exceptions, the methodologies used for protein crystallization, most of those described above, have been devised for the growth of crystals from very small volumes of mother liquor. Indeed, reducing the amount of material needed has been a principal objective in X-ray crystallography laboratories for some time. Crystals can now be obtained from only 2 to 4 μL of solution, and entire screening matrixes can be explored using only 2 to 5 mg of protein or nucleic acid. Two major applications of protein crystallization, for formulations, and for downstream processing of macromolecules do not, however, benefit from reduction in scale.

On the contrary, they profit from very large volumes and high yields of the crystalline product. As a consequence, the microtechniques described earlier may represent model approaches, but they are, in themselves, quite inappropriate to bulk processes. Large-scale, industrial crystallizations are, at least currently, limited to methods based essentially on batch crystallization in combination with manipulation of pH and temperature. In addition, many chemical compounds that might be used as precipitants or additives in laboratory settings cannot be used in many cases in bulk processes because of expense, hazard, or the creation of unacceptable by-products. Although this might appear severely limiting, remarkable success has been achieved using only these simple tools. The large number of biological macromolecules in crystalline form and produced in kilogram (or larger) amounts available on the commercial market attests to this. The restriction on techniques is compensated to a great extent in industrial processes by other factors. For example, the objective protein is usually in relatively high concentration and comprises a significant fraction of the total protein present. It is usually produced by genetically engineered or overproducing strains of simple microorganisms; hence, it does not suffer from many of the sources of microheterogeneity, that afflict many natural products. In addition, crystals produced in large-scale processes are usually very small crystals of marginal perfection that would be unsuitable for analytical work. Such crystals may, however, be entirely appropriate for the intended purpose of purification and product formulation.

PROTEINS CRYSTALLIZATION

One of the most useful techniques for producing a supersaturated solution is adjustment of pH to values where the protein is substantially less soluble. This may be done in the presence of a variety of precipitants so that a spectrum of possibilities are created whereby crystals might form. The gradual alteration of pH is particularly subtle because it may be accomplished by a variety of gentle approaches that do not otherwise perturb the system. Although micro- dialysis is probably equally suitable, more success with pH has been achieved by the vapor diffusion method, using "sitting" microdroplets on spot plates or in one of the plastic plates available for protein crystallization. The ambient salt, effector, or buffer conditions are established before dispensing the microdroplets in the depressions on the plate. The pH is then slowly raised or lowered by adding a small amount of volatile acid or base to the reservoir. Diffusion of the acid or base then occurs from reservoir to sample, just as for a volatile precipitant. If the pH is to be raised, for example, a small drop of concentrated ammonium hydroxide can be added to the reservoir; a drop of acetic acid may be used to lower it. The pH can also be gradually lowered over a period of days by simply placing a tiny chip of dry ice in the reservoir. The liberated CO_2 diffuses and dissolves in the mother liquor to form weak carbonic acid. When a specific pH end point is required, the mother liquor maybe buffered at that point and then moved significantly away by addition of acid or base.

Table 10.2 Sources of microheterogeneity

Sources of microheterogeneity
Presence, absence, or variation in a bound prosthetic group, substrate, coenzyme, inhibitor, or metal ion
Variation in the length or composition of the carbohydrate moiety of a glycoprotein
Proteolytic modification of the protein during the course of isolation or crystallization
Oxidation of sulfhydryl groups during isolation
Reaction with heavy metal ions during isolation or storage
Presence, absence, or variation in post-translational side-chain modifications such as methylation, amidiation, and phosphorylation
Microheterogeneity in the amino or carboxy terminus or modification of termini
Variation in the aggregation or oligomer state of the protein caused by association or dissociation
Conformational instability caused by the dynamic nature of the molecule
Microheterogeneity caused by the dynamic nature of the molecule
Partial denaturation of sample
Genetically different animals, plants, or microorganisms that make up the source of protein preparations
Bound lipid, nucleic acid, or carbohydrate material or substances such as detergents used in the isolation

The microdroplets of mother liquor may then be returned to the buffer point by addition of an appropriate volatile acid or base to the reservoir. Individual proteins display distinct solubilities at varying levels of salt concentration and can be selectively separated from solution as amorphous precipitate or crystals. At very low ionic strengths where salting-in occurs, the solubility of the protein increases with the ionic strength because of a decrease in its activity coefficient. The salting-out effect, which Hofmeister suggested to be caused by a reduction of the chemical activity of water by salt, occurs as the ionic strength increases. Because ionic strength is a function of the second power of the valence, multivalent ions are most efficient. In the salting-out region of a protein's solubility profile, the logarithm of protein solubility as a function of the ionic strength is linear and can be expressed by

$$\log S = \beta - K_s\,(I/2)$$

where I is the ionic strength in moles per kilogram of water, S is the solubility of the protein in grams per kilogram of water, and K_s is a constant. K_s serves as a measure of the slope of the solubility extrapolated to zero ionic strength and is the logarithm of the solubility at that point.

In principle, the precipitation or crystallization of a protein is effected by manipulating the factors that appear in the above equation. One can crystallize a macromolecule by increasing the ionic strength at constant pH and temperature a K_s fractionation procedure, or it can be brought about at constant ionic strength by manipulating the pH and temperature, which in most cases strongly influence solubility. With all techniques, the objective is to produce a supersaturated protein solution by very gradual and nonperturbing adjustment of conditions. In practice, a solubility minimum is determined for the protein at some specific pH, temperature, and ionic strength. The protein solution is set initially at the requisite salt concentration for this solubility minimum, but either the pH or temperature is set initially to maintain protein solubility. The temperature is then slowly relaxed to minimize protein solubility, or the pH is gradually equilibrated via the liquid or vapor phase to an appropriate value. In the case of temperature shifts, the protein is generally raised to a higher temperature, as, for example, with squash globulin, insulin, or excelsin, a process that increases solubility.

The solution is then cooled slowly to where the protein is no longer soluble. This can be achieved by placing the warm solution in an insulated or thermally controlled container or by placing the sample in a Dewar flask. The approach to supersaturation thus occurs at an almost imperceptible rate. Exactly the same technique can be applied if the protein is more soluble at cold than warm temperature. Under these circumstances, the protein solution is cooled under established ionic strength conditions and then allowed to warm slowly. This approach was used, for example, to grow crystals of DNase by Kunitz and is the basis of Jakoby's sequential extraction procedure using decreasing concentrations of ammonium sulfate. The use of pH gradients in space and time has been successful in many investigations; they can be used in association with direct liquid dialysis or with vapor diffusion methods as well. They can, furthermore, be used in conjunction with a host of different precipitating agents. Temperature can be used in an analogous manner.

Proteins and nucleic acids are structurally flexible and are believed to alternate between several conformational states. In particular, they may assume substantially different conformations when they have bound coenzyme, substrate, or some other ligand. As a consequence, a protein with bound effector may exhibit different solubility properties than the nonliganded form. In addition, if multiple conformational states are available, the presence of effector may be used to select for only one of these, thereby imposing a degree of conformity of structure that would otherwise be absent. The addition of ligands can sometimes be used to induce supersaturation and crystallization in those cases where its binding to the protein produces solubility differences under a given set of ambient conditions. The effector may be slowly and gently combined with the protein, for example, by dialysis, so that a resulting complex becomes supersaturated.

The addition of ligands, substrates, and other small molecules has seen widespread use in protein crystallography, because it provides useful alternatives if the apoprotein itself cannot be crystallized. No matter what the ultimate objective of macromolecular crystallization is, it is first necessary to (i) identify a set of physical and chemical conditions that yield some kind of crystals, (ii) optimize those conditions, and (iii) reproducibly apply the conditions according to reliable procedures that will consistently produce the requisite crystals. Thus, the initial step consists of screening large numbers of possible precipitants, buffers, effectors, pHs, temperatures, etc. The second is a careful screening using a finely sampled matrix of conditions focused on those that yielded crystals, and the third is on devising reproducible procedures for application. Clearly, the first stage is the greatest hurdle that must be overcome, but it is not necessarily the step demanding the most effort. Table 10.3 is included here, not to alarm or intimidate the novice, but to alert those interested in macromolecular crystal growth of the complexities and the many factors that may have an impact on a crystal growth investigation. Fortunately, only a few of these factors may be simultaneously operative for a particular problem; the trick is to discover which they are for your specific case.

Table 10.3 Factors that do or could affect protein crystal growth

Factors
pH and buffer
Ionic strength
Temperature and temperature fluctuations
Concentration and nature of precipitant
Concentration of macromolecule
Purity of macromolecules
Additives, effectors, and ligands
Organism source ofmacromolecule
Substrates, coenzymes, inhibitors
Reducing or oxidizing environment
Metal and other specific ions
Rate of equilibration and rate of growth
Surfactants or detergents
Gravity, convection, and sedimentation
Vibrations and sound
Volume of crystallization sample
Presence of amorphous or particulate material
Surfaces of crystallization vessels
Proteolysis
Contamination by microbes
Pressure
Electric and magnetic fields
Handling by investigator and cleanliness
Viscosity of mother liquor
Heterogeneous or epitaxial nucleating agents

How does one begin the process, that is, how do you discover those initial conditions that yield crystals and provide starting points? Over the years, a vast number of reports and observations entered the literature, without much form or even clear purpose. Attempts were made to sort and classify; to shape ad hoc experiments into more rational approaches; to gather, collate, and systematize the experiences of thousands of investigators into rationally constructed databases; and finally to extract from those bases arrays of conditions most likely to yield protein and nucleic acid crystals. These sets of conditions form the foundation for the crystal screens, now commercially available, that have virtually revolutionized the initial search for crystallization conditions. Although some important macromolecules continue to slip through these still imperfect nets, the number of successes is impressive. Improved screening techniques alone are to a great extent responsible for the explosion of new crystals, grown not only by crystallographers, but by the myriad of biochemists and molecular biologists whose need for structure is acute.

The search for novel, and perhaps even better, precipitants and conditions for crystal growth continues, and we can reasonably expect even more useful developments in this area. With arrays of high-probability conditions that need be tested against no more than a few milligrams of protein or nucleic acid, attention has focused on the macromolecules themselves. Three principles have clearly come forward: purity, stability, and solubility. Failure in crystallization can probably be attributed, in general, to one or more of these. Homogeneity has always been recognized as important, though not always essential. The trick is to recognize when it is the crucial factor and direct one's efforts there. Important advances have been made with regard to purification. Not only are there now more powerful techniques and instruments available, but crystallography has benefited to an extraordinary degree from the refinements of recombinant DNA technology and the development of high-yield expression systems.

With the introduction of endogenous purification tools such as the histidine tag or glutathione sulfur transferase (GST) chimers, even greater availability of ultrapure crystallizable protein seems on the horizon. Recombinant systems provide us not only with pure, native proteins, but a way to introduce mutations and thereby create almost limitless alternatives should the native molecule prove unyielding. They offer systematic approaches as well to the intelligent engineering of solubility properties and stability. Chemical and conformational stability of the target macromolecule, too, has long been recognized and given serious attention. Ligands are used to lock discrete conformations, disulfides introduced by mutation or chemical modifications are made to reduce mobility, proteins and ribosomes from thermophilic organisms are used for their natural stability, and proteolytic or mutational truncations are made to remove naturally flexible domains or elements. We are only beginning to understand how to manipulate conformation to enhance crystallization properties, and this area also promises much for the future, particularly for challenging topics such as antibodies, multiple domain enzymes, and multimolecular complexes.

Perhaps the principle least recognized but now receiving the most attention is solubility, meaning here not only dissolution of macromolecules in a liquid, but in a disaggregated, monodisperse molecular form. Indeed, we have come to recognize aggregation and polydispersity as a principal reason why many macromolecules cannot be crystallized. This includes not only membrane and lipophilic proteins but macromolecules that have a strong propensity to aggregate, such as antibodies, hormone polypeptides, and proteins whose purpose is ordered aggregation, such as those from viral capsids. Extending observations that nonionic detergents promote the growth of hydrophilic as well as membrane proteins, light-scattering studies have confirmed the salient importance of monodispersity in macromolecular crystallization. The current consensus among those involved in crystallization research is that if by some means monodispersity can be ensured, then there is better than an 80% chance that crystals will be obtained.

To promote monodispersity, the common approaches are to incorporate detergents in crystallization trials or to crystallize at low protein concentrations, at higher temperatures, or in the presence of chaotropic agents. This is an area that continues to be under active investigation. Additional factors have emerged from recent crystal growth research and from the empirical results generated by the current revolution in crystallography. Among them are the importance of surfaces and their promotion of heterogeneous nucleation, the value of unique environments such as microgravity, gels, thin capillaries, the broad range of polymeric precipitants useful in growing crystals, and new approaches to solving the problem of persistent microcrystals.

CRYSTALLIZATION

Crystallization is a mass transfer separation process commonly used in many chemical industries. It permits highly selective purification of a material from an impure mixture, and the crystalline state of the final product is frequently advantageous. Whereas bulk crystallization has been applied for the purification of many small organic biological products, including antibiotics and amino and organic acids, and has played an important role in chiral resolutions of organic materials, until recently it has been applied only rarely to the bulk purification of proteins. It is well known that proteins, despite their large size, can be crystallized. Nevertheless, a survey of the vast protein crystallization literature, almost entirely devoted to producing crystals for X-ray studies, reveals that protein crystallization has been perceived generally to be tedious, difficult, and disrupted by the presence of impurities.

Table 10.4 The conflicting crystallization needs of X-ray crystallographers and bulk protein manufacturers

	X-ray crystallography	*Bulk manufacture*
Aim of crystallization	Protein structure determination	Protein purification
Protein used	Usually novel	Usually well characterized
Protein quantity available	Usually limited (mg)	Large (grams or more)
Number of crystals desired	One	Unlimited
Crystal size	Generally >300 *μ*m	25 *μ*m or more
Lattice order	Very highly ordered	Ordered
Solution purity	Preferably free of impurities and protein heterogeneity	Usually contains many impurities

These perceptions have perhaps arisen because of the unique requirements of protein X-ray crystallography, which remains the primary method for the determination of protein structure. The significant differences highlight the fact that many of the well-known issues that bedevil the production of protein crystals for X-ray crystallography are not germane to the bulk crystallization of proteins for manufacturing uses. In fact, the bulk crystallization of many proteins is found to be reasonably straightforward, similar in nature to the crystallization of inorganic molecules, and capable of achieving high degrees of purification at good yield and large throughput. This article will focus on existing knowledge relevant to the bulk crystallization of proteins from solution and largely bypass the more specialized field of protein crystallization for X-ray crystallography, for which excellent reviews already exist. Useful general reference books on bulk crystallization of materials include those by Randolph and Larson and Mullin.

METHODS FOR OBTAINING PROTEIN CRYSTAL

In addition to the beauty of the crystalline form, protein crystals share many features in common with smaller inorganic and organic molecule crystals. They consist of highly ordered molecular arrays with distinct crystal faces. Proteins usually exhibit polymorphism, leading to several crystallographic forms, depending on various factors including the nature of the precipitant or the temperature. Obtaining the most appropriate crystal form is often important for process manufacture, because "chunkier" crystal forms are easier to wash, store, and transport without excessive crystal damage and fragmentation. Needle forms, which are difficult to process, occur commonly in industrial practice. Transition boundaries between different polymorphic forms must be established experimentally and may be affected by many factors, including temperature, pH, and precipitant type and concentration. A common example is the tetragonal-orthorhombic transition for lysozyme, which occurs at about 25°C. Further, the crystals may undergo significant habit change in the presence of impurities or under varying conditions and exhibit dendritic morphologies during very high crystal growth rates.

Table 10.5 Polymorphic forms of proteins

Protein	*Crystal form*	*Condition*
Lysozyme	Tetragonal, orthorhombic	Temperature
Lysozyme	Monoclinic; triclinic; tetragonal	Nature of salt
Cytochrome *c*2	Hexagonal; triclinic; tetragonal	Ionic strength, pH
Ovalbumin	Monoclinic; triclinic	pH

A significant feature of all protein crystals is their large degree of incorporated solvent, which is typically 50% v/v, but varies generally between 30 and 80%. This has the practical result of making protein crystals unstable when exposed to air, resulting in their rapid collapse to an amorphous state. Consequently, protein crystals are usually maintained as a liquid crystal slurry.

SOLUBILITY

Crystallization processes depend on the production of a supersaturated solution that leads to the formation of the crystalline solid phase, consisting primarily of solute. Consequently, an accurate protein solubility data set determined for the conditions chosen for the crystallization process is essential. Two processes are fundamental to crystallization: nucleation and subsequent crystal growth by addition of additional solute to the nucleated crystal. The extent of supersaturation is the primary determinant of both the yield of crystalline solute from the solution and the rate of crystal growth.

Solubility Curve

Under any set of conditions, there is a unique protein solubility given by the solubility curve. This curve represents the thermo-dynamic boundary at which there is stable coexistence of solid and dissolved protein forms. Below the solubility line, solution conditions thermodynamically favor the maintenance of a single, homogeneous liquid phase, and protein crystals added to this solution will dissolve. Solution conditions that lie above the solubility curve represent supersaturated conditions under which the formation of a protein-rich solid phase and a saturated liquid phase are favored. The solubility data represent the lower limit of the crystallization process in that once the protein concentration in solution declines to this concentration, the system is at equilibrium and no further crystallization is possible.

Metastable supersaturation region

The supersaturation region can be further divided into a metastable region and a nucleation region. The metastable region represents a region of low supersaturation in which nucleation does not occur spontaneously within times useful for crystallization processes, despite the existence of supersaturation. However, if crystals are already present, crystal growth will occur. This is the preferred region for the operation of industrial crystallization processes, since it provides the greatest degree of control over the crystal number in the crystallizer and, hence, crystal size distribution and quality. The width of the metastable region is of great significance for crystallization processes and appears to be reasonable generous for proteins, although the available data are extremely limited.

Table 10.6 Typical size and metastable regions for various molecular species

Molecular category	*Size of metastable region (c/c^*)*
Ionic, inorganic salts	≅ 1.1 or less
Small organics (e.g., sugars)	≅ 5–30
Proteins	≅ 1–20
Large proteins (MW> 200 kDa)	≅ 1.1 or less

When crystallization processes are operated in the metastable region, the crystallization process must be initiated by providing crystal nuclei to the crystallizer (seeding), so that crystallization can begin. The number, mean size, and size distribution of these seeds are variables that permit the quality of the final crystal product and the process crystallization time to be manipulated.

Nucleation supersaturation region

In some crystallization processes, seeding is not preferred, usually because of requirements for aseptic operation (e.g., pharmaceutical preparations) or when a very small (<30 μm diameter) final mean crystal size is desired. In these cases, batch crystallization processes must be initiated in the nucleation region. In this region of higher supersaturation, spontaneous production of large numbers of tiny crystal nuclei will occur after an initial induction time. The control of such processes can be difficult, because rates of crystal nucleation and growth are usually very high at the large supersaturation typical of this region. The nucleation region can be further subdivided into a primary nucleation region, in which nucleation will occur spontaneously from the supersaturated solution (heterogeneous nucleation if active foreign nuclei, such as dust, are present, or homogeneous nucleation if they are not) or secondary region (at lower supersaturation) if crystals are already present. In industrial crystallization, secondary nucleation predominates. However, the threshold between these regions is fuzzy.

Solid phase

At very high supersaturation, conditions will favor the production of a disordered solid phase (precipitation), which usually forms rapidly on nucleation. Such operation generates amorphous protein precipitates, which lack the purity and structural properties of protein crystals. Precipitation processes have been widely used for protein recovery and are covered elsewhere in this encyclopedia. Recent dynamic light scattering (DLS) studies of supersaturated protein solutions have suggested that there is a significant difference in the manner in which homogeneous nucleation occurs for crystallization and precipitation processes.

Factors Influencing the Solubility of Protein

Proteins vary widely in their solubility in aqueous solutions. Most commercially manufactured commodity proteins exhibit high solubilities (40 to 100 mg/mL) in aqueous buffers, but more esoteric proteins can have unusual behavior. For example, some proteins (prolamins) are more soluble in aqueous ethanol than in water. Accurate measurement of protein solubility is required for proper design of crystallization processes. This is frequently a tedious process, made difficult by the challenge of obtaining sufficient quantities of pure protein material and the common contamination of the material with heterogeneous forms of the protein, whose effect on the solubility is unclear. For real crystallization processes, the effect of impurities present, often in significant concentration, on protein solubility must not be overlooked. Issues and protocols for protein solubility determination have been discussed by Ries-Kautt and Ducruix. Recently, a useful apparatus for the rapid measurement of protein solubility using only small quantities of protein has been developed in which solutions near the solubility limit are equilibrated with small volume crystal beds of the protein. Protein solubility is affected by many factors, and the major factors are discussed below.

Effect of pH

The pH of solution usually has a marked influence on protein solubility. The solubility often exhibits a minimum at the isoelectric point (IEP), with a sharp increase in solubility within small pH changes either side of the IEP. In concentrated protein solutions, typical of those necessary for protein crystallization, this generalization is fallible. For example, glucose isomerase has an IEP of 4.0, but has minimum solubility in ammonium sulfate at about pH 7.

Effect of temperature

Proteins exhibit a wide range of solubility responses to temperature. The solubility of some proteins is largely unaffected by temperature; others, such as glucose isomerase and lysozyme, show marked increase with temperature; and others exhibit a retrograde response. The two common crystal forms of lysozyme have been noted to have a markedly different solubility response to temperature.

Effect of ionic strength

Increasing ionic strength decreases globular protein solubility, although diverse effects occur, including regions of low ionic strength ($I < 0.15$ M) at which salting-in phenomena are commonly observed. In some cases, a solubility minimum with concentration has been observed for some protein–salt combinations, particularly for divalent salts. Protein solubility during precipitation processes has often been analyzed by the Cohn plot in which protein solubility is graphed as a logarithmic function of ionic strength. It has been frequently noted, however, that protein solubility in crystal suspensions is lower than for amorphous precipitate suspensions and that the Cohn relationship exhibits poor fit for the former.

Nature of salt

The varying ability of different salts to alter protein solubility is well known and is described by the Hofmeister series (lyotropic series), which ranks ions according to their salting-in or salting-out effect. However, care is needed in applying the series, as shown for lysozyme.

Role of nonelectrolytes

A wide variety of nonelectrolytes can influence protein solubility, usually in a complex manner. Organic solvents (ethanol, acetone), polyethylene glycol (PEG), sugars, and even amino acids have been used to salt-out proteins. Where waste disposal of inorganic salt-rich mother liquors is difficult or expensive, nonelectrolyte nonsolvents may be a good alternative. The complex behavior of these systems is illustrated well by ethanol, which salts-out at low temperatures and low concentrations, but salts-in at all concentrations as temperature rises. These effects do not depend on the dielectric constant (polarity) of the solvent, but on the unusual properties of ethanol and water mixtures.

CONDITIONS FOR PROTEIN CRYSTALLIZATION

Obtaining the appropriate conditions for protein crystallization is more challenging than for many small organic and inorganic molecules, because relatively little is known about protein solubilities, often only small quantities of the protein are available early in process development, protein crystallization even under optimal conditions can be slow (days, months), and the use of a precipitant, rather than temperature, to create supersaturation adds an extra degree of freedom in locating conditions for good crystallization. Because X-ray crystallography has until recently been the sole method for protein structure determination, immense effort has been put into the development of protocols for quickly finding suitable conditions for protein crystallization at the microscale. Typical techniques used are well known and include dialysis, vapor diffusion, and batch crystallization methods.

Various statistical search strategies, such as sparse matrix screens using commercially available solution kits, and robotic technology for repetitive microscale tests are used to reduce the amount of material and time required and optimize the likelihood of finding a suitable set of conditions. This body of work exists as an accessible database and serves as an excellent resource for quickly narrowing down suitable systems for the bulk crystallization of proteins. Unfortunately, the transient solution environment typical of microscale systems means that additional tests are required to define crystallization conditions exactly.

Process engineering issues can then be applied as an additional set of criteria. Precipitants used for the crystallization will need to be available in bulk at low cost and, for some sites, waste disposal concerns may impose constraints on the use of concentrated salt solutions. In this vein, the use of biodegradable alternative precipitants has been investigated.

Table 10.7 Values of B_{22} corresponding to crystallization conditions

Protein	*Precipitant and pH conditions*	$B_{22} \times 10^{-4}$ *mol/mL/g²*
Lysozyme	2% NaCl, pH 4.6	–3.0
Canavalin	0.7% NaCl, pH 7	–0.8
Concanavalin A	1 M $(NH_4)_2SO_4$, pH 7	–2.5
Concanavalin A	0.1 M NaCl, pH 6	–1.9
BSA	52% satd $(NH_4)_2SO_4$, pH 6.2	–2.0
Ovostatin	7.5% PEG 8000, pH 7.5, 20°C	–7.1
Ribonuclease A	50% n-propanol, pH 5, 24°C	–4.1
α-Chymotrypsin	10% PEG 3350, pH 4.6	–8.4
STMV	12.5% SAS, pH 6.5	–1.8
Ovalbumin	43% SAS, pH 5.4	–6.1

Both DLS and fluorescence spectroscopy techniques have been applied recently to attempts to differentiate conditions leading to precipitation, or crystallization. Although changes in solute structure are seen, the predictive ability of these techniques remains elusive. An alternative approach for predicting optimal solution conditions for protein crystallization is to measure the osmotic second virial coefficient (B_{22}) of the protein in the crystallization mixture, but at dilute (<8 g/L) protein concentrations.

It was discovered that for the proteins tested there exists a crystallization slot characterized by B_{22} values in the range of –1 to -8×10^{-4} mol/mL/g². At more negative values, the solution was found to promote the formation of amorphous protein aggregates, whereas at positive values, the protein preferred to remain in solution. Although independent experimental validation of this approach has not yet been published, subsequent work by Wilson to extend the method to more proteins and by others investigating the application of lattice model or colloidal theory to describe protein interaction in concentrated solutions has supported the validity of this approach. Measurements of B_{22} may provide a faster approach to optimal conditions than do microscale tests, which require several days for the result to be known.

INDUSTRIAL CRYSTALLIZATION

The knowledge of rates of protein crystal growth is of primary importance in industrial crystallization processes, but except for lysozyme, few systematic studies have been published. Most kinetic studies have been performed using static crystal measurements, the most common method of which involves nucleating several crystals in a growth cell located on a microscope and following the increase in crystal dimension with time as supersaturated solution is pumped gently around the crystals. These are useful for fundamental studies of growth mechanisms, but have a number of disadvantages for process design:

1. The growth rate is typically expressed as a face (e.g., two-dimensional increase) growth rate whose applicability to three-dimensional crystal growth in suspended crystal slurries in industrial crystallizers, in which the hydrodynamic conditions may be vastly different, has not been validated.
2. Because crystals grow at different rates, a large number of crystals must be followed to obtain a reasonable approximation of mean population growth rates.

Nevertheless, the method has benefits in that data can be obtained on individual crystals. Few kinetic studies of protein crystallization have been performed in bulk crystallizers in which the growth of a large crystal population can be measured under conditions more typical of industrial conditions.

Rates of Crystallization

Supersaturation

Chief among the factors determining the rate of crystal growth is the degree of supersaturation, which defines the driving force for crystallization. Supersaturation (s) can be defined a number of ways and is typically defined in the manner most convenient for the user. The relative supersation (σ) = s/c^* is commonly used. In industrial practice, it is convenient to express the relationship between the growth rate (G, μm/h) and supersaturation (s) as a power law:

$$G = ks^n$$

where k is the nth-order growth rate constant and n is the kinetics order. From the studies performed on protein growth kinetics, a range of dependencies of growth rate on supersaturation have been reported, but most have revealed a second-order dependence. This may be interpreted as surface integration kinetics control of the rate of crystal growth. In isolated cases, a first-order dependence was measured, for example, for concanavalin A, which is more indicative of a diffusion-controlled growth mechanism. Protein crystal growth rate appears to be independent of crystal size. In the case of lysozyme, which has been extensively studied, the usual growth rates experienced in static cell studies suggest that about two to three layers of molecule are added to the crystal face every second, amounting to a face growth rate of 0.1 to 1 nm/s (0.4–4 μm/h).

Recent work has proposed that protein crystal growth occurs by the addition of oligomers of lysozyme into the crystal, rather than by monomer incorporation. There is a large body of supporting evidence for this view, both with lysozyme and for other proteins, although this finding remains controversial. A comparison of protein growth rates with those of other commonly crystallized materials reveals that proteins are slower to crystallize relative to most smaller molecules for a given relative supersaturation. Nevertheless, known protein crystal growth kinetics are comparable to those of aluminum trihydroxide, which is crystallized industrially on a very large scale. Furthermore, the metastable zone available for protein crystallization (e.g., the values of supersaturation in which it is convenient to operate a crystallizer) is many fold more than that possible in many small molecule crystallizations. Most commercial protein crystallization processes are batch operations. During isothermal batch crystallizations, the supersaturation will fall rapidly from an initially high value to much lower values, especially because supersaturation is typically achieved by the addition of precipitant. In these cases, it may be uneconomical to wait for the system to reach equilibrium before terminating the crystallization. Although low supersaturation is undesirable because it will result in long crystallization times, excessive supersaturation is also undesirable and generally results in any of a number of inconvenient features, including greater inclusion of impurities in the crystal lattice, unusual and dendritic crystal forms that are hard to process, and amorphous precipitation. Where the solubility of the protein being crystallized exhibits sensitivity to pH, temperature, or other factors, the supersaturation may be maintained at a moderate value by changing these factors as crystallization occurs, provided that there are no deleterious results such as decreased rate, change in crystal form, denaturation, etc. This can only be determined by thorough solubility studies.

Effect of temperature and pH

The effect of temperature and pH on the kinetics of protein crystallization have been little studied. There are frequent comments in the literature that different pH and temperature conditions, or even changes in the primary structure of the protein, altered the rate of crystallization or made crystallization easier. In most cases, there are insufficient data to confirm that the experimental variable (pH, temp, etc.) had a direct influence on the rate of crystallization, rather than an indirect influence through some other effect, such as an altered protein solubility, which in turn would result in increased supersaturation for the same initial protein concentration and hence a faster crystallization rate. The pH has been observed to have a marked effect on the crystallization rate, independent of its effect on solubility, in the case of ovalbumin.

An increase in pH at values up to 1 pH unit higher than the isoelectric point (pH 4.6) resulted in an increase in the rate of crystallization of up to 10-fold. Whether this trend continued at larger deviations from the IEP was not studied. Reports on the effect of temperature are rare, but the probable dependence of the rate of protein crystallization on kinetic mechanisms implies that the rate should increase with temperature at a given supersaturation. Although the results above are too limited to generalize to all commercially produced proteins, it is interesting to note that conditions that have been observed to enhance crystallization rate (increased temperature, pH values away from the IEP) also correspond to conditions of greater protein solubility. Consequently, a compromise is necessary between high crystallization rates and the need for high recovery of protein as crystal.

Effect of impurities

The presence of some impurities in a crystallizing solution is known to result in depressed rates of crystal growth for many molecules. In most cases, such poisoning is difficult to prove unequivocally. The impurities likely to be present in protein solutions undergoing crystallization can be arbitrarily divided into three categories:

1. Nonprotein impurities
2. Protein contaminants not derived from the product protein
3. Heterogeneous forms of the product protein

The term *pure* typically means crystals devoid of the latter two categories, because large quantities of the precipitant inevitably will be present in the solvent channels within the protein crystal. In fact, it is exactly this property that makes protein crystals amenable to X-ray crystallography.

Role of nonprotein impurities

The rate of protein crystallization appears to be remarkably unaffected by the presence of small compounds, such as salts, precipitants, sugars, and pigments, compared to small molecule crystallization processes, although they may have a pronounced effect on solubility of the protein. This resilience may be because of the comparatively small number of contact points between adjacent protein molecules in the crystal, which makes disruption of the ordered lattice by small molecules more unlikely compared to the tightly packed lattices of salts and other small molecules. This is convenient for protein bioprocessing. However, polysaccharides derived from microbial cell wall or components of the media are commonly present in fermentation broths and have been implicated as causing difficulties in bulk crystallization processes, particularly affecting the crystal morphology and properties of the final product. This problem was resolved by the use of carbohydrate-hydrolyzing enzymes.

Role of contaminating proteins

The presence of contaminating proteins has often been considered a primary reason for not adopting crystallization as a protein purification process. In fact, in the few cases studied to date, the presence of contaminating proteins had little or no deleterious effect on the rate of crystallization. The rate of crystallization of ovalbumin was unaffected by the presence of up to 0.22- mol fraction of total protein as lysozyme and conalbumin; similarly, the rate of lysozyme crystallization was unaffected by the presence of up to 5% w/w ovalbumin. However, the literature on this issue is scarce.

Role of heterogeneous forms of the product protein

An important issue is the effect of microheterogeneous forms of the product protein on rates of crystallization. There is mounting evidence that even small changes in the primary sequence of a polypeptide may lead to large alterations in its crystallizability. Unfortunately, the term *crystallizability* is vague and may refer to an improved propensity for nucleation, altered solubility under the conditions used for crystallization, or an actual enhancement in the rate of growth of the crystal face for a given supersaturation. Discerning between these effects requires extensive characterization of protein behavior, which typically is not performed. Nevertheless, numerous reports of protein variants that have either improved crystallizability or that inhibit crystallization of the preferred protein exist.

Thus, variants of recombinant human insulin were generated by site-directed mutagenesis that exhibited improved crystallizability under physiological conditions, whereas for ovalbumin, the subtilisin-treated variant, plakalbumin (which possesses 6 fewer amino acid residues out of the original 385), is reported to exhibit greatly improved crystallizability. This opens the possibility of engineering protein variants with improved crystallization characteristics to improve properties germane to their final use or to reduce processing cost. Some forms of heterogeneity, however, may also be responsible for poisoning rates of protein crystallization. This has been reported for lysozyme in mixtures of that protein from different sources. This adverse effect of variant forms is more likely, because they would share a high degree of homology with the native protein and might be expected to integrate into the lattice in such a way as to poison or disrupt further growth. The ability of crystallization process to select for crystals of only one microheterogeneous form of a protein is unclear.

Other operating factors

The effect of a number of other operating variables on the rate of protein crystallization have not been investigated. It would be expected that the precipitant used, or its concentration, would alter the rate of crystallization. However, data are scarce. Certainly, the effect of salt concentration (ammonium sulfate) on the rate of ovalbumin crystallization was negligible, independent of its effect on protein solubility, which is large. Agitation, or a convective environment, has been argued to have a deleterious effect on crystal growth in small-scale studies performed on microscope cells, although other studies have contradicted these findings. Nevertheless, some agitation is essential in industrial crystallizers to prevent the crystals settling in the vessel. A more serious effect of agitation is crystal breakage.

Protein crystals appear to be relatively fragile, probably because of the small density of intermolecular contact points between neighboring molecules in the crystal lattice. Consequently, rough handling of the crystals should be minimized. High pressure (up to 3,000 bar) was reported to increase the crystallization rate of glucose isomerase relative to that of solutions at atmospheric pressure; however, it is more likely that the enhanced rates observed were the result of higher rates of nucleation at high pressure, which would generate larger crystal surface areas for growth. This latter effect would be advantageous for the production of large numbers of small ($<10\ \mu m$) protein crystals in the absence of seeding, because induction times for nucleation were also reduced by high pressure.

Growth Rate Dispersion (GRD)

Growth rate dispersion (GRD) refers to the existence of a population of crystal growth rates under identical crystallization conditions. The primary cause of GRD remains unclear, but it is now accepted that crystals, under conditions of constant supersaturation, will grow at different, but constant, growth rates. The most significant effect of GRD is to broaden the crystal size distribution with time, which is generally undesirable for most industrial applications. The existence of GRD in protein crystallization is largely unstudied, but appears to be small relative to crystallizations that exhibit the effect strongly, such as sugars.

SELECTIVE PURIFICATION

Crystallization is capable of highly selective purification of most compounds from impure mixtures. Protein crystals sought for X-ray crystallography need to be highly pure, because the crystal is required for structure determination. This requirement has commonly led to the comment that protein solutions must be highly pure to attain crystallization. However, for bulk crystallization purposes, this requirement is not relevant. Because of the highly open structure of protein crystals, small impurities, such as salts, sugars, etc., are likely to be present in the final crystals to a similar degree as in solution, although protein crystals appear to exclude the pigments and colorants found routinely in fermentation broths. Successful protein crystallization has been routinely obtained in the presence of other contaminating proteins with degrees of purification similar to those achieved by ion exchange. Most of the carryover of contaminating protein is in the liquid adhering to the exterior surface of the crystals. Studies showed that ovalbumin crystals (MW 45,000 Da) produced in bulk crystallization were substantially free of both smaller and larger protein contaminants, which comprised 22 mol % of the total protein initially present in solution, after a single wash of the crystals. Similar results have been observed for lysozyme.

APPLICATIONS OF CRYSTALLIZATION

Despite the ubiquity of crystallization for the structural determination of proteins, actual process applications of crystallization as a technology for the purification and recovery of proteins remained relatively rare until the last few years. An increasing number of proteins are now produced in multikilogram or multiton quantities by crystallization from clarified fermentation broth. These include insulin, glucose isomerase, asparaginase, subtilisin, lipases, thermolysin, and penicillin acylase. The first major use of crystallization was in the purification of insulin, which was crystallized as a zinc complex from salt solution. This technology was developed in the 1950s and simulated the natural process found for storage of insulin as zinc hexameric crystalline arrays in the pancreas. More recently, the ability of insulin to be crystallized has been used in slow release forms to achieve a more sustained titer of the hormone in the blood, due presumably to the regulation of insulin adsorption by slow dissolution kinetics at physiological pH.

In the majority of cases, halide salts are preferred for producing the supersaturation needed to obtain crystallization, although a recent report suggests that the use of PEG may greatly enhance the process at lower salt contents. Crystallization is usually performed using clarified fermentation broth that has been concentrated normally by ultrafiltration so that supersaturation is attained on addition of precipitant. The crystallization pH is generally adjusted to values at or near the isoelectric point of the protein to take advantage of the lower solubility at these pH values. Alkaline proteases, such as those from *Bacillus* species (e.g., subtilisin) and lysozyme, are exceptions. These enzymes are typically crystallized at lower pH values to minimize autolysis during the process. Process temperatures vary widely. Temperature change plays a minor role compared to precipitant addition for obtaining supersaturation. This is usually the result of the constraints imposed on process temperature by protein stability requirements or the relative insensitivity of protein solubility to temperature, particularly at high salt concentrations. Typically, the crystals produced in commercial crystallization processes are small (30 to 100 μm in diameter) to aid their more rapid dissolution, in either product applications or subsequent purification steps. By the end of the crystallization process, the crystal content of the liquor can be up to 20 to 30% v/v. The crystalline product can be separated from the mother liquor by settling of the denser crystal phase and decantation of the mother liquor from the crystals, or by usual solid–liquid separation technologies such as microfiltration, pressure filtration, or centrifugation. Washing to remove the mother liquor from the crystals improves crystal purity and the overall purification ability of the process.

Generally, the crystals must be maintained as a concentrated slurry in a suitable salt solution to avoid redis solution or drying out and reversion to an amorphous state. A novel approach to protein crystallization, which overcomes the problem of maintaining the crystal form, is the subsequent cross-linking of the crystals by, for example, glutaraldehyde, which renders the crystals insoluble in aqueous solutions. This process has been used to develop mechanically robust and stable enzyme crystals called CLECs (cross-linked enzyme crystals) suitable for repeated use and easy separation from the reaction liquor in organic syntheses. A further advantage of CLECs is their enhanced stability relative to the noncrystalline form and even to the un-cross-linked crystal form of the enzyme under adverse conditions, such as low water environments, elevated temperatures or the presence of proteases, which makes them of value in the synthesis of many chirally pure drugs and chemical intermediates. A recent example comprising a lipase revealed the half-life of the soluble enzyme to be 5 h compared to more than 13 days for the cross- linked crystal form.

CONCLUDING REMARKS

In addition to its critical importance for protein structure determination by X-ray crystallography, protein crystallization is now emerging as a useful bulk technology for protein purification. It is particularly well suited to commodity proteins where bioprocessing costs must be contained in the face of demands for improved product quality. Protein crystallization offers potentially high degrees of purification at any scale in reasonable process times and cost. Nevertheless, the gaps in knowledge concerning the application of protein crystallization at process scale remain large, despite its successful use in the production of commercially significant proteins. Protein crystals also offer exciting prospects as novel forms of catalysts in controlled- release formulations and as biomaterials. The sensitivity of protein crystallizability to even single amino acid changes in the polypeptide chain further suggests that fruitful results can be obtained using tools of modern molecular biology and protein engineering, already well established in the production of commodity proteins. Perhaps what is needed most, however, is the realization that crystallization, long overlooked in production processes for proteins and far more elegant than its wilder sibling—precipitation—deserves to be considered alongside the many other operations commonly selected for protein purification.

11

EFFICIENT CLEANING METHODS

The importance of adequate cleaning in a biological process cannot be overemphasized. The required stringency of cleaning and cleaning validation is best evaluated by performing a risk analysis. The risks associated with the use of improperly cleaned equipment and materials (such as filters and chromatography media) vary with the product source, intended use of end product, the position of the equipment and materials in the processing train, the phase of development of the process, and the type of unit operation. A further consideration is whether the equipment or facility is used for more than one product. Robust cleaning methods, suitable cleaning agents, and cleaning validation ensure that a product with the desired properties is manufactured without contamination from processing additives, well-defined impurities, unexpected contaminants, and previously run product.

AVOIDING CONTAMINATION

For certain complex product sources such as plasma, the potential safety risk from inadequate cleaning is clearly very high. The impurities in the source material are not fully characterized, making evaluation of cleaning efficiency more difficult. For other products, such as secreted products from a mammalian cell culture such as chinese hamster ovary (CHO) cells, the impurities are clearly defined, cleaning efficacy is more readily assessed, and the risks are fewer than with products derived from multiple sources such as blood donors. Robust cleaning procedures are also required for synthetic processes such as those used to produce oligonucleotides used for antisense therapy and synthetic peptides, but the cleaning methods and assessment of their efficacy may be even simpler.

If the end product is intended for use as a therapeutic agent delivered in a large dose for an extended time in a healthy individual, the criticality of cleaning is obvious. For a product intended for in vitro diagnostic use, the potential contamination that can arise from inadequate cleaning can result in an incorrect diagnosis and even result in a dangerous treatment for a patient. For industrial enzymes, contamination from previously run product may or may not have a detrimental effect on product performance, but even in this case the limits of cleanliness should be defined.

PROCESSING SCHEME

A typical processing scheme will consist of several unit operations. The closer one gets to final product, the more critical the cleaning becomes. For example, it has been suggested by some regulators during oral presentations that firms using contract manufacturers for formulation and filling consider supplying their own dedicated equipment to ensure no risk of contamination from other products caused by insufficient cleaning. But even up-stream in fermentation or synthetic processes, the lack of adequate cleaning can lead to contamination of the next batch with unknown impurities that adulterate the product in such a way that its safety, potency, and efficacy are compromised.

Table 11.1. Unit operations in a typical biotechnology process

Unit operations
Cell culture or fermentation Isolation
Clarification
Purification
Formulation or final filling

PHARMACEUTICAL IMPORTANCE

From research to full-scale manufacturing, proper cleaning is essential for reproducible processes. Not surprisingly, there is a great rush to get a product into the clinic to demonstrate efficacy. In this rush, cleaning is sometimes overlooked, resulting in laboratory data that are unreliable. For example, when a researcher uses a chromatography column for developing a purification process and fails to adequately clean precipitated impurities, the purification process in the next run is likely to yield a product that does not represent what is intended. Good manufacturing practices (GMP) are required for all clinical trials. This requirement is also applied in many nations other than the United States. Although cleaning validation might not be complete at this stage, routine cleaning is essential, and it is advantageous to have a cleaning validation plan. Cleaning and cleaning validation are essential in the manufacture of marketed products and are part of good manufacturing practices.

COMMERCIAL VALUE

In the past, most facilities used for manufacturing therapeutics were dedicated to one product or a family of products from the same source. For complex biologicals, such as plasma products and traditional vaccines, this is still the case. For these biologics, starting materials, and even the final products, may not be sufficiently well defined to evaluate the effectiveness of the cleaning regimes in removing all traces of previously run product and processing materials. Advances in analytical technologies, however, are allowing firms to better assess cleaning effectiveness, especially for well-characterized products. Products that are considered well characterized include therapeutic DNA plasmids, therapeutic synthetic peptides of 40 or fewer amino acids, monoclonal antibodies for in vivo use, and therapeutic recombinant DNA-derived products. Today, there is a trend to reduce processing costs by employing contract manufacturers to produce clinical trial materials, thereby avoiding the construction of costly facilities for products that may not ultimately obtain regulatory approval. There are also many firms that campaign products (produce one at a time) to fully utilize space, equipment, and personnel. Regulators carefully scrutinize cleaning and cleaning validation in such facilities to ensure that there is no cross-contamination of products. Cleaning validation using sensitive, specific, and nonspecific analytical methods is essential for such operations.

METHOD OF UNIT OPERATIONS

Each unit operation has somewhat different cleaning requirements, which as previously stated are dependent on the nature of the product source and its intended use, its position in the process train, and whether it is used for more than one product. Cleaning is not an issue to be addressed after a process is finalized; it should be designed into the process. In fact, the selection of the most suitable equipment and materials may be dependent on their ability to be cleaned. Records kept during research and development may be valuable in the design of a cleaning process. For example, a researcher may find that a standard cleaning procedure recommended by an equipment vendor is not sufficiently harsh for a particular product feed-stream.

Although each feedstream and each unit operation have unique features that must be considered before the most efficient cleaning protocol can be designed, there are many commonalities that can used to minimize development time. References such as regulatory guidelines and PDA's book on cleaning should be consulted. Keeping up with the latest technical and regulatory publications and comments at meetings and workshops is essential. In the design of the cleaning regime for each unit operation, critical factors to consider include the type of cleaning method, compatibility, and detection methods. Cleaning may be automated or manual. Automated cleaning can provide more consistent results. Manual cleaning, on the other hand, is clearly much more difficult to validate, and operators must be certified to perform the task. Cleaning-in-place (CIP) is preferred, but for some components, cleaning-out-of-place (COP) is necessary. Cleaning may be accomplished by chemical or physical methods, or, most likely, by a combination of both.

Chemical Cleaning

When chemical methods are used, the contact time, temperature, and cleaning agent concentration must be specified. The order of adding cleaning agents and rinsing agents may be critical. Additionally, if the chemical agent is flowing through the equipment, then flow must be specified. It is essential to ensure that the cleaning agent is compatible not only with the specified equipment, but also with any ancillary components that might be exposed to the agent. The nature of the soil must also be considered. Whereas hot alkali and detergents might work for one unit operation, for another they may actually exacerbate the cleaning problem by causing deterioration of the equipment or aggregation of impurities. Organic solvents used for removal of lipids may extract potentially harmful chemicals from some equipment components in a given unit operation.

If a packed chromatography column containing a soluble hydrophobic protein is cleaned by treatment with a salt solution, the protein is likely to precipitate and actually leave behind a larger residue. For filters and chromatography media, compatibility with the cleaning agent is of particular concern. For example, when incompatible cleaning conditions are used, an immunoadsorbent might leach significant amounts of antibody, leading to reduced capacity and hence inconsistent results in the next batch.

Table 11.2 Common cleaning mechanisms and cleaning agents

Mechanism	*Agent*
Dissolve	Water, alkali, acid
Saponify	Alkali
Degradation of proteins	Alkali, acid
Wetting	Surfactants
Emulsification, suspension	Phosphates, surfactants
Sequester	Chelating agents

Physical Methods of Cleaning

Physical methods include high velocity flow, jet sprays, and agitation. If physical methods are used, it is essential to ensure compatibility with the equipment. Physical methods commonly used for cleaning tanks and fermentation vessels (e.g., spray balls) are clearly not suitable for any packed chromatography columns or thin piping.

Physicochemical Methods of Cleaning

A combination of physical and chemical methods is usually used. Mechanical forces remove gross soils and deliver cleaning solutions to large areas, but chemicals (such as alkali, acid, and detergents) are generally required as well. According to Rohsner and Serve, cleaning processes require energy as well as chemical and physical reactions. Cleaning mechanisms are further addressed by LeBlanc et al. Some development work is necessary to put into place the most efficient and cost-effective cleaning methods for each unit operation. It is important to recognize that the nature of the soil may change somewhat upon scale-up of a process, necessitating a modification of the cleaning procedure. This is most common in the earlier processing steps such as fermentation. Ranges of acceptable conditions should be established and validated to accommodate the worst-case soil.

Fermentation technique

At this early step in the process, the nature of the soil to be cleaned is not well defined, but it may contain substances such as lipids; soluble and precipitated proteins; endotoxins (if Gram-negative bacterial fermentation is used); small, soluble highly charged molecules; large DNA; particulate cell debris; and even whole cells. The amount of each substance to be removed depends on the fermentation or cell culture process. Filamentous fungi generally require harsh cleaning, vessels used for yeasts and bacteria are more easily cleaned, and animal cell cultures in which the products are secreted require even less stringent conditions. Assessing the cleanliness at this early stage requires the use of nonspecific assays such as total organic carbon, and in some cases specific assays are also used. If the product is secreted and very few cells die during the process, the substances to be removed from the bioreactor may be limited, and relatively mild cleaning procedures might be very effective.

However, if the cells die and lyse at unknown rates in a process that is not well controlled, there will be more debris and released cellular components to remove. The heat used to kill cells may also cause aggregation and precipitation, making the cleaning more difficult. The type of cell culture media (i.e., serum, serum-free, defined) must also be considered. In addition to designing the cleaning procedure to accommodate the type of cells or microorganisms, it is necessary to evaluate the bioreactor design and ensure that the CIP mechanisms are suitable. CIP system design and operation for bioreactors have been addressed by Chisti and Moo-Young.

Physical removal of cells

Cell removal and clarification most commonly use filtration and centrifugation operations. Techniques such as expanded bed absorption, two-phase separations, and precipitation are also used. Each one of these methods may require its own unique cleaning procedures. Cleaning must ensure removal of cells, cell membranes, cell walls, nucleic acids, proteins, lipids, etc.

Method of filtration

In filtration operations, one must consider the membranes as well as the hardware. Chemical cleaning is most common, and the cleaning agents must be compatible with both the membrane and the wetted components of the hardware. In addition to the piping and membranes, wetted components may include spacers, gaskets, valves, and seals. Adhesives may also be present. The cleaning agent should not damage any of these over time. Some membranes are incompatible with cleaning agents such as sodium hydroxide; some may be cleaned with mild detergents and then sanitized with bleach. Furthermore, some filter materials have pressure and temperature requirements that must be considered when designing the cleaning protocol. And as with other unit operations, the number of uses of wetted components may need to be defined to ensure that with extended exposure to cleaning agents extractables are not released and that the membrane performance is not changed. The stringency of cleaning required is dependent on the level of gel polarization and membrane fouling. Brose and Waibel have addressed the adsorption of proteins to microfiltration capsules. The cleanliness of tangential flow membranes is often determined by measuring the flux of clean water through the system at a standard transmembrane pressure. Guidance and cautions for performing this measurement have been described by Michaels. Other methods of assessing cleanliness include total organic carbon, physical inspection, and comparing the specifications of a water-for-injection (WFI) rinse.

Physical separation

Because centrifuges represent a significant capital cost, they are frequently used for more than one product. Cleaning validation is essential to prevent cross-contamination. Although continuous desludging disk centrifuges can be designed so that they can be cleaned in place, those that do not desludge continuously cannot. The bowl in tubular bowl centrifuges must usually be cleaned out of place after disassembly. Aerosols generated during centrifugation may require further evaluation of the equipment area to be cleaned. For low-volume products, cleaning can be minimized by using disposable self- contained tubes. This approach may be preferable, for example, when centrifugation is used to prepare multiple products over a relatively short period of time, such as for monoclonal antibodies used for in vitro diagnostics.

Method of phase partitioning

With the exception of aerosol generation, phase partitioning and precipitation present the same general cleaning issues as found in tanks and centrifuges. Equipment design plays a major role in determining how well these pieces of equipment can be cleaned.

Expanded bed adsorption technique

Expanded bed adsorption and the use of large (on the order of 300 μm) beads are two techniques in which product isolation is achieved by binding to chromatographic media. Unlike standard chromatography, particulates and even whole cells are applied to the chromatography media. Cleaning is essential to ensure that the equipment does not retain soils from previous runs and that media can be reused. Chromatography media, whether used for isolation or purification, have a high degree of surface area that allows for adsorption and even entrapment of impurities. Even though chromatography media are dedicated to one product, cleaning must address the specific feed-stream—both impurities and product—and its interaction with the media. Furthermore, the equipment must be designed to ensure that cleaning procedures will remove all traces of previously run product and that no residual live organisms remain.

In a study to evaluate the effectiveness of a cleaning and sanitizing method, a challenge test was performed by adding an *Escherichia coli* homogenate containing approximately 10^9 colony forming units (cfu) per milliliter to two *streamline* expanded bed columns. One column contained 250 mL of *streamline* sp and the other 250 mL of *streamline deae*. Sodium chloride cleaning solutions of increasing concentrations from 0.1 to 0.5 M were added, and the reduction of living microorganisms measured. The relative reduction in living cells is approximately 10^5. The increase in the number of cells in the eluate from the *deae* illustrates the necessity to understand the nature of the impurity to be removed and its interaction with the component being cleaned. In this case, the increase occurs when the increased sodium chloride concentration releases negatively charged *E. coli* cells from the positively charged *deae*.

Robust chromatographic process

A robust chromatographic process provides the purity required for proteins, peptides, oligonucleotides, plasmids, viral vectors, etc. But it is essential to select equipment and media that can be cleaned in place. Most of today's chromatographic media can tolerate rather harsh conditions, such as exposure to sodium hydroxide. In affinity chromatography, the ligand usually determines the cleaning solutions that can be used. Protein A, for example, has been shown to withstand even sodium hydroxide. On the other hand, with immobilized monoclonal antibodies, it is often necessary to protect the column by using only clarified, partially purified feed.

The cost of preparing a monoclonal antibody for use as an immunoadsorbent used in the purification of a therapeutic product is extraordinarily high, because the immobilized antibody must have the same purity as the therapeutic itself. One example of the highly successful use of such a column is described by Feldman et al. for the purification of factor VIII from human plasma. As with other unit operations, specific impurities must be addressed in designing the cleaning regimes to ensure that there is no residue carried over from batch to batch of product. Clearly, efficient cleaning is even more important the closer one gets to final product purity. All cleaning procedures, contact time, and temperature must be specified. To maintain column lifetime, the least harsh method that provides a safety window should be implemented. From a typical bacterial fermentation, one might have to remove soluble and precipitated proteins, hydrophobic proteins and lipids, nucleic acids, viruses, and endotoxins.

Hydrophobic interaction chromatography (HIC)

Soluble proteins are the easiest to remove. For ion exchange, gel filtration, and most affinity chromatography media, a high salt wash (e.g., 1 to 2 M NaCl) is most common. For hydrophobic interaction chromatography (HIC), low-ionic-strength buffers or water are usually sufficient. In some cases, a detergent must be used, but this approach is more costly and requires validation of detergent removal. Precipitated proteins frequently require up to 1 M sodium hydroxide or 1 M acetic acid. This is often followed by a water rinse and then a salt wash (except for HIC). Like precipitated proteins, hydrophobic proteins usually require sodium hydroxide.

Other approaches include the use of ethanol, detergents, and a combination of cleaning agents. For example, ethanol (20%) in 1 M sodium hydroxide has been used at 40°C for 30 to 60 minutes to remove host cell proteins from an ion exchanger. For lipids, very little information is available. Commonly used solubilizing agents may destroy some chromatography media. Detergents and organic solvents (i.e., ethanol and acetonitrile) have been used with some success, but it is necessary to be aware that explosion-proof areas may be required for organic solvents. For those already working in such an area, this does not present a problem, but for firms used to working only in aqueous conditions, this may increase costs significantly. Furthermore, solvent disposal or recycling must be implemented. Removal of lipids before chromatography is strongly recommended to alleviate these concerns and added expenses.

DNA clearance

Nucleic acids bind tightly to anion exchangers. Some earlier work to evaluate their removal was performed by loading radiolabeled calf thymus DNA onto a DEAE Sepharos Fast Flow column. Combinations of sodium chloride and sodium hydroxide were insufficient to totally remove all of the DNA. More recently, the ability of different cleaning agents to remove nucleic acids from a quaternary amine anion exchanger was tested using three different samples. It is interesting to observe that for the monoclonal antibody, DNase was required to thoroughly remove all traces of nucleic acids.

Table 11.3 Mass balance of calf thymus DNA clearance from Q-sepharose fast flow

Fraction tested	*μg DNA*	*% of total DNA*
Starting material	699.75	100
Buffer Awash	0.79	0.11
100% B eluate	253.80	36.27
Acid wash	0.64	0.09
1 M NaOH/1 M NaCl	449.60	64.25
Water wash	3.92	0.56
Total DNA eluted	708.75	101.2

Viruses

Regulatory agencies have expressed concern about the retention of viruses on columns. As stated at the International Conference on Harmonization Guideline on Viral Safety, "Assurance should be provided that any virus potentially retained by the production system would be adequately destroyed or removed prior to reuse of the system. For example, such evidence may be provided by demonstrating that the cleaning and regeneration procedures do inactivate or remove virus".

Role of endotoxins

Like nucleic acids, *endotoxins* are highly negatively charged and may bind tenaciously to anion exchangers. Sodium hydroxide has been shown to be very effective in inactivating endotoxins. Although a contact time of up to 5 hours with 1 M NaOH is indicated here to inactivate endotoxin, the added action of flow may enhance removal. In a study on a chromatography system challenged with a solution containing 1,200 endotoxin units (EU) per milliliter, it was found that circulating 1 M NaOH for 60 minutes was sufficient to inactivate endotoxin in the both the system and the packed column of a quaternary amine anion exchanger that was placed in line. In addition to the *Limulus* amebocyte lysate (LAL) test for endotoxin, a total organic carbon assay was used to analyze the rinse fluids for any carbon-containing residues remaining from the endotoxin. No residue was detected.

Cleaning and sanitization methods

The ability to remove microorganisms from chromatography media is an issue that should be addressed in development. Depending on the microorganisms, they may produce endotoxins, enterotoxins, spores, and proteases—all of which could adulterate a product. Sanitization of chromatography columns is readily accomplished with sodium hydroxide. Combining cleaning and sanitization is the most cost-effective approach. In many cases, addition of sodium chloride to a sodium hydroxide solution provides an efficient, simple method that is easy to validate for removal of bioburden (tested by routine monitoring) and impurities from the feedstream. The importance of compatibility of the wetted materials with such combinations should be considered at the design phase. Cleaning issues in formulation and filling operations are addressed in depth elsewhere.

Some relevant texts include those by Olson and Groves and Cole. A recent publication that addresses reducing pyrogens in cleanroom wiping materials provides valuable information on cleaning.

Table 11.4. Mass balance of total DNA clearance from a monoclonal antibody on Q-sepharose fast flow

Fraction tested	*μg DNA*	*% of total DNA*
Total DNa in MAb	481.50	100
Flow through	0.31	0.06
Wash	0.12	0.02
Peak 1	10.00	2.08
Peak 2	270.68	56.21
100% B wash	6.08	1.26
Acid wash	4.00	0.83
1 M NaOH/1 M NaCl	64.00	13.29
2 M NaOH	3.21	0.67
3 M NaCl	0.65	0.13
Water wash	0.29	0.06
Subtotal	359.34	74.61
Dnase 1 treatment	125.00	25.96
Total DNA eluted	485.9	100.6

TECHNIQUES FOR EVALUATING CLEANLINESS

Several methods for evaluating cleanliness were discussed previously in conjunction with the different unit operations. There are three basic methods to evaluate cleanliness: visual inspection, assaying rinse water, and assaying surfaces by swabbing. The nature of the unit operation and equipment design frequently determine which method will be most appropriate. A combination of methods is usually necessary for validation. Visual inspection may be quite useful as a qualitative assessment of cleanliness for equipment such as tanks. As noted by several regulators, "you can always see the ring around the bathtub." However, more sensitive, quantitative methods are also required. For multiproduct facilities, it is absolutely necessary to use product-specific assays. Product-specific assays are also used for dedicated facilities. Rinse water can be collected after cleaning and examined for residues.

In many cases, the quality of the rinse water is compared to WFI that has not been run through the component being tested. Other tests commonly run on rinse water include LAL for endotoxin, total organic carbon (TOC), protein assays, sodium dodecylsulfate–polyacrylamide gel electrophoresis (SDS-PAGE), spectrophotometric analysis, and product-specific assays such as ELISAs. Swabbing is performed by using non-carbon-containing materials that do not interfere with subsequent testing but from which swabbed soils can be extracted. The swabs themselves should not generate particles, background carbon, or nonvolatile residues.

Swabbing of surfaces allows evaluation of the cleanliness of areas in which non-water-soluble residues might remain. However, there may be components or areas of systems that are hard to reach by swabbing. In this case, a soil can be applied to pieces of the equipment, called coupons, that are cut out or provided by a vendor. The soil is allowed to dry, and then the coupon is swabbed after cleaning. Coupons can be useful in designing the cleaning protocol. The coupon can be placed inside equipment and cleaned during a normal cleaning cycle. Sometimes, the coupon and the soil are heated to simulate a difficult cleaning problem. Cleaning methods can be tested on the baked-on soil. After the swabbing is performed, the swab is extracted, and the extract is analyzed by assays such as TOC, HPLC, and ELISA. A survey of five U.S. biotechnology companies revealed which assays are commonly performed and how they are used at different stages of manufacturing. The report presents the purpose of each test, detection limits, acceptance criteria, and methodology in a concise tabular form.

LIMITATIONS

How clean is clean? As analytical methods become more sensitive, the likelihood of finding something is increased. Attempts to define acceptance limits of cleanliness have been made by Fourman and Mullen and Zeller. Fourman and Mullen suggested that "No more than 10 ppm of any product should appear in another product" and "No more than 0.001 dose of any product will appear in the maximum daily dose of another product." But clearly, a risk analysis should be performed to assess the potential hazard for each product. Whereas for some therapeutics these low levels suggested by Fourman and Mullen may be accepted by regulatory authorities, for many multiproduct facilities, "not detectable" may be required. On the other hand, for diagnostic products, the real issue should be whether the product is consistent and whether the carryover affects its accuracy. The acceptance criteria for upstream unit operations such as fermentation may not be as stringent as those for final purification and processing steps. Typical acceptance criteria for CIP of fermenters has been discussed by Naglak et al. In addition to setting specifications for residual impurities derived from the product feedstream and final product, it is necessary to consider the removal of cleaning agents. Analytical strategies for organic cleaning agent residues have been addressed by Gavlick et al. The questions: "What is the level of detergent residue that would be acceptable to the Food and Drug Administration (FDA)? What is the basis for arriving at this level, if any?" were answered by the FDA's Office of Compliance as follows: "FDA has repeatedly stated that it is the firm's responsibility to establish acceptance limits and be prepared to provide the basis for those limits to FDA. Thus, there is no fixed standard for levels of detergent residue. Any residues must not adversely alter drug product safety, efficacy, quality, or stability".

GUIDES FOR CLEANING

Although it is not a regulatory requirement, a master plan for cleaning validation can be quite valuable. The master plan will provide a time frame for performing the work, prioritizing tasks, designating responsibilities, describing protocols, and listing all facility, equipment, and process validation requirements. Although the time frame will most likely change, the plan will ultimately save time and may prevent regulatory delays by ensuring that no critical items are overlooked. A final validation report should be signed by management. Agalloco described a series of questions that provide insight into cleaning validation: What items are being removed? How are samples taken? What analytical tests are utilized? When shall cleaning validation be performed? Which physical parameters shall be evaluated? How much residual will be permitted? These questions should be addressed during the preparation of the master plan. Two cleaning validation guides written by the U.S. FDA are very useful.

Although the intent of these guides is to provide FDA investigators with information that results in consistency in inspections, there are many relevant points that should be considered by any firm that has to perform cleaning validation. For example, in the mid-Atlantic version, it is stated that operators performing cleaning operations should be aware of problems and have special training, and the length of time between the end of processing and each cleaning step should be controlled. The documents further address issues such as equipment design, documentation, analytical methods, sampling, and establishment of limits.

Although no one approach is required, the guidance provided in these documents should be followed or a reason why it was not adhered to provided. One of the more difficult techniques to validate is swabbing. Some of the variables include operator swabbing technique, extraction efficiency from the swab, and recovery of residue from different wetted materials. In spite of its difficulties, this technique provides valuable information that cannot be obtained from visual inspection and rinse water analysis.

Total protein analysis and total carbon analysis of swab samples for the cleaning validation of bioprocess equipment have been addressed by Strege et al. Use of this technique for validating detergent residues has also been presented. Most cleaning validation is performed at pilot or full scale, but for certain unit operations such as chromatography, the parameters that are to be validated can often be established with scaled-down models using production feedstream.

Seely et al. scaled down a chromatography system and used cleaning conditions that were harsher than those used in normal production to determine if residues remained on the packed column. At the same time, production columns were also evaluated for cleaning efficacy by collecting relevant product fractions from blank runs and assaying them for UV280 absorbing material and endotoxin. In addition, reversed-phase HPLC, ELISA, and silver-stained SDS-PAGE assays were performed. No endotoxin or UV280 absorbing material was found.

A trace amount of the product was found, but the level (about 0.3% of the amount normally collected) was thought not to represent a significant cycle-to-cycle carryover risk to the final product. Furthermore, the eluate after storage was evaluated by TOC for evidence of further cleaning effects of the storage solution.

Once cleaning is validated, it is much more convenient to monitor key process conditions such as cleaning agent concentration, pH, conductivity, temperature, contact time, and flow if applicable. Revalidation of cleaning may have to be performed when there is a change in a process or equipment, after maintenance or shutdown periods, or after area contamination. Each firm should establish the frequency of cleaning revalidation.

EXAMPLES

Some examples illustrate the critical issues in cleaning and cleaning validation.

CIP Validation in a Bioreactor

In a 1994 paper, Geigert et al. presented a case study on the CIP of a bioreactor. They addressed five key questions: What components need to be cleaned from the bioreactor? What is a scientifically sound and appropriate sampling plan? How will the test samples be handled prior to analysis? Are the test methods chosen scientifically sound and appropriate for the intended purpose? What is clean? The components to be cleaned included active protein, host cell components, media ingredients, and cleaning agents. The sampling plan included swabbing of a few select surface sites, performing a worst-case cleaning challenge using artificial soil or reducing cleaning parameters, and evaluating liquid from the rinse.

Both final rinse and earlier rinse samples were taken to show that the cleaning process was effective. Samples were evaluated for their stability during storage and handling. As an example, it was noted that hydrophobic proteins may adhere to certain plastic storage containers. Several assays, both specific and nonspecific, were evaluated. Three assays were selected: SDS-PAGE, a bioassay, and pH. TOC was not yet set up at the facility; SDS-PAGE and the bioassay had already been validated by quality control and only needed to be further validated for the limit of detection for the cleaning studies. The absence of cleaning agent was determined by pH, and the absence of final product by the bioassay. Four protein products were analyzed by SDS-PAGE with silver staining. The limit of detection varied from 1 ppm to 10 ppm.

The bioassay was performed for the three products that might be campaigned in a single bioreactor. Two products were from yeast; one was from *E. coli*. The two yeast products could be detected at levels as low as 4 ppb, but the *E. coli* product, which was very hydrophobic, aggregated. This study illustrates that the test procedures to evaluate cleanliness are critical in supporting the validation work. By using validated test methods, the firm was able to provide a high degree of assurance that a 1,600-L bioreactor could be cleaned sufficiently between product campaigns, thereby avoiding the additional costs of purchasing costly dedicated equipment.

Pilot-scale Chromatography

In this example, a firm had used a pilot-scale chromatography unit for development work in the purification of monoclonal antibodies (IgA and IgM). To avoid the expense of purchasing a new unit, the firm decided to use this system for the manufacture of clinical trial material. This required extensive cleaning validation. Because the unit contained tubings and threaded fittings, it was decided that replacing low-cost wetted components was less expensive than validating their cleaning. After the new components were installed, the cleaning of the system was validated by demonstrating absence of carryover on previously exposed sites and by using new surface sites to establish a baseline for future cleaning validation. Samples were taken by swabbing and collecting rinse water, and samples were analyzed by both TOC and ELISAs. The IgM ELISA has a limit of detection of 25 ng/mL, and the IgA ELISA has a limit of 4 ng/mL. The results showed that the 16 used surfaces sampled were below the detection levels of the ELISAs.

Multiproduct Pharmaceutical Contract

A multiproduct biopharmaceutical contract manufacturing facility must address cleaning and cleaning validation to satisfy its own requirements as well as the requirements of their customers and regulators. Sherwood has presented a description of CIP experiences in such a facility. The facility manufacturers monoclonal antibodies and recombinant proteins by mammalian cell culture. The author points out that "Each company must determine their own cleaning policies and acceptance criteria based upon knowledge of the equipment and processes employed." Visual inspection, final rinse water, and swabbing of surface samples were used to evaluate CIP of fermenters and purification holding vessels. Because visual inspection is subjective, two observers are always used. For the rinse water samples, sample containers are single use, and sample storage temperatures and duration were validated.

The following tests were performed on rinse water: pH and conductivity, nitrate and protein concentration, bioburden, TOC, endotoxin, and an enzyme immunoassay (EIA). The EIA was only performed for the purification holding vessels. In general, nonspecific assays are used for screening both product and manufacturing materials in fermentation, whereas in the purification area more specific assays are used. For the swab samples, TOC and protein concentration were performed for the fermentation, and in the purification area, the EIA was added. Acceptance criteria are more difficult to establish in such a facility because the final product dose is not known. Acceptance limits have to be based on analytical method performance criteria, pharmacopeial requirements, and historical data. The usefulness of performing cleaning validation was illustrated by the performance qualification of the CIP of the fermenter vessel. In one run, a sight glass swab showed a high TOC result. After investigation, a manual cleaning of the sight glass was implemented. For the initial performance qualification, three consecutive CIP runs were monitored and found to meet acceptance criteria. On an annual basis, one CIP run is fully monitored. The author noted that ongoing monitoring in such a contract manufacturing facility may also require final rinse and swab sampling.

12

ENVIRONMENTAL CLEANING

Waste air or waste gases of industrial processes may contain volatile organic or inorganic odorous or toxic compounds. These gases are not to be released to the environment, according to national or European legislation, which defines the allowances of the individual compound to be released. Environmental biotechnology including waste gas treatment was first developed in countries using the German language and in The Netherlands. In these countries, waste gas treatment was applied by the industry for elimination of odor emitted from wastewater treatment plants, slaughterhouses, and such facilities. For more than two decades, plants have been operated with increasing efficiency, and their application has been extended to the chemical, pharmaceutical, and food industries. Initially, most information has evolved from these European countries; experience has been compiled and merged into guidelines for design and application of industrial biofilters, trickle-bed reactors, and bioscrubbers. These guidelines have been published by the Verein Deutscher Ingenieure in German and in English.

The developing technology was subsequently recognized on other continents. Previous articles on biological waste gas cleaning have covered different processes, mass transfer, biological, chemical, and olfactory aspects, and costs, and the reader is referred to comprehensive reviews with detailed references therein. High concentrations of organic or inorganic pollutants are currently eliminated from waste gases by sorptive processes or thermal of catalytic oxidation. Low pollutant concentrations and odorous compounds are also being traditionally eliminated by biological processes. Plant investment and operating costs of biological waste gas cleaning processes are significantly less than alternative physicochemical processes. This fact is the major driving force for academic and industrial research to develop biological processes and to extend these for elimination of high concentrations of pollutants from waste gases. Presently, the three types of plants mentioned are currently used for biological treatment of different qualities of waste gases, which later are treated in detail. Biofilters require a substantial area and are filled to a low height with organic matter such as compost or chopped wood, which serves in part as a source for minerals for the attached microorganisms. Biofilters are suited for cleaning low concentrations of pollutants. Trickle-bed reactors require less area, are operated as towers, and are filled with incompressible porous minerals as air–liquid contactor and as support for microbial growth.

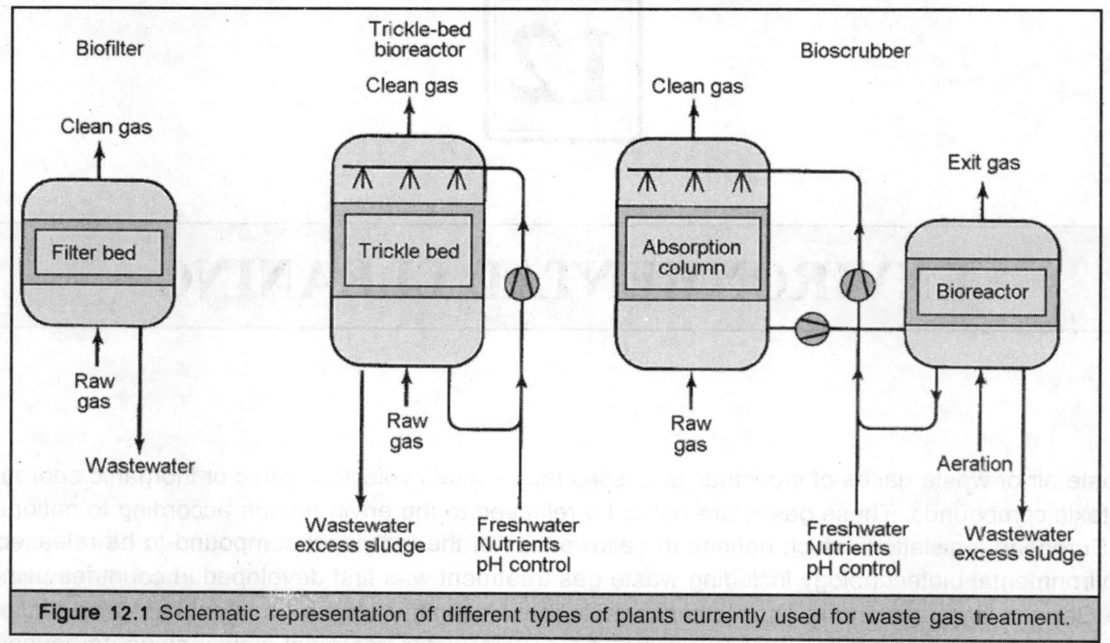

Figure 12.1 Schematic representation of different types of plants currently used for waste gas treatment.

Bioscrubbers are generally composed of two units—one for scrubbing the pollutants from the gas phase and the other as fermentor for microbial degradation of the pollutants comparable to activated sludge. Microorganisms, bacteria or fungi, use the organic or inorganic compounds for energy transformation and for biomass formation. Biological waste gas cleaning is an aerobic process and is brought about by aerobic microorganisms either suspended in water or attached to some support material as a biofilm.

Process development is required to establish and to maintain optimum conditions for these microorganisms to achieve an optimal elimination of pollutants from air. Microorganisms require different elements to build up biomass, which is composed of proteins, sugar polymers, fatty acids, nucleic acids, and so forth. The energy to build up the respective monomers and to polymerize these is derived from the biochemical oxidation of volatile organic matter to carbon dioxide or from gaseous inorganic reduced compounds. Because the microorganisms are present in the aqueous phase, the organic or inorganic substrates must be transported from the gas to the liquid phase. Because biological waste gas cleaning is also an aerobic process, oxygen has to be transported to the liquid film and products have to be transported either to the gas or to the liquid phase.

The degree of elimination of a pollutant and the selection of the process is determined by different factors: (i) the quality of the waste gas determined by the chemical nature of the pollutant, composition of different pollutants, pollutant concentrations, pollutant metabolic accessibility, and the volume of the gas emitted per unit of time; (ii) the biomass and its physiological state determined by the rate of conversion of the substrates; (iii) the availability of the substrates determined by the mass transfer of the pollutants and of oxygen from the gas phase to the biofilm and the export of degradation products therefrom (mass transfer depends on the process selected [e.g., biofilter, trickle-bed reactor, or bioscrubber] and on the physicochemical nature of the pollutants to be eliminated); and (iv) costs of investment and of operation for waste gas cleaning determine the selection of each of these factors by the impact on overall productivity, reliability, and thus profitability.

BIOLOGICAL TREATMENT

Industrial waste gas cleaning systems are currently operated under nonaxenic conditions and, as a result, contain different populations of bacteria, fungi, and protozoa. Such populations may establish randomly by means of the content of bacteria and bacterial and fungal spores in the surrounding air. To rapidly establish a microbial population in the reactors, these are often inoculated with suspensions of wastewater treatment plants. Because the composition of the waste gas determines the development of microorganisms best fitted to respective conditions, initial efficiency of a plant may be suboptimal. Also, because microorganisms are the crucial phase in biological treatment, they are emitted by the stream of the clean gas that is the subject of hygienic concern.

Production of Biofilms

Initial colonization of the microbes is the result of adsorption onto the surface of the support material, which occurs almost instantly. Secondary colonization is characterized by the formation of filaments and of extracellular polysaccharides, which together form the biofilm. The kinetics of biofilm formation, its composition, changes, properties, and impact on biocorrosion have been reviewed previously. The composition of the species changes with duration of operation. Moreover, under steady-state conditions of the plant, the biofilm represents an ecosystem with grazing protozoa digesting free-living bacteria. Protozoan predation has resulted in a low and constant level of bacteria, which, however, may not be detrimental to the efficiency of a plant but increase stability of trickle-bed reactors, for example.

Enhancement of carbon mineralization up to 22 to 80% and decrease in pressure drop by preventing the extent of clogging have been reported from laboratory waste gas treatment systems. Moreover, most protozoa are not attached to surfaces and could be purged out by flushing the different systems. The use of pure bacterial cultures with defined catabolic properties with the ability to degrade xenobiotic or toxic compounds at known rates has been suggested to be advantageous with respect to waste gas cleaning. Several bacterial species specialized to degrade different solvents have been adsorbed on activated carbon. With suitable bacterial strains, 90% of the solvents could be eliminated from the gas phase, and mass transfer to the biofilm appeared to be limiting the efficiency of the laboratory trickle-bed reactor. No reports are presently available for industrial application of monocultures.

Health Hazards

Health hazards for personnel or the surrounding population emerging from waste gas treatment plants have not been reported. However, different bacterial species so far identified from various waste gas treatment plants have been mainly related to the degradative capability of the respective facility. Among these, species of the gens *Pseudomonas* have been identified that are highly versatile in utilizing aromatics and other volatiles. One concern is the release of potentially pathogenic microbes to the environment with the exit gas stream. Some studies have been devoted to this fact. The transport of microbial germs, mainly bacteria and to a smaller extent spores of molds, in the outlet gas of the different full-scale biofilters varied between 10^3 and 10^4 cells/m^3, which is of the same order of magnitude as indoor air. Search for fungal spores, the potential causative agent of mycogenic allergies, revealed a minor reduction of spores emitted from clean gas as compared to raw gas from moist biofilters, whereas emission from a dry biofilter temporarily exceeded that determined from raw gas by more than three orders of magnitude.

Identification of microorganisms is laborious, time consuming, and costly, and most reports consider the absorption of pollutants rather than the emission of microbes. The technology to detect and to identify specific bacteria within multispecies biofilms, however, is presently available and is used to describe microbial ecosystems. Nucleic acid hybridization techniques and comparative sequence analyses are increasingly being applied to environmental microbiology. More recently, fluorescent dye-labeled oligonucleotide probes have been used for identification of single cells.

MICROBIAL DEGRADATION

Organic or inorganic pollutants of waste gases are substrates for microorganisms. Organic carbon sources have a dual function as substrate for the living cell. Dissimilation of the carbon source characterizes its quality as an energy source to donate electrons for aerobic respiration, which yields the energy required for growth and maintenance of the cell. Carbon is also assimilated by the cell for growth to build up cell constituents. Carbon dioxide and water are the products of the dissimilatory aerobic respiration, and biomass is the product of the resulting assimilation. The quality of a given carbon source as an energy source depends on its degree of reduction and on the type of metabolism, as is shown for dissimilation of one-carbon compounds such as methane, methanol, and formaldehyde. These compounds are oxidized to carbon dioxide, water, and the reductant. Moreover, substitutions of hydrogen by halogens such as methylenechloride cause the release of hydrochloric acid. Degradation of methylmercaptane results in the release of hydrogen sulfide and that of methylamine results in the release of ammonia.

The by-product hydrogen sulfide is subsequently oxidized by thiobacteria to sulfuric acid, and ammonia is oxidized by nitrifying bacteria to nitric acid, both resulting in acidification of the environment. Therefore, the nature of the pollutant to be mineralized by the microorganisms has a significant impact on the pH of the microenvironment within the plant and on pH-dependent microbial degradative activity. Acidification as a result of the elemental composition of the pollutants as well as the composition and distribution of the different microbial species may require mixing, washing, and pH control of the biofilm attached to the support material or being suspended in the aqueous phase. The different principles of waste gas treatment processes such as biofilter, trickle-bed reactor, or bioscrubber differ in efficiency in mass transfer, capacity to harbor biomass, and costs. Thus, knowledge of the quality of the waste gas with respect to its composition or load is essential for selection of the process. One crucial criterion in microbial transformations is the overall rate of degradation. The mean residence time of a gas passing certain industrial waste gas treatment plants is occasionally less than a second. In this period of time, mass transfer and degradation of the pollutant must occur—and it does.

Table 12.1 Respiration and mineralization of differently reduced and substituted one-carbon compounds, nitrification, and sulfur oxidation

Respiration
$4[H]^a + O_2 \rightarrow H_2O$
Mineralization
$CH_4 + O_2 \rightarrow CO_2 + 4[H]$
$CH_3\text{-}OH + H_2O \rightarrow CO_2 + 6[H]$
$CH_3\text{-}NH_2 + 2H_2O \rightarrow CO_2 + NH_3 + 6[H]$
$CH_3\text{-}SH + 2H_2O \rightarrow CO_2 + H_2S + 6[H]$
$CH_2\text{-}Cl_2 + 2H_2O \rightarrow CO_2 + 2HCl + 4[H]$
$CH_2O + H_2O \rightarrow CO_2 + 4[H]$
Nitrification
$NH_3 + 2H_2O \rightarrow HNO_2 + 6[H]$
$HNO_2 + H_2O \rightarrow HNO_3 + 2[H]$
Sulfur oxidation
$H_2S + 4H_2O \rightarrow H_2SO_4 + 8[H]$

ELEMENTAL COMPOSITION

The composition of the waste gas usually contains few elements—carbon, hydrogen, or oxygen as in the case of acetone—and other elements (N, P, S, Mg, K, Fe, etc.) are also required to build up biomass. Formation of biomass with constant elemental composition is a function of the organic substrate when each element required for growth is supplied by the environment and expressed as the cellular yield ($Y_{X/S}$) (gram dry cell weight per gram of substrate). In biofilters, some minerals may originate from the support material such as wood, heather, or compost, whereas in other waste gas cleaning plants minerals may be supplied from the added water. Mostly, there will be a severe imbalance in supply with elements required for microbial growth, whether the major pollutant was a hydrocarbon, solvent, hydrogen sulfide, ammonia, or methylen chloride. Imbalance in nutrient supply may originate from the surplus of the carbon source or from the limited supply of minerals, both resulting in two different consequences. First, imbalance of the elemental composition of the biomass may result from storage of different compounds.

Excess phosphate supply to certain bacteria results in accumulation of intracellular polyphosphate; excess carbon and energy source may result in either intracellular accumulation of polyhydroxy fatty acids or the formation of extracellular polysaccharides and biofilms. Second, the growth rate may be affected and the requirements for maintenance energy may increase significantly. Maintenance energy is defined as the amount of substrate required by the cell to maintain vital cell functions without growth. The concept of maintenance requirements was first described by Pirt and was later extended by Neijssel and Tempest with the definition of overflow metabolism.

$$1/Y = m/\mu + {}^{1}/Y_{max} \qquad ...(1)$$

where Y is the cellular yield, Y_{max} the maximum cellular yield, μ the specific growth rate (1/h), and m the maintenance coefficient (g substrate/g dry cell weight h^{-1}).

Overflow metabolism may result in two different phenomena: (i) the uncoupling of energy transformation of respiration leading to a low cellular yield and higher rate of oxidation of the substrate to carbon dioxide and (ii) the formation of polymeric or monomeric by-products. The applicability of these academic findings to waste gas cleaning systems has been subsequently proven successful in laboratory plants. However, defined conditions of specific nutrient limitations are susceptible to changing concentrations of volatile organic carbon in the raw gas, which changes the balance in nutrient supply, and thus conditions for optimal mineralization to carbon dioxide and water are difficult to maintain.

COMPONENTS OF WASTE GAS

Waste gas originates from very different industries, from air exchange of buildings, and from various processes. More than 500 biofilters and more than 200 bioscrubbers are currently operated in Germany and in The Netherlands, although outside Europe this technology is less often used. Most of these facilities emit gases at high flow rates of 1,000 to 120,000 m^3/h but with low concentrations of volatile organic and inorganic compounds. Odor removal is quantified by olfactometry, which is costly and requires reliable calibration as evident from an inter-laboratory comparison of odor measurements of *n*-butanol. Recent developments aim at on-line sensors for the quantification of odors.

Odor threshold concentrations to be recognized by humans depend on the respective compound and composition of the volatiles. Some threshold concentrations for sensing the odor of chemicals such as dimethylethylamine, phenol, or volatiles of defined composition have been correlated to total organic carbon determined by flame ionization detection, which is a reliable and available technology for continuous exit gas analysis. Threshold concentrations for odor sensing vary from 0.1 to 4.9 mg organic carbon/m^3 for defined chemicals and from 0.1 to 0.5 mg organic carbon/m^3 for waste gas from food processing and other industries. Usually, the degree of reduction of odor intensity is more effective than the elimination of volatile organic carbon. Flame ionization detection of volatiles allows quantification of organic carbon but does not differentiate between different components of the waste gas. Thus, the degree of elimination of individual compounds cannot be differentiated.

To follow the elimination of different components, identification by gas chromatography is required. This analysis has been used to follow degradation of different solvents in waste gases. Corrosion of the detecting devices may be observed on on-line quantification of haloorganic compounds such as chloromethanes, chlorobenzenes, chloroethanes, or fluorinated or bromated compounds by flame ionization detection because of formation of the respective hydrohalic acids. Therefore, sampling, gas chromatography, and different detection systems may be advisable.

Table 12.2 Compilation of some industries applying biological waste gas cleaning

Biofilter	*Trickle-bed reactor*	*Bioscrubber*
Animal rendering	Chemical industry	Aluminum foundry
Bone processing	Concrete slap production	Animal rendering
Chemical industry	Fibrous skin production	Blood processing
Coal mine exit gas	Grease melting plant	Chipboard production
Dump gas	Plastic production	Egg farms
Feces drying	Rayon production	Grease melting plant
Fish smoking	Sewage treatment plants	Grinding wheel production
Fish processing		Printing plant
Foundries		Tobacco industry
Fragrance production		Polyester production
Food production		
Pig farms		
Slaughterhouse		
Tobacco processing		
Varnish production		

TRANSFER AT COMMERCIAL LEVEL

The principles of mass transfer were first elaborated to understand and to design chemical reaction processes and have been described elsewhere. However, these principles can be transduced also to biochemical and biological reactions where substrates must be delivered to the microbial cell and metabolic products have to be removed therefrom. This is especially important when a substrate is supplied from the gas phase to the aqueous phase, which is the environment for microbial life and activity. The principles of mass transfer are applicable to the different processes treated here. However, the differences in design of these plants are based on different principles of gas-liquid contactors for transfer of volatile chemicals of different properties such as hydrophilic or hydrophobic and soluble or insoluble compounds. Therefore, the principles of the transfer of substrates of different physicochemical properties from the gas phase to the biofilm are discussed first, and subsequently mass transfer in biofilters, trickle-bed reactors, and bioscrubbers.

Total Mass Transfer

Total mass transfer rate (J) from the gas to the liquid phase is driven by the difference in concentration of a given component in equilibrium with the gas phase (c_G) and the actual concentration in liquid phase or the biofilm (c_B), the specific interfacial area (a), and a correlation factor designated mass transfer coefficient (h) as described by equation 2:

$$J = \beta a(c_G - c_B) \qquad ...(2)$$

The specific interfacial area is generally defined as the quotient of the total area for mass transfer and the total reactor volume. The mass transfer coefficient is influenced by the different steps for mass transfer from the gas phase to the liquid phase and to the biofilm. These steps are the superpositions of convection and diffusion in both fluid phases and the absorptive transition of the component to the liquid phase. The equilibrium in absorption at the transition phase is expressed by Henry's law, equation 3:

$$c_G = Hc_F \qquad ...(3)$$

where c_G is the partial pressure of the component in the gas and c_F of that in the fluid. A similar correlation holds for the phase-interface liquid or biofilm. During steady-state operation of a plant, mass transfer rate (J) is in equilibrium with the substrate consumption rate (J_R) of the biomass as expressed in equation 4:

$$J = J_R = K_R a c_L^r \qquad ...(4)$$

where K_R is the reaction rate of the biomass, a the specific area, c_B the concentration of the component in the biofilm, and r the order of the reaction. Mostly, biological reactions are of zero order when these are independent of the substrate concentration and of first order when the substrate concentration determines the reaction rate. The overall reaction rate is determined by the different steps just mentioned. However, in special cases one step may dominate the process, for example, when the rate of absorption of a pollutant is low as given for chemicals with Henry coefficients >1 MPa or low biological degradation rates. Parameters affecting the capacity of waste gas cleaning facilities are discussed next.

Significance

Several examples have been described that demonstrate the significance of single steps on the overall process. The degradative activity of a given biofilm may be limited by suboptimal physiological conditions, which are discussed here, and the mass transfer rates. These conditions apply to the organic or inorganic electron donor for energy transformation, in our case the volatile pollutants, mass transfer rate of the electron acceptor oxygen, the low rate of pollutant degradation by the biofilm, or a combination of these caused by local differences within the reactor.

1. Limitation by mass transfer of oxygen from the gas to the liquid phase may be given for a pollutant highly soluble in water, easily transferred from the gas into the biofilm, and rapidly degraded therein. Although the oxygen partial pressure of air is 21 kPa, it is barely soluble in water (7.5 mg of oxygen/L at 30°C), and thus its mass transfer rate competes with that of low concentrations of the pollutants.
2. Limitation by mass transfer of the pollutant from the gas to the liquid phase may result from rapid degradation of a pollutant by the biomass. In this case, mass transfer can be enhanced by increase in superficial gas velocity or by reduction of the irrigation rate. Both operations decrease the thickness of the boundary layers δ_G and δ_F.

3. Limitation by microbial degradation may result either from low degradative activity or from low biomass. Both affect the efficiency of waste gas cleaning similarly. Low degradative activity is given for toxic or xenobiotic pollutants and other persistent chemicals not accessible to the common microbial metabolism. Low biomass may result from insufficient supply of mineral nutrients, nonphysiological pH, and support material inadequate to allow sufficient biomass attachment.

4. Different limitations, as have been described, maybe present in one system at different locations. In a system with a high load of a pollutant, insufficient degradation may take place at the entrance of the reactor where the mass transfer rate is higher than the degradation rate. As the pollutant is removed from the gas stream while passing through the reactor, the concentration in the gas phase decreases. Finally, a limiting mass transfer rate may result as compared to the potential microbial degradation rate.

The efficiency of elimination of a pollutant is clearly dependent on its solubility and mass transfer characteristics. Consequently elimination of a pollutant from the gas phase is linked to the Henry coefficient of the respective compound whether it is eliminated by physicochemical processes or microbial degradation. The differences in the processes of waste gas cleaning beneficial or deterious to the aspects as stated are discussed in the respective sections.

BASIC CHARACTERISTICS

For a given waste gas to be treated, the appropriate type of plant has to be selected, and its size has to be estimated. Therefore, basic characteristics are crucial for plant design. Available area and costs are criteria for choosing between biofilter and trickle-bed reactor, while the hydropathic nature of a pollutant is crucial to decide for or against a bioscrubber. For elimination of pollutants with Henry coefficients >1 MPa, bioscrubbers are usually not chosen. In general, the mean residence time of the gas and the fluid is crucial for the three types of plants. The residence time is a measure to define the time given for contact of the gas, the liquid, and the biomass to allow mass transfer. The mean residence time (t_m) of a gas or fluid in a reactor with a given volume (V_R) at a given flow rate (V^*) is ex-pressed in equation 5:

$$t_m = V_R/V^* \quad \text{...(5)}$$

The inverse relation of this mean residence time of the gas defines the volumetric flow rate (τ_V), as expressed in equation 6:

$$\tau_V = V^*/V_R \quad \text{...(6)}$$

The relation of the flow rate of a gas or fluid to the area of the cross section of the plant (A_R) is defined as area load (τ_A) as expressed in equation 7:

$$\tau_A = V^*/A_R = u_0 \quad \text{...(7)}$$

where u_0 is the superficial velocity of a gas or fluid, which is the velocity with which they pass the free cross section of the reactor.

In waste gas cleaning, two criteria are important with respect to a pollutant: its concentration and its amount. The amount of a pollutant fed to a plant of a defined volume per unit of time is given by its concentration per volume of gas (c_{in}) and by the gas flow rate, and is expressed as volumetric load (V_L) as shown in equation 8:

$$V_L = (V^*/V_R)c_{in} \quad ...(8)$$

The pollutant is metabolized in the plant by the biomass and eliminated from the gas phase. Generally, the elimination is equivalent to the degradation of the pollutant, and this term is subsequently used. The overall degradation is defined as volumetric degradation rate (V_D), as expressed in equation 9:

$$V_D = (V^*/V_R)(c_{in} - c_{ex}) \quad ...(9)$$

The difference in concentration of a pollutant in the feed and the exit gas (c_{ex}) related to the feed concentration characterizes the efficiency of a plant, as expressed in equation 10:

$$\eta = (c_{in} - c_{ex})/c_{in} \quad ...(10)$$

The efficiency of a plant is determined by various parameters such as the chemical nature of the pollutant, biomass degradative activity, and gas or fluid flow rate. The latter is determined by the resistances within biofilters or trickle-bed reactors given by the type of filling, its density, porosity, and height of the package resulting in pressure loss, and may be a significant cost factor for operation. The power (P) required to achieve a certain gas flow rate (V^*) will compensate for the overall pressure loss (Δp) across the reaction zone of a plant, as expressed in equation 11:

$$P = \Delta p V^* \quad ...(11)$$

Other pressure losses resulting from such components as pipings, nozzles, and orifices are not specific to waste gas cleaning facilities and are not treated here.

The porosity (ε) is a crucial parameter of pressure loss for biofilters or trickle-bed reactors, and is defined by the interstitial volume of a rector. This factor is given by the difference in volume of the empty reactor (V_R) and the volume of the particles (V_P) related to the total reactor volume, as expressed in equation 12:

$$\varepsilon = (V_R - V_P)/V_R \quad ...(12)$$

CATEGORIES OF WASTE GAS TREATMENT PLANTS

As mentioned, three categories of plants are currently used for treatment of industrial waste gases of different origin and composition: biofilter, trickle-bed reactor, and bioscrubber. Their principles, industrial sizes, capacities, and advantages or disadvantages are outlined next. Certain characteristics increase with the plant used. The applied volumetric flow rate increases roughly 10-fold as the order of biofilters, trickle-bed reactors, and bioscrubbers. This order reflects not only an inverse relationship of efficiency and size of the plants but also their increasing complexities.

Construction of Biofilters

Different designs are currently applied for the construction of biofilters: single bed, multiple bed, and closed bed, either requiring land or installed on top of a roof. Biofilters require an area of 10 to more than 1,000 m^2 filled to a height of 0.5–1.6 m with different materials as support for microbes and for fine distribution of the gas. The filters are operated at rates of 100–120,000 m^3/h.

The gas flow rate related to the filter volume varies from 10 to 580 $m^3\ m^{-3}\ h^{-1}$. The superficial gas velocity of biofilters varies from 0.1 to 8 cm/s. The mean residence time within a filter bed of 1-m height is rather long and amounts to about 10 to 700 s. The filling material may be compost, peat, heather, chopped wood, bark, or disrupted root wood to which microorganisms are attached. Biofilters are suited to eliminate odors and volatile organic or inorganic compounds at rather low concentration, as detailed next.

The area of the biofilter contains a wall of wood or concrete and a system for distribution of the main stream of the gas. It is practical for the biofilter to be vented from the bottom, that the orifices or slits allow constant gas flow rates over the filter area, and—for larger plants—to allow machines to enter the filter for exchange of the filling. A waste gas may contain dust particles that are removed by a deduster before feeding into the biofilter. The gas passed through the biofilter should either be humidified to about 95% or the filter should be irrigated to ensure a moist environment as required for microbial activity. The concentration or the composition of the pollutants may vary with the operation of the emitting source, resulting in changes of load. Moderately higher concentrations can be temporarily reduced and a portion of the compounds stored in the filling material according to its adsorptive properties.

The compounds are subsequently released therefrom for mineralization by the biomass. Thus, the efficiency of the biofilter can be maintained. The filling material supplies the attached microorganisms with the minerals necessary for growth. The degradative capacity of the biofilters is limited by the low supply of minerals, which in turn causes a low amount of active biomass present on the support material, its limited buffering capacity for protons or volatiles, and the lack of control of the microbial environment with respect to pH. The filling material is of natural organic matter and subject to microbial degradation at different rates according to the structure, surface, and density. Moreover, environmental conditions may add to the rate of decomposition and loss of mechanical properties of the filling material.

Decomposition of the filling material results in settling, reduction in pore size, and increase in pressure loss and consequently in increased energy requirements for air compression. Moreover, settling of the filling may result in channeling wherein the gas stream circumvents but does not penetrate portions of lower porosity within the filter bed. This in turn causes reduction of the residence time of the gas and reduction in efficiency of pollutant elimination. Reduction in efficiency of the plant requires exchange of the decomposed filling material, which usually occurs after 1-3 years of continuous operation depending on filling texture, mechanical properties, and degradability and the load of the waste gas to be treated.

Elimination by Trickle-Bed Reactor

Pollutants with low solubility in water may be eliminated from waste gases by trickle-bed reactors. The advantage of operation of a trickle-bed reactor over biofilters is not only in the increased mass transfer but also in the ease of controlling physiological conditions such as pH and the addition of minerals in the trickling fluid in the bottom of the column. Moreover, trickle-bed reactors require less land than biofilters and are operated in single or multiple stages. They are built as columns filled with incompressible and mostly inert support material for gas-liquid contact and attachment of biomass. The bed heights vary from 1 to 3 m, the reactor volumes from 2 to 160 m^3, and the flow rates from about 3,000 to 85,000 m^3/h. The mean residence time of the gas within the columns is generally less than 4 s, which is considerably shorter than reported for biofilters, and the superficial gas velocities vary from 0.1 to 1.6 m/s. The volumetric gas flow rates are rather divergent and vary from 250 to 4,700 $m^3\ m^{-3}\ h^{-1}$. The inert support material may be slag, natural porous stones, coke, charcoal, granular clay, ceramics, sintered glass, or plastics and other material proven to be suitable in absorption technology. The filling is trickled from top, and the surface area of the filling provides a major basis for mass transfer from the gas to the fluid phase.

Sufficient empirical data are available for choosing geometry, size, and filling for various industrial applications. Moreover, based on the geometry of the fillings, the total surface area of the trickle bed can be calculated. The thickness of the fluid film is proportional to the irrigation rate and inversely related to the total surface area. This in turn allows calculating the mass transfer rate for a given pollutant. Also, total biomass can be deduced from the mean thickness of the biofilm. Trickle-bed reactors may be seeded with microorganisms by spraying activated sludge from wastewater treatment plants. The microbes will be selected and the population of the biofilm will be established according to the given pollutants. A major resistance for mass transfer from the gas phase to the biofilm is the thickness of the fluid film, determined by the rate of irrigation and rate and direction of the gas flow.

Therefore, the irrigation rate should be reduced to generate thin films. Depending on the waste gas to be treated, the rates vary from 0.04 to 50 $m^3\ m^{-2}\ h^{-1}$ as reported from laboratory and industrial reactors. However, industrial plants are preferentially operated at $<0.5\ m^3\ m^{-2}\ h^{-1}$. To enhance mass transfer and to reduce diffusional resistance by the liquid film, intermittent irrigation has been applied for elimination of pollutants by a laboratory trickle-bed reactor, which still ensured sufficient moisture for the biofilm. Mass transfer may also be increased by increasing the surface area as given with support material of high specific surface. Application of such material results in lower interstitial volume, lower bed volume, and a smaller plant size. However, the thickness of the biofilm will increase with time of operation of the trickle-bed reactor, decreasing or abolishing the small interstitial volumes, which results in increase in pressure loss, clogging, and channel formation. The mean residence time is reduced and the efficiency of the plant declines in eliminating pollutants from the gas stream. The otherwise favorable characteristics of trickle-bed reactors with respect to mass transfer and control of circulation have led to the concept of removing excess biomass by various means: increased irrigation rate to wash off the biofilm, periodic pulses of air pressure after flooding, and movement of the trickle bed itself to shear off the biofilm.

In laboratory systems this procedure of biomass removal has led to continuous operation for more than 2 years with constant degrees of elimination of pollutants. Research data now suggest that the degradative capacities of trickle-bed reactors can be better described mathematically than biofilters and allow conclusions for growth of biomass and of operation of trickle-bed reactors in industrial scale. Reduction of biomass as a result of a biological equilibrium of biomass formation and biomass degradation has been reported from a pilot trickle-bed reactor after long operation. In this system, elimination of the acidogenic pollutant dichloromethane was reported, involving conversion into biomass, whereas clogging as a result of biomass degradation was not observed.

Dichloromethane is a substrate with low yield in energy transformation and maintenance requirements.

Selection of Bioscrubbers

The decision to select bioscrubbers for waste gas cleaning is based on the physicochemical properties of the pollutants and the availability of space for placing the plant. Bioscrubbers are usually smaller than biofilters and consist mostly, but not exclusively, of two units, one for absorption of the pollutant from the gas phase and the other as a fermentor for their mineralization and growth of microorganisms. Bioscrubbers are usually inoculated with activated sludge, and the microbial population will develop according to the volatiles supplied with the gas stream. Waste gas is passed through a spray tower, bubble column, absorber column, or other gas-liquid contactors for absorption of the pollutants. Industrial bioscrubbers presently treat 9,000 to 150,000 m^3 waste gas per hour. Scrubber units are built in heights of 2-4 m and volumes of 10-100 m^3 and exhibit volumetric gas flow rates of 1,000-3,600 $m^3\ m^{-3}\ h^{-1}$. Bioscrubbers are usually operated with superficial gas velocities of 0.8–2.0 m/s and with mean residence times of <3 s. The specific flow rate of the liquid phase to be circulated per area of the absorber unit ranges from 12 to 40 $m^3\ m^{-2}\ h^{-1}$.

Elimination of hydrophobic pollutants from the gas phase in single-unit bioscrubbers is usually not advisable because of the high volumes of water to be moved. However, single-unit bioscrubbers have been successfully applied to clean industrial volatile pollutants with Henry coefficients of more than 1 MPa. Moreover, new developments in part operating in an industrial scale have been reported for application of bioscrubbers that combine both units in one apparatus. These systems consist of special devices for gas or liquid dispersion, special orifices at the bottom of bubble columns that contain the dispersed microorganisms, or gassed closed-loop airlift reactors, which, however, await industrial application. The physiological conditions for the microorganisms such as pH, salinity, and oxygen partial pressure can be well controlled in bioscrubbers, as similarly described for trickle-bed reactors. Also, mineral salts nutrients can be added to the medium to allow microbial biomass to grow.

The advantage of bioscrubbers over trickle-bed reactors is in the ease of controlling the environment for microorganisms and mechanically improving mass transfer rates by movement of the liquid according to the need to absorb pollutants from the gas phase. On the other hand, elimination of hydrophobic compounds with low solubilities in water requires movement and dispersion of large volumes of liquids. Bioscrubbers are an interesting alternative to other technologies that have been outlined with respect to the ease of operation, environmental control, and adjustment to changing conditions in the quality of the feed waste gas. Thus, bioscrubbers can be applied for elimination of hydrophobic pollutants. A variation of the process uses silicon oil for absorption of hydrophobic volatiles. Problems, however, may arise from disposal of silicon oil-containing sludge.

APPLICATIONS

Application of biological treatment has been proven useful to economically clean waste gases of different sources. Presently, waste gases with concentrations of less than 1 g of organic carbon/

m³ or of inorganic compounds are treated by biological processes. At these concentrations, biological processes are economically advantageous over alternative processes such as condensation, thermal or catalytic oxidation, or adsorptive and desorptive technologies. Performance of membrane bioreactors for waste gas treatment is presently being examined at a laboratory scale.

Table 12.3 Processes and ranges of concentrations commonly used to eliminate hydrocarbons from waste gases

Process	*Concentration (g organic carbon/m³ waste gas)*
Condensation	>250
Membrane separation	>20
Thermal oxidation	6–20
Catalytic oxidation	1–10
Absorption	1–50
Adsorption	1–25
Biological waste gas treatment	<1
Biological treatment with biomass removal	1–10

Reactor performance, however, is hampered by unstable biofilms or by clogging of the liquid channels from excess biomass formation. On solution of these problems, membrane bioreactors are expected to be useful tools for treatment of poorly water-soluble pollutants such as highly chlorinated hydrocarbons. In a comparative study, costs for investment and for operation were related to treatment by biofilters. Costs of investment include those for the plant and for measuring and controlling devices. Costs of operation include consumables, energy requirements, and costs for personnel. From this study it is evident that bioscrubbing is more costly with respect to investment and operation. At present, data on the economy of trickle-bed reactors are not available. However, biofiltration and bioscrubbing are significantly less costly than catalytic oxidation or adsorption.

Table 12.4 Relative costs of investment and of operation for different waste gas treatment plants

Type of process	*Investment costs (%)*	*Annual costs of operation (%)*
Biofilters	100	100
Bioscrubbers	158	196
Catalytic oxidation	182	288
Adsorption	236	365

Adsorptive processes aim at the recovery of solvents, for example, whereas thermal or catalytic oxidation aims at the elimination of the pollutants. To make use of the low costs of biological processes for elimination of a pollutant, it is anticipated to extend this technology to waste gases with organic loads significantly larger than 1 g organic carbon/m³. Two alternatives are presently considered. First, classical engineering processes as specified are applied to eliminate high concentrations or peak concentrations toxic to microorganisms and to combine these with conventional biological processes. This concept was successfully applied recently for elimination of solvents from waste gas of the lacquer industry. Second, the inexpensive biological processes are to be extended to treat waste gases with high organic load. The conversion of this concept requires addition of mineral nutrients, which results in formation of high amounts of biomass. Removal of the excess biomass is required to prevent undesired consequences such as clogging in trickle-bed reactors or increase in salt concentration in bioscrubbers or both. However, at laboratory scale high concentrations of biomass allowed the complete (100%) elimination of as much as 4.2 g toluene/m^3 or 8.7 g *n*-butanol/m^3 from the gas stream.

CONCLUDING REMARKS

Biological waste gas treatment is less costly and requires less maintenance than conventional physicochemical, thermal, or catalytic processes. However, problems may arise from the chemical nature of the compounds to be eliminated from waste gases and from the material of which parts of the plant are constructed, such as the walls of the reactor and such parts as piping, orifices, and nozzles. The waste gases to be treated may contain compounds that are hardly or very slowly biodegradable, such as benzene and xenobiotics, and other compounds toxic to different cell functions, such as solvents, cyanic acid, and hydrogen sulfide. However, successful operation of laboratory-scale waste gas treatment plants to eliminate such compounds point to the applicability of biological treatment to this category of pollutant also. Different compounds may be converted to organic acids by incomplete oxidation when the mass transfer of oxygen is insufficient. Acidification,

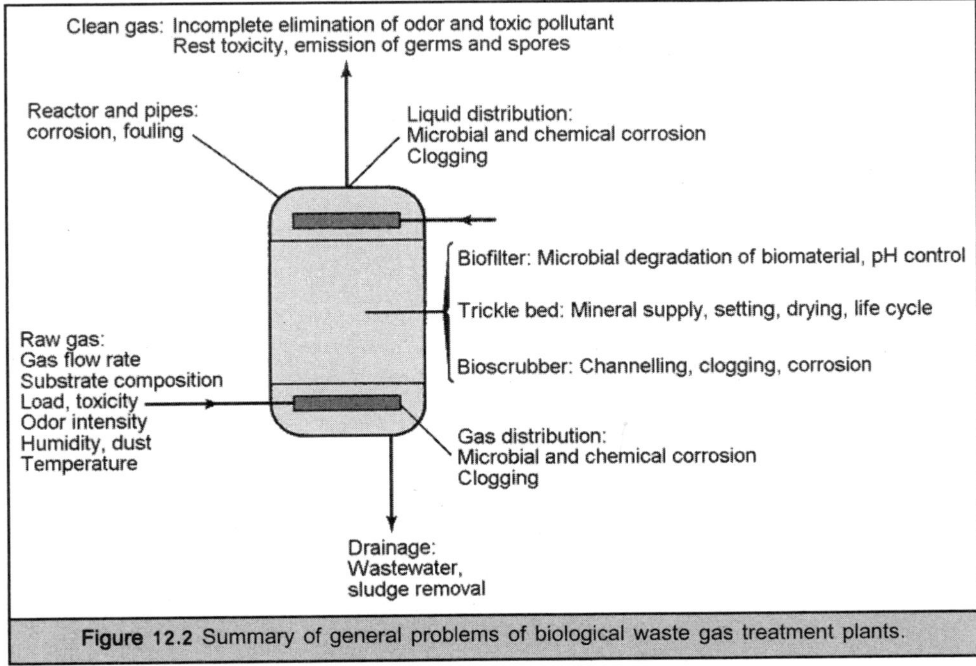

Figure 12.2 Summary of general problems of biological waste gas treatment plants.

however, not only results in decrease of pH and an unfavorable change of the physiological conditions for the microorganisms but also in corrosion of the material of the plant.

The waste gas of rendering plants contains volatile organic sulfur compounds converted to sulfuric acid; this causes severe corrosion of concrete supports of the gas distribution devices of the filter bed. Therefore, such devices have to be exchanged for such components as wooden beams. The same applies for other acidogenic compounds of waste gases. Therefore, the appropriate material for piping and such should be selected according to the expected composition of the waste gas for the three types of plants. Trickle-bed reactors may especially suffer from clogging and channeling by excess formation of biomass or slime with high loads of the waste gas or extended operation, as was outlined. Therefore, altered processes are presently being examined to extend the application of biological treatment of waste gases loaded with compounds that are slowly degraded or that contain high loads of organic and inorganic pollutants. This application in turn will require devices for the control of the environment for the microorganisms and the removal of excess biomass, and recent developments have been devoted to this need.

13

INDUSTRIAL WASTEWATER TREATMENT

The treatment of municipal and some industrial wastewater generally uses a combination of primary, secondary, and tertiary treatment. Primary treatment simply screens and settles large particulates and skims off floating greases and oils. Secondary treatment biologically removes organic carbon, and in newer plants, soluble nitrogen or phosphates. Tertiary treatment attempts to limit the microorganisms and other pathogens in the treated water by membrane filtration or deep-bed filters and some form of disinfection using chlorine, ozone, or ultraviolet light. The secondary or biological treatment of wastewater uses a number of key processes, unit operations in the process engineering terminology. These include primarily processes for biological reactions and processes for settling and clarification. Chemical precipitation to remove phosphorous is sometimes used just before the settling and clarification. There are many flowsheet configurations, often of a proprietary nature, that attempt to obtain the optimal combination of anaerobic, anoxic, and aerobic conditions for the microorganisms.

OBJECTIVES OF STUDY

The goal of the operation of a wastewater treatment plant is to satisfy the effluent requirements at all times at minimum cost. The effluent quality and a good performance has to be guaranteed *consistently,* which will have a great impact on instrumentation, control, and automation. Effluent requirements are usually formulated in terms of daily, weekly, or annual averages of suspended solids, biological oxygen demand (BOD), total nitrogen, and total phosphorus. To comply with the limits of the effluent concentrations, certain safety margins are required. The goal of automatic control is to keep the effluent quality consistently good, despite disturbances. This allows the margins to be smaller, thus saving costs. The capital costs are usually the dominant part of the total cost. Traditionally, many plants have been designed with large tank volumes to meet the varying loads to the system. The total cost can be decreased, if the balance between design and operational costs is systematically considered. There are many components of the operational costs, such as consumption of electric energy (for pumping, aeration, etc.), chemicals for phosphorus removal, or polymers for sludge. Obviously, labor costs are important. The balancing of costs is, of course, an important reason for control. Considering the increasing demand of consistent operation and effluent water quality around the clock, however, will probably be the main driving force for plant control and automation. Driving forces for instrumentation, control, and automation (ICA) in a large number of countries were discussed at an IAWQ symposium in 1993. The major incentives were categorized as

1. Effluent quality standards
2. Economy
3. Increasing plant complexity
4. Available tools (sensors, computers, models)

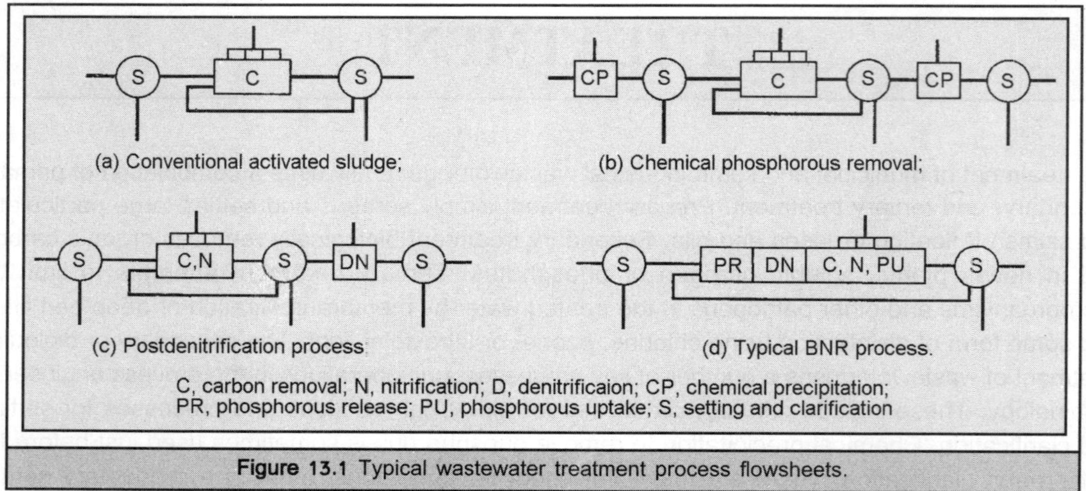

Figure 13.1 Typical wastewater treatment process flowsheets.

In some countries, there is a water shortage, which in itself is a driving force for improved wastewater treatment, such as for recycling. In many industrial countries there is today an elaborate requirement for nutrient removal. This has led to the need for quality control based on advanced on-line instrumentation. With more complex processes, the needs for energy conservation and consistent operation are apparent. Ammonium and nitrate on-line instrumentation in the activated sludge was developed and documented in the late 1970s. At that period, the on-line measurements

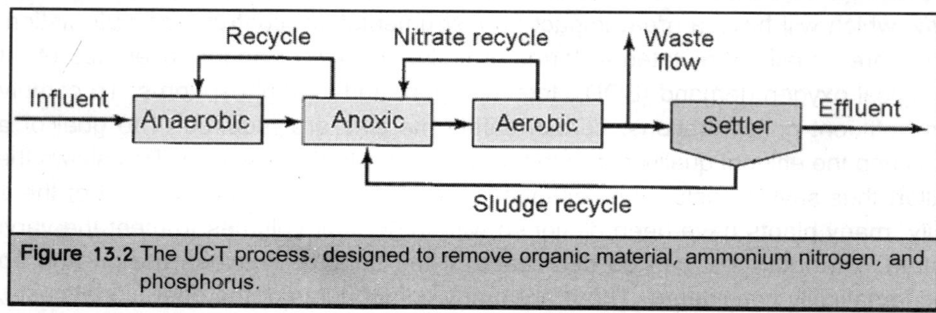

Figure 13.2 The UCT process, designed to remove organic material, ammonium nitrogen, and phosphorus.

were still limiting the applicability. Very large efforts have been spent on the development of N and P analysers, and there has been a lot of progress in the instrumentation technology.

In 1990s, the control is extended to include biological P removal in many plants. Today, commercial applications of nutrient removal control based on advanced instrumentation are common. A typical representative layout of nutrient removal plants is the University of Cape Town (UCT) activated sludge plant.

Strategies

The essence of the control problem in wastewater treatment is related to disturbances. They are critical to the approach to control and have some specific features:

1. The daily volume of wastewater treated can be huge.
2. The disturbances in the influent are enormous compared to most industries.
3. The influent must be accepted and treated; no returning it to the supplier.

Because the disturbances to a wastewater treatment system are significant, the plant is for all practical reasons never in steady state. The disturbances are related to flow rate changes as well as concentration and composition variations. The disturbances are not only related to the influent flow characteristics, but may be generated internally in the plant because of operational procedures. For example, poor pumping operations contribute to hydraulic disturbances. Clarifier upsets are often caused by inadequate pump control. A sudden flow increase caused by a pump start will influence the flow propagation along the plant within a fraction of an hour. A proper settler operation requires that the influent pumping has to be smooth.

Variable speed pumping is a proved technology today and is a crucial component to create a smooth hydraulic operation of the plant. If primary pumps or other recycle pumps are operated on/off in too-large steps, the resulting solids-liquid separation will suffer, as will the effluent water quality. Concentration disturbances are also generated within the plant itself. For example, sludge thickener overflow is usually returned to the plant inlet. This overflow is usually operated with on/off pumping without considering the influent load. Because the recirculated flow may have a large concentration, it will make a significant contribution to the overall load, both in terms of hydraulics and substrate loading.

It is usually impossible to remove the source of the disturbance, even if this is an important option, especially for industrial effluents, where production control in the industry may be improved. Sometimes the magnitude of the disturbance can be reduced *before* it reaches the plant. An integrated sewer-treatment plant control can attenuate hydraulic disturbances, and some waters can be pretreated in order to avoid harmful effects on the plant, such as neutralization of wastes with high or low pH.

The control can also compensate for effects *within* the process. A common example is dissolved oxygen control. All these disturbances will more or less influence the effluent water quality. An ultimate operational goal is to minimize the influence of disturbances and to maintain consistently a high effluent quality.

IMPLEMENT CONTROL

It is far from straightforward to implement control in an activated sludge system; some of the reasons are listed here:

1. The process has significant nonlinearities limiting the usefulness of simple controllers.
2. There is a very wide range of response times, both a problem and a blessing.
3. There are many interactions within the process with major recycles.
4. There are many external couplings from the sewer to the receiving waters.
5. Designers often leave too few manipulated variables and too little control authority.
6. The concentrations of nutrients (pollutants) are very small, even challenging sensors.
7. Sensors do not exist for many states and are expensive for others.
8. The value of the product in the marketplace is remarkably low.

Other attributes why control is far from straightforward include:

1. The microorganisms change their behavior and their population distributions.
2. The separation of the effluent from the biomass is challenging and easily disturbed.
3. Future effluent standards will be tight and based on spot checks.

Disturbance rejection is a major goal of activated sludge control. Large disturbances must be attenuated, and we must do it as early in the process as we can. This will ease the task of controllers later in the process.

The formulation of the goal of a wastewater treatment operation is much more demanding than to define the effluent standards. When a goal or a performance criterion is defined, then the necessary control actions can be analyzed. There are several goals of wastewater treatment, and one has to recognize the different levels and interest groups. The specific goals are set so that they contribute to meeting the more general corporate and community goals. They include at least

1. To meet effluent discharge requirements.
2. To achieve good disturbance rejection.
3. To optimize operation to minimize the operating cost.

The first goal of course includes keeping the plant running. This is the major task for many operations, at least in terms of manpower. It is not a trivial task to maintain advanced instrumentation and equipment, to make sure it runs smoothly, and to detect any faults in its operation. Much of this is traditional automation and process control and is not specific to running wastewater treatment systems. The environment may be very specific and harsh for much equipment, but that is also the case in other industries as well, so from a strict control point of view the problems are traditional. However, keeping the plant running and simply recording effluent quality is too often taken as the only goal. This has been and is still largely because of the lack of process knowledge on the part of plant operators. Too many operators of plants rely on the consultant for expertise in both design and operation. To satisfy the effluent quality requirements, it is crucial to ask: is this task solved primarily by design or do instrumentation, control, and automation play significant roles in this part? The traditional answer among designers used to be *no*. When only carbonaceous removal was considered, the plant was considered self-regulating and could take care of load disturbances and still produce a satisfactory effluent.

The traditional method was large volumes, excess aeration, and chemical dosage. Consequently, a lot of plants have been oversized. No incentive was given for improved operation, because operational costs were not compared with investment costs. Often they were met from different financial sources. Because of the significant load variations and disturbances typical for wastewater treatment systems, the plant is hardly ever in steady state. Despite the wide swings of load, the plant still has to *consistently* produce a satisfactory effluent. If the volumes of the tanks are not oversized, this means that some control system has to minimize the influence of disturbances and ensure that the operation is satisfactory.

Pressures to fully utilize volumes and the complexities of nutrient removal are making process control essential. This of course includes disturbance attenuation. Minimizing the operating cost and maximizing the use of available capacity is becoming necessary. Here, control and optimization are the key tools. Costs to be minimized include electrical energy to produce air, chemicals, pumping, and carbon addition.

VARIABLE CONTROLLING METHODS

There are quite a few variables to manipulate in an activated sludge process. Still, the possibilities to control the plant in a flexible way are quite limited. In many plants, however, there may be a potential to make a better use of the manipulated variables that are available. We categorize them in the following groups:

1. Hydraulic, including sludge inventory variables and recirculations.
2. Additions of chemicals or carbon sources.
3. Air or oxygen supply.
4. Pretreatment of influent wastewater.

There are several other manipulated variables in a plant that are related to the equipment and to basic control loops in the process, such as local flow controllers, level controllers, and so on. They are not included in this discussion. A majority of the manipulated variables are meant to change the hydraulic pattern in the plant. The different flow rates will influence the retention times in the different units. Moreover, the clarification and thickening processes are particularly sensitive to the *rate* of change. The hydraulic flows also determine the interaction between different unit processes. We classify the hydraulic manipulated variables into four major groups:

1. Variables controlling the influent flow rate.
2. Variables controlling the sludge inventory and its distribution, including return sludge flow rate, waste sludge flow rate and step feed control.
3. Internal recirculations within the activated sludge system.
4. External recycle streams, influencing the interactions between sludge and liquid processes.

Sequential batch reactors (SBR) offer other control possibilities because of the flexibility associated with working in time rather than in space. Some of the features of SBRs are described in some detail in the work by Irvine-Ketchum. The Danish Bio-Denitro and Bio-Denipho processes are not the traditional fill-and-draw process but are alternating processes with continuous discharge. Some of the control variables are sometimes called *selection mechanisms* for the viable bacteria in the system. They are classified in the groups:

1. Electron acceptors (e.g. oxygen or nitrate).
2. Substrate.
3. Thickening or clarification properties.
4. Temperature.
5. Growth rate.
6. Free swimming organisms.

Not all of these can be manipulated. Only the electron acceptors (mostly oxygen) and (to some extent) the thickening and clarification properties can be controlled (by chemical precipitation). Moreover, substrate can be added (e.g. carbon sources).

MANIPULATED VARIABLE CONTROLS

Out of the many manipulated variables, we will consider a few. Probably the most important is the control of dissolved oxygen (DO) by using the air supply. Other important measures are the sludge inventory, controlled by the return sludge flow rate, and the waste sludge flow rate.

Control for Aerator System

The DO supply is a key control variable for the aerator system, both for carbonaceous and nitrogenous organisms. It influences the operating costs significantly, so there is an incentive to minimize the air consumption. From a microbiology point of view the choice of a proper DO set point is crucial, and too little DO will limit the growth of organisms. From a process control point of view, DO control is known and proved technology. As long as the air supply is variable, the control of DO is quite straightforward.

The tuning of the controller has to be different at different loads because of the inherent nonlinearities, and this can be compensated for by self-tuning regulators, prove in full scale for a long time. Another way is demonstrated in pilot scale, where a nonlinear controller can be used to operated over a wide range of loads. The oxygen transfer rate determines how fast gaseous oxygen is transferred into DO. This in turn decides that the DO concentration can be manipulated by air flow changes within fractions of an hour. Changes in the oxygen demand, however, may influence the DO concentration within minutes. Not only the total amount of oxygen but its distribution in space and time are essential for the operation. Some of the considerations are:

1. *Total air supply.* The total demand of air is basically determined by the total mass of viable organisms in the system multiplied with the biodegradable substrate concentration. The air supply control has an essential impact on the energy cost for the plant.

2. *DO set points*. It is important to be able to select the DO set point. However, little is known about the proper choice of it. Generally, under closed loop DO control, it is made constant. There is, however, some evidence that a time variable set point may be advantageous.
3. *DO spatial distribution (profile)*. It is a great advantage to be able to individually change the air flow rates along the aerator. This makes it possible to obtain an accurate air flow spatial distribution.
4. *Pure oxygen addition*. Some aeration systems are supplied with pure oxygen instead of air. Pure oxygen is now and then used in conventional air systems to add extra DO to the system under extreme load situations.
5. *Anoxic/aerobic zones*. Because loading conditions may change significantly on a weekly or longer basis, it is of interest to vary the size of the anoxic reactor volume and change the relative retention times between nitrification and denitrification. Such a control is, of course, easy to realize in SBR systems.

Given an adequate controller, it is straightforward to change the DO set point. However, to know which set point is the best from a microbiological point of view is a much more intricate question, and much remains to be studied before the ideal DO control system has been found. Consider an example of a compromise between competing goals within the same plant. The anoxic unit in the activated sludge system, is used to denitrify the nitrate, that is, to reduce the nitrate to free nitrogen by the oxidation of organic carbon.

The dominating organisms in the process are the heterotrophs, a proportion of which reduce nitrate in an (ideally) oxygen-free environment. Therefore, the DO level has to be kept at a minimum. If not, the organisms will use oxygen instead of nitrate as the electron acceptor, and the zone will work at least partly like an aerobic reactor. There are several disturbances that may violate the anoxic condition. The nitrate recycle stream may have a high DO concentration that will rapidly influence the DO concentration of the anoxic zone, thus the denitrification (DN) rate. It is apparent that from a DN point of view that the DO concentration at the outlet of the aerator has to be kept as low as possible, so that no more oxygen than necessary is recirculated. However, for the nitrification taking place in the aerator, a high DO concentration is required. Consequently, the DO concentration at the aerator outlet has to be a compromise between the needs of nitrification and denitrification.

Sludge Inventory Control

The sludge inventory can be controlled primarily by three manipulated variables: the waste-sludge flow rate, the return-sludge flow rate, and the step-feed flow rates. Manipuiation of the waste-sludge flow rate is used to control the total inventory of sludge in the process. Because the total inventory is a function of the total growth rate of organisms, it is used to control the sludge retention time, or the sludge age. It will influence the system in a time scale of several days or weeks. Manipulation of the return-sludge flow rate is used to distribute the sludge between the aeration basins and the settler units or between the acidogenic and methanogenic reactors in two-stage anaerobic systems. Recycle from the settling stage is an important variable for obtaining the right operating point in the reactors, but seldom useful for the control on an hour-to-hour basis. Some systems are supplied with several feeding points for the return sludge. This has a potential for sludge redistribution for certain loads, such as toxic loading.

A *combination* of different recycle may be important. In systems with chemical precipitation, sludge from the secondary settler may be combined with chemical sludge from a postprecipitation settler unit. In that way, the flocculant properties may be influenced, and the chemicals better utilized for phosphorus removal. By controlling the step feed in an activated sludge plant, the sludge within the aeration basin can be redistributed, given the proper amount of time. As a special case of step-feed control, one will obtain a *contact stabilization* structure.

METHODS OF MEASUREMENT

Measurement technology is crucial for adequate operation. In particular, on-line analysis is of key interest to control the effluent water quality. It is of great importance to extract maximum information from available instrumentation. Therefore, detection and diagnosis, based on combination of measurements, are models that make up a basis for successful plant operation.

Clinical Method

Sensor technology is crucial for the success of control and operation, and great progress has been made over the last few years. Measurement for control is not necessarily the same as measurements for quality check or for regulatory agencies. The sampling rates have to be consistent with the plant dynamics, and the location of some of the sensors has to be chosen with care, for example, DO probes. Important types of measurements include:

1. Flow rates in different plant units.
2. Air flow rates and air pressure.
3. Sludge flow rates.
4. Temperature in water basins and in digesters.
5. Gas (e.g., methane) flow rates and temperatures.
6. Alkalinity.
7. pH.
8. DO in different locations.
9. Sludge levels.
10. Suspended solids in different places. BOD, COD, TOC.
11. Phosphorus fractions.
12. Nitrogen fractions.
13. Respiration rate.

There are many instruments that satisfactorily measure flows, levels, pressures, dissolved oxygen in liquids and suspended solids, and dry solids concentrations in sludge. Even if an instrument is deemed to be reliable, proper maintenance is still required. Maintenance requirements will vary greatly between different applications. The development of respirometry has been significant. Within IAWO, task group prepared a scientific and technical report during 1997. Also, advanced N and P analyzers are becoming common in the advanced operation and control of nutrient removal plants.

Mathematical Model

It is clear that more information than the primary sensor signals is required in order to get good operation. The idea of estimation is to mix sensor information with mathematical models. In this way, variables that are not measurable may be reconstructed by combining a mathematical model with sensor information about some other variable. Consider the respiration rate (or the oxygen uptake rate) in an aerator. It can be measured by using respirometers. Alternatively, it can be obtained from measuring the DO concentration and calculation of the DO mass balance. Consider the DO mass balance in a complete mix aerator:

$$\frac{dc}{dt} = d_1 \cdot c_i - d \cdot c + k_L a \cdot (c^s - c) - R \quad \text{...(1)}$$

where c is the DO concentration; c_i the influent and c^s the saturation DO concentrations, respectively; $k_L a$ the oxygen transfer rate, and R the oxygen uptake rate. The parameters d and d_i are dilution terms. If the oxygen transfer rate $k_L a$ is known, then R can be easily calculated from the DO measurements and the mass balance (equation 1). Usually, however, $k_L a$ must be determined, because it varies slowly with time. Therefore, both the oxygen transfer rate and the respiration rate R to be determined simultaneously from measurements of the DO, preferably under closed loop DO control. This is not trivial, because there is an identifiability problem. A positive change in R can be explained by a similar positive change in $k_L a$ and no unique solution is found. However, certain measures can be taken in order to find the parameters uniquely, so that the respiration is found under DO control.

CONCLUDING REMARKS

Because of the combined development of instrumentation, computer technology, operating technology, and the understanding of plant dynamics, the goal is today a plant-wide control. Recognizing the interaction between various unit processes, the overall control of a plant tries to make maximum use of volumes, energy, and other resources. Because biological nutrient removal (BNR) will become more common, plant complexity will grow. This may be a major driving force for more instrumentation, control, and automation. In a BNR process, there is a delicate balance between carbon, nitrogen, and phosphorus removal. Some important couplings are:

1. The phosphorus release has to take place in an anaerobic environment—no oxygen or nitrate can be present. If oxygen-rich water is recirculated, it may impede the P release. If nitrate is allowed to enter the anaerobic zone, the bio-P process is also inhibited.
2. Denitrification takes place in an anoxic environment. The various couplings to other unit processes were discussed earlier.
3. Both the phosphorus release and the denitrification require a certain carbon concentration. At low loads this may not be sufficient, and an extra carbon source has to be added.

It will become more and more important to understand the impact of load variations. Furthermore, the increased quantities of sludge makes it necessary to develop better control methods for sludge treatment and disposal processes. The drive for a sustainable environment is an incentive to make better use of automation in order to reach such goals. An efficient operation of a plant has to consider not only the couplings between various unit processes within the plant, but couplings to external processes. Realizing that sewer operation must be coordinated with plant control, there has been a strong development of combined sewer and plant control. Disturbances can be detected and controlled as early as possible.

14

BIOLOGICAL WASTEWATER TREATMENT

Activated sludge for biological wastewater treatment has been broadly available for a long time. But, water pollution and eutrophication in rivers lakes, and inland seas have resulted from insufficiently treated water. To protect and control these occurrences, it is critical to consider biological oxygen demand (BOD) and nitrogen-phosphorus (N-P) removal in the development of anaerobic–aerobic wastewater treatment systems. Increasing volumes of wastewater, combined with limited space availability and progressively tightening standards and quality control, require the development of new intensive biotechnologies for wastewater treatment. Recently, there has been a significant increase in the use of immobilization processes for municipal and industrial wastewater treatment. Immobilization processes offer several advantages because of higher volumetric load, increased process stability, compactness, and so on. The successful application of an immobilized system as biocatalyst relies on the proper choice of the immobilizing materials and reactor type. The most important requirements for wastewater treatment are as follows:

1. High cell mass loading capacity.
2. Simple nontoxic immobilization procedure.
3. High surface-area-to-volume ratio.
4. Mechanical stability.
5. Chemical stability.
6. Reusability.
7. Suitable for conventional reactor systems.
8. Economic feasibility.

An ideal system that possesses all these optimal properties does not exist, and a suitable compromise has to be found. For wastewater treatment, bioreactors with immobilized microbes have been examined as a combination of activated sludge process for BOD and nitrogen removal and as an anaerobic process for methane production. But, the attempted immobilization to an anaerobic process is scarcely used in a practical scale because the handling of microbes is difficult at the immobilization and the effects of immobilization are not clear.

The immobilized microbe treatment process clearly is an effective method of aerobic wastewater treatment. On a laboratory scale, the immobilization of microorganisms has also been applied to the removal of specific compounds such as phenol and its derivatives, coloring matter, and other inorganic compounds such as cadmium, uranium, and hydrogen sulfide. Consequently, they can maintain higher biomass concentrations, and a high degree of organic carbon or nitrogen removal, or both, is achieved by these compact systems. Moreover, minimized production of excess sludge is expected by maintaining a low food-to-microorganism ratio.

TYPES OF WHOLE CELL OF IMMOBILIZATION

Whole-cell immobilization is always used in wastewater treatment and may be defined as the physical confinement or localization of intact cells to a certain defined region of space with the preservation of some decomposition activity. Accordingly, the immobilization technique using biocatalysts is developed to reuse the catalytic activity and for continuous operation in a bioreactor. The immobilization of microbes for wastewater treatment is broadly classified into four major categories: adsorption of attachment to a surface, entrapment within a porous matrix, containment behind a barrier (semipermeable membrane and microcapsule), and self-aggregation. These techniques can be applied to all wastewater treatments that use a viable cell system.

Role of Adsorbent

Attaching cells to a surface is a very popular immobilization technique because it is simple and fast. A suitable adsorbent for attachment should display a high affinity toward microbial cells and yet cause minimal denaturation. A surface-attached process is the immobilization method in which cells are bound to a surface, regardless of the type of binding. Microbial cell walls carry a charge that depends on the pH of wastewater and other environmental conditions. Biofilms resulting from the growth of cells on the surfaces of inorganics such sand or rock are present in conventional industrial processes, and the self-immobilization of cells is a widespread process. But in recent years, intensive treating processes have indicated a need for the development of immobilization for a given process. The strength with which the cells are bonded to the support varies, with cell type, support type, and the variation of operation condition. The industrial use of surface-attached cell systems (biofilm) has been extensively examined in the field of wastewater treatment. For strengthening the attachment, porous supports such as porous glass are used. However, in the fluidized-bed type bioreactors for wastewater treatment, inorganic materials are generally not used because of the difficulty in controlling their specific gravity, brittleness, and shortcoming of abrasion-resistance.

Entrapped Technique

The second immobilization technique is entrapment within a porous matrix that is synthesized in situ around the cells. This method is rapidly coming into widespread use. The cells are entrapped in a matrix that protects them from the shear field outside the particles. A high degree of cell viability is retained in most entrapment methods. These systems are therefore useful for a variety of growth-related processes.

An advantage of the preformed support particles is that they are more resistant to the compression and disintegration in fluidized-bed bioreactors. Cell growth is controlled by the continuous flow of wastewater past the particles. A variety of compounds can be used for cell entrapment, with minimal loss of viability. Natural polymers used for immobilization include polysaccharides, such as alginate, carrageenan, cellulose, and agar, and polypeptides, such as gelatin and collagen. These natural polymers are compatible with microorganisms, but they are not suitable for wastewater treatment because of their assimilation by microorganisms in activated sludge.

Table 14.1 Environmental application of gel-entrapped microorganisms

Substrate	*Microorganism*	*Entrapment method*
Ammonia	*Nitrosomonas europaea*	Ca-alginate
		κ-carrageenan
	Nitrosomonas europaea	Polyelectrolyte
	Paracoccus denitrificans	Complex-stabilized Ca-alginate
	Thiosphaera pantotropha	Agarose
	Scenedesmus quadricauda	κ-Carrageenan
Dairy effluents	*Candida pseudotropicalis*	Alginate
Nitrate	*Anasistis nidulans*	Collagen
	Denitrifying sludge	PVA
	Pseudomonas denitrificans	Ca-alginate
	Thiobacilus denitrificans	Agar/agarose
Nitrite	*Nitrobacter agilis*	Ca-alginate
		Carrageenan
	Nitrobacter sp.	PVA
Nitrogen	*Anabaena* CH3	Ca-alginate
	Nitrifiers	Polyethyleneglycol resin
Phenol	*Candida tropicalis*	Alginate
	Pseudomonas sp.	Ca-alginate, polyacrylamide hydrazide PVA-boric acid-Ca-alginate

Accordingly, only synthetic polymers, such as polyacrylamide, poly(vinyl alcohol), poly(ethylene glycol), and polypropylene, and hydrophilic photo-cross-linkable resin can be applied to the immobilization of biomass in terms of mechanical strength and durability that are required for industrial usage. For successful immobilization, the hydrogel must be effective in cell viability and function (biocompatible). The method of entrapment can reduce the culture viability. The entrapment in a polyacrylamide gel destroys viability in 50 to 90% of the cells by the toxic acrylamide monomers.

Containment Behind a Barrier Technique

The third major category of cell immobilization is containment behind a barrier. When cell separation from the effluent is required, these systems are highly useful. The barrier that immobilizes cells can be as simple as the liquid–liquid phase interface between two miscible fluids. Mohan and Li immobilized the denitrifying bacteria, *Micrococcus denitrificans,* by emulsifying the cell suspension with surfactant into a hydrocarbon solvent. Cells in the surfactant-enclosed aqueous droplets did not leak from aqueous phase through the organic phase to the cell-free aqueous phase surrounding the droplets. Because the cells are retained behind a semipermeable wall or membrane, this type of system is very similar to microcapsulation in an engineering sense. Nutrients (mainly BOD components) are supplied to, and products are removed from, the cell mass by diffusion. In wastewater treatment, the barrier loses its separation activity during long operations.

Cellular Flocculation

Cells naturally aggregate or flocculate and produce extracellular polysaccharides. The large size of the aggregates makes their use possible in reactors designed for immobilized cells, such as packed-bed or fluidized-bed reactors. Artificial flocculating agents or cross-linking agents may be added to enhance the process of aggregation for cells that do not naturally flocculate. Similar mechanisms govern both artificially and naturally induced flocculation. The durability of the aggregate will depend on its hydrodynamic interactions with the environment. Many cells produce extracellular polysaccharides, some of which can act to promote the cohesion of a biofilm. These substances might be important determinants of the diffusive resistance in these films.

INDUSTRIAL IMMOBILIZED TECHNOLOGY

Much research has focused on industrial applications using immobilized cell technology for wastewater treatment. In an industry setting, entrapment and adsorption are the predominant techniques used for wastewater treatment. The nitrifying process is described in this section, examples include entrapment using PBG gels and adsorption using photo-cross-linked resins.

Entrapment Technique

The entrapment of microbes is the most popular immobilization technique for living cells in wastewater treatment. The BOD sensor was developed to abbreviate the long measurement time by entrapping the activated sludge. In wastewater treatment, immobilization processes should be mild to minimize the damage to the living cells. Entrapping materials must also have proper permeability to allow sufficient diffusion and transport of oxygen and essential nutrients, metabolic waste, and secretory products across the hydrogel network at supplemental culture. Considering the use of synthetic polymers serving as the synthetic extracellular matrixes for cells, it is important to understand the transport properties through these gels to predict whether nutrients can diffuse into the matrix. Supplemented culture media provide the cells with essential nutrients for growth and development as well as a means of oxygen and metabolic product transport while in vitro. Traditionally, mass transfer behavior in the gel matrix has often been recognized as a rate-limiting factor in an entrapping cell system. Some of the immobilizing matrixes are highly hydrated, allowing almost unhindered diffusion for products and substrates with molecular weights less than 10,000 Da. The effective diffusivity of oxygen, glucose, and various ions in gel has been studied by several groups. For industrial applications, microbes must be cultivated for entrapment and storage to keep the activity of immobilized microbes after immobilization.

Mass transport

Mass transport through gels is governed by the size and shape of the pores and solute molecules as well as by the nature of the interaction between solute molecule and gel material. The value of the equilibrium coefficient reflects the relative affinity of the solute for the gel relative to its affinity for the surrounding medium. The species concentration in the gel at equilibrium is obtained by mass balance using the known values for the reservoir and gel volumes. The value of K, the partition coefficient, is calculated by the following equation:

$$K = C^{gel}/C^{bath} \quad \ldots(1)$$

The average pore diameters of gels were estimated from the diffusion of solutes and the hydrodynamic diameter of solute.

Biomass decay is an important parameter in the dynamics of the immobilized microbe system. *Nitrosomonas europaea* cells entrapped in a gel can lose their viability during cultivation; they die first in the center of a gel bead but also will die in the outer layers at low substrate loadings. If the substrate concentration is raised, the number of viable microbes near the surface of the gel increases again. The fluorescein diacetate and lissamine-green staining techniques were useful in distinguishing between viable and dead cells in gel beads. Successive nitrification and denitrification are effective biological processes for the removal of nitrogen in the form of ammonia, nitrite, and nitrate from aqueous waste. In this biological nitrogen-removal system, nitrification is generally the rate-limiting step because of the extremely low growth rate of obligate autotrophic nitrifying bacteria.

The use of immobilized cells in nitrification overcomes this disadvantage and shortens greatly the hydraulic retention time (HRT) in continuous nitrification. In a synthetic resin gel such as polyacrylamide, the gel strength decreased about 50 to 60% in the presence of microbes. But the density of microbes in the gel did not influence its strength. The immobilization of nitrifying bacteria by using urethane prepolymers of the hydrophilic type exhibited a drastic improvement in the activity yield. These immobilized cells were prepared by entrapment using alginate, polyelectrolite complex, polyethyleneimine, photo-cross-linkable resin, poly(vinyl alcohol), polyethylene, and poly(ethylene glycol) (PEG).

Role of polyethylene-glycol (PEG)

The PEG gel was prepared by radical reaction under existing microbes to exhibit about 3 kg/cm^2 of gel strength. These gels are nontoxic for microbial cells and show viability. Furthermore, this method is easy industrialized for a large-scale production. Microorganisms such as nitrifying bacteria grow near the surface of gels (about 60-μm depth). After about a month's cultivation, they have constant activity. The rate of oxygen consumption the parameter of activity, is 60 mg O_2/L gel/h at initially and 800 mg O_2/L gel/h at the end of precultivation. At the steady state, the concentration of microbes in the gel carrier is increased to contain about 10^{10} cells per milliliter of carrier. On the other hand, the concentration of nitrifying microbes in the activated sludge is usually 10^3 to 10^5 cells/mL.

Nitrification and denitrification of entrapped microbes were scarcely influenced by the operation temperature, and the activity of nitrification was about three times higher than that of the activated sludge, the reaction rates are exhibited on the basis of suspended solid for comparing the activity of immobilization with that of activated sludge. The nitrification rate of precultured immobilized gels was about 100 mg N/L gel/h at 13°C. Accordingly, the system using entrapped microbes showed a high denitrification rate owing to high BOD loading resulting from short retention time in the nitrification tank.

Regular microbial wastewater treatment

It is composed of an initial and final sedimentation tank, a denitrification tank, and a nitrification tank. The volume ratio of denitrification to nitrification is 1.5/1.0, and the gel separator (the net of 1.5-mm mesh) in the nitrification tank is installed for preventing gel bead washout. The mesh size of the gel separator should be half the gel diameter for sufficient separation. Wastewater treatment was recycled between the denitrification tank and the nitrification tank. Retention times are 4.8 and 3.2 h for denitrification and nitrification, respectively. The average total nitrogen removal is represented as

$$\text{Total N removal (\%)} = R/(R + 1) \times 100 \qquad \text{...(2)}$$

with R as the recycling ratio (–). The total N removal is increased according to the enlargement of the recycling ratio, but the dissolved oxygen in the denitrification tank is also increased. Consequently, recycling is generally operated at a 1:2 recycling ratio to maintain the anaerobic condition of denitrification and compensate for the increased power cost in nitrification. Gel beads are fluidized by the air blowing at the bottom of nitrification tank.

The packing ratio of entrapped gels in the reactor is usually 5:15 volumetric percent depending on the treatment. In the entrapment method, gels are added at 7.5 volumetric percent to the nitrification tank. The floating carriers are distributed homogeneously. In the entrapment method, specialized equipment and sufficient time are needed to prepare an adequate number of gels for large-scale wastewater treatment. Even in winter nitrification was thoroughly carried out, and BOD as well as suspended solids in the effluent were lower than 10 mg/L. This method needs about 8 h of retention time to eliminate 70% of nitrogen in wastewater.

After continuous operation for 2 y, it was concluded that gel beads must have above 1.5 kg/cm^2 compressive strength; therefore, the initial compressive strength of gel absent from microbial cells needs to be over 2.7 kg/cm^2. For the nitrification, photo-cross-linkable resin was used to obtain an activity of 1.5 kg of N/m^3 gel/day for continuous wastewater treatment. On the other hand, the drop in denitrifying activity at the immobilization was restored in the medium containing nitrite or nitrate with sodium acetate as a hydrogen donor.

Commercial Utilization

Systems in which cells are immobilized by adsorption are popular because of the ease of this type of immobilization. The strength with which the cells are bonded to the support varies with cell type and support type. The adsorption method differs from the entrapment method, advantages include easy gel formation (no consideration of microbes at cross-linking reaction), mass transfer of gel barrier, good storage stability before usage, and low transportation cost of gels at the industrial scale. A disadvantage is the long induction period for steady wastewater treatment.

Table 14.2 Operation condition of plant

Item	*Run 1*	*Run 2*	*Run 3*
Operation period	18	11	18
Infuluent (m^3/day)	2,720	2,280	1,350–3,150
Retention time (h)	6.6	7.9	5.7–13.3
Aeration (Nm^3/h)	610	670	800
Recycling ratio (-)	2.8	2.7	2.0–7.0
MLSS (mg/L)	2,800	1,840	1,880
BOD-SS loading (kg/kg day)	0.070	0.149	0.0965–0.302
BOD loading (kg/m^3 day)	0.196	0.275	0.182–0.570
Total N and SS loading (kg/kg day)	0.0285	0.0535	0.0226–0.0772
Total N loading (kg/m^3 day)	0.0793	0.0980	0.0426–0.146
Recycling ratio of sludge (%)	64	64	47–1 09
Temp. of water (°C)	23.5	14.0	14.2
Total SRT (day)	14.6	9.0	8.8
Aerobic SRT (day)	5.8	3.6	3.5

The growth rate of nitrifying bacteria is about 10 times less than that of general microbes. The attachment of biomass on the gel surface occurs during the start-up phase. After this phase, the reactor is operated continuously at different hydraulic and surface loading rates. Accordingly, the gels must have high strength and good cell attachment on their surfaces. A positively charged macroporous cellulose carrier is used for immobilization of nitrifying bacteria by the adsorption method. This carrier promotes stable adhesion of bacteria onto the carrier walls and allows transfer of nutrients into the carrier. A porous cellulose carrier was produced from cellulose with a polyethyleneimine coating to form a stable cross-linked structure. Similarly, polyurethane foam and other carriers have been used for wastewater treatment.

Importance of photo-cross-linked gels

To use immobilized microbes in wastewater treatment, the preparative method, especially the support material used for the preparation, must be selected from the standpoint of its economy and feasibility on a large-scale operation. For these reasons, photo-cross-linkable resins are promising support materials for the immobilization of microbes. Although this immobilization process may have various conceivable features, its features are summarized as follows:

1. The immobilization procedure (gel formation) is simple because of ultraviolet irradiation.
2. The gel materials have great mechanical strength.
3. The surface of gel beads can be regulated by the properties of photo-cross-linkable resin.
4. The manufacture of photo-cross-linkable resins and the process of gel formation can be easily industrialized.

The photo-cross-linkable resins usually have a main chain of PEG, a mixture of PEG and poly(propylene glycol) (PPG), or poly(vinyl alcohol) (PVA) with photo-cross-linkable ethylenic double bonds. By irradiation of ethylenic double bonds with UV eight, polymerization is induced, forming three-dimensionally-cross-linked gels.

Recyclic lines

The sludge used is taken from the recycle line of an activated sludge treatment plant treating wastewater, because organic nitrogen and ammonia nitrogen are present in wastewater. Acclimatization of the sludge is carried out in a batch reactor. The carrier elements are made of hydrophilic photo-cross-linkable resin with a 4-mm bead diameter. Because the nitrifying bacteria do not flocculate differently from the activated sludge, it is difficult to attach the nitrifying bacteria on the surface of the carrier. The adsorption rate of nitrifying bacteria is maximum at pH 4.8, and nonbiological factors are important at the initial stage of attachment. Fluorescence studies of the adsorbing surface clarified that the affinity to water and ζ-potentials of the surfaces are important factors for adsorption.

Table 14.3 Comparison of PEG-PPG/PVA type gels and PVA type gels

Species of gels	*PEG-PPG/PVA type*	*PVA type*
DO (mg/L)	2.8	6.0
NH_3-N value added (kg/m^3 day)	0.29	0.21
Nitrification velocity (kg/m^3 day)	0.21	0.15
Cells attached (mg/L)	6,920	1,620

Among the synthetic polymers, photo-cross-linkable resin has largely resolved the problems with inorganic carriers described in the Introduction. Its surface structure maybe suitable for attachment of microbes, controlling the properties of photo-cross-linkable resins. Several factors affect cell affinity and behavior on hydrogel: the general chemistry of the monomers and the cross-links, hydrophilic and hydrophobic properties, and surface properties involving charged functional chemical groups. Physical properties of the hydrogel also govern the adhesion affinity; therefore, altering pore size and network structure can convert cell adhesion as well as morphology and function. The attachment of biomass on the charged gel surface occurs during the start-up phase.

Designing gel carriers using photo-cross-linkable resin has resulted in good cell growth on their surface with good mechanical strength.

The surface of PEG-PPG-type hydrophilic resin is most hydrophobic and has an excellent affinity to microbes. When observing the surface of PEG-PPG-type photo-cross-linked resin and the results of contact angle, the phase separation of the gel in which hydrophobic domains of about 1-μm diameter separated on the hydrophilic phase of PEG was seen on SEM of the surface and with the optical microscope. From these results, the surface of hydrophobic domains formed on hydrophilic phase, the so-called macrophase separation of hydrophilic gel, may be best for the attachment and growth of nitrifying microbes. The structure of the gel bead is controlled with a combination of PEG and PPG, PVA type, and PEG type. Beads with a diameter of about 4 mm and compressive strength of more than 20 kg/cm^2 can be produced by immobilization.

Continuous treatment of wastewater

Continuous wastewater treatment can be carried out by a process using gel for microbial adsorption. Removal performance of the pilot plant with a retention time of 8 to 9 h in the bioreactor is obtained as follows. The average total N of the influent is 31.8 mg/L, and that of the effluent total N is 8.3 mg/L. The average P of the influent is 2.57 mg/L, and that of the effluent total P is 0.60 mg/L. The nitrification rate of immobilized gels is of zero-order reaction in which the produced NO_z-N increases directly with the reaction time, and the nitrification rate is about 2.5 kg NO_k/m^2 gel/day.

Table 14.4 Operation conditiond of pilot plant

Item	*Step 1*	*Step 2*	*Step 3*
Influent flow rate (m^3/d)	25.0	25.0	28.6
HRT (h)	8.7	9.0	8.0
Anaerobic tank (m^3)	1.0	1.5	1.5
SRT (day)	16	11–14	8–11
Aerobic SRT (day)	5.7	3.7–4.7	2.7–3.7
Gel diameter (mm)	5.0	5.0	4.2
Charge ratio (%)	20	20	16 -n
DO (mg/L)	1.8–5.0	2.0–5.8	2.8–7.8
Influent BOD (mg/L)	49–151	45–110	48–118
Total N (mg/L)	14–34	13–47	16–45
Total P (mg/L)	1.2–2.9	2.0–5.8	2.8–7.8

In the relationship between the dissolved oxygen (DO) and the nitrification rate coefficient (K_M) of gels, K_M largely depended on the DO. It is considered that the oxygen diffusion through the biofilm on the surface is the rate-determining step for nitrification. The phosphorus uptake increased in proportion to the denitrified nitrogen, and the weight ratio of phosphorus to nitrogen is about

0.75. From these facts, nitrifying microbes are suggested to be phosphorus-accumulating microorganisms. Since 1997, a denitrification plant using these gels (nitrifying tank volume, 10,000 m^2) has been operating in Tokyo.

CONCLUDING REMARKS

The development of processes using immobilized microorganisms has made it possible to eliminate BOD and nitrogen in half the time (6 to 8 h), compared with conventional activated sludge. In addition, these processes maintained constantly high activity during winter. The treatment cost containing the depreciation is 80 to 90% of the activated sludge. To gain the advantages of immobilization, reductions are needed in the immobilized gel costs, particularly for large-scale operations, which include raw materials costs, immobilization facility, transportation cost, storage cost, and so on. Wastewater treatment processes using immobilized cells are still under development and will be further improved in the future. Particularly, it is expected that carrier gels will regulate their structure and form a particular reaction site in gels or on the surface of gels independent of the surroundings.

15

INDUSTRIAL APPLICATIONS OF ENZYMES

The use of enzymes as purposefully added ingredients to laundry products dates back to 1913 when Otto Rohm patented the use of crude pancreatic extracts in laundry presoak compositions to improve the removal of biological stains. It wasn't until the early 1960s, however, with the isolation and commercial production of alkaline active proteases, that enzymes became widely used in laundry products. By 1969, at least 50% of European detergents contained a protease. Within 6 years of the introduction of alkaline proteases in detergents, 80% of all European laundry detergents contained enzymes. Today, the penetration of enzymes into laundry detergents across the globe is quite high. Most premium and lower-tier laundry products contain at least one enzyme, usually a protease, and many are formulated with up to four different enzyme activities—protease, amylase, lipase, and cellulase. Enzymes in laundry and cleaning products account for about 40% of the more than 1 billion dollar industrial enzyme market, with detergent proteases accounting for about 50% of the volume. In the United States, the detergent sector is the largest market for enzymes, accounting for about 20% of total enzyme demand and more than 25% of industrial enzyme sales.

The move to compact detergents, beginning in 1987, helped raise the demand for detergent enzymes, and global enzyme sales doubled between 1989 and 1994. Four factors are responsible for the increased reliance of detergent manufacturers on enzyme technology. First, they provide consumer-recognizable cleaning benefits. Historically, proteases and amylase have been used to improve the removal of proteinaceous and starch-based soils, respectively. More recently, detergent-compatible lipases have been introduced to help clean greasy and oily stains arising from foods and body soils. Second, some enzymes enable the detergent formulator to add completely new performance benefits.

Cellulases are the best and most recent example where the development of cost-effective, alkaline-active cellulases has enabled the delivery of consumer-valued benefits such as color maintenance and fabric restoration. Third, enzymes are highly weight efficient. The ability to act catalytically imparts a space efficiency to enzymes that is highly desirable to the formulator of compact detergents. The fourth factor contributing to the increased use of enzymes in detergents is that in recent years the performance/cost ratio for enzymes has improved significantly as a result of the availability of more efficient enzymes and the industry trend toward reduced pricing. Current market trends and consumer needs are influencing the development of enzymes for detergent applications, with the emphasis on enzymes that have improved performance/cost ratios, increased cold water activity, and improved compatibility to other detergent ingredients. In addition, enzyme suppliers and detergent manufacturers are actively pursuing the development of new enzyme activities that address the consumer-expressed need for improved cleaning, fabric care, and antimicrobial benefits. For applications beyond laundry and cleaning products, enzymes with low allergenicity are being developed to minimize worker and consumer safety concerns.

MARKET VALUE OF ENZYMES

As noted in the "Introduction", enzymes have long been of interest to the detergent industry for their ability to aid in the removal of soils and stains as well as to deliver unique benefits that cannot otherwise be obtained with conventional detergent technologies. As a result, the detergent enzyme market has grown nearly 10-fold during the past 20 years. In 1997, sales of detergent enzymes totaled about $550 million. U.S. demand accounted for about $160 million and is expected to grow to $260 million by 2001. Through the 1980s and early 1990s the market was split between five major suppliers: Novo-Nordisk A/S of Denmark with the dominant market share (>55%); Gist-Brocades N.V. in The Netherlands; Genencor International in the United States; Solvay in Belgium; and; Showa-Denko in Japan.

These suppliers marketed a full range enzymes for liquid and powder detergents. Beginning in 1995, however, there was considerable rationalization in the detergent enzyme market owing to the relatively high cost of manufacturing coupled with increased pressure from detergent manufacturers to drive down raw material costs. Genencor International purchased the detergent enzyme businesses of Gist-Brocades and Solvay, and Novo-Nordisk acquired Showa-Denko's detergent enzyme business. Today, Novo and Genencor are the main suppliers of detergent enzymes, supplying more than 95% of the global market.

HYDROLASES IN DETERGENT

The majority of enzymes used in detergents today are hydrolases; the catalytic event involves addition of water across a chemical bond. In addition, most detergent enzymes exhibit *endo* specificity, that is, they preferentially attack internal bonds as opposed to terminal linkages. Before the conversion of the detergent market to compact product forms, the use of enzymes in detergents was limited primarily to proteases.

The market conversion to compacts, which require high levels of cleaning performance from low product dosages, fueled the interest in formulating new enzyme activities that can provide a more robust cleaning profile. This, coupled with recent advances in genetic and protein engineering, has led to new classes of enzymes being available for detergents, including amylases, lipases, cellulases, and peroxidases. Each of these general enzyme classes is discussed in detail in the following sections.

Proteolytic Enzymes

Most powder and liquid laundry detergents on the market today, both low density and compact, contain a protease. Protease enzymes enhance the cleaning of protein-based soils, such as grass and blood, by catalyzing the breakdown of the constituent proteins in these soils through hydrolysis of the amide bonds between individual amino acids. Proteases serve a multifunctional role in the overall cleaning process. They boost the efficacy of surfactant and dispersant systems by degrading proteinaceous soils into smaller, more dispersible fragments, thereby directly facilitating the soil removal process. In addition, they decrease the redeposition of proteins, especially hydrophobic proteins such as those found in blood, resulting in improved overall whiteness. The plateau observed at higher enzyme concentrations is typical of most detergency evaluations of enzymes. Typically, detergent formulators dose enzymes at the breaking point of such curves in order to obtain optimum cleaning performance at lowest possible cost.

The hemoglobin in the blood stain has been imaged with a colloidal-gold-labeled antibody marker to help visualize remaining protein. There is a considerable amount of residual blood protein left on the fibers after washing. There is a marked reduction in the amount of residual blood protein consistent with the visual observation of enhanced stain removal. Number of different commercial detergent proteases are currently being marketed. However, they are all closely related, nonspecific endopeptidases, with molecular weights around 27 kDa and highly homologous primary amino acid sequences. Current commercial detergent proteases are derived from various species of *Bacillus subtilis* and are characterized by a common catalytic triad consisting of serine, aspartic acid, and histidine. The serine hydroxyl group functions as a nucleophile, attacking the peptide amide bond to form a tetrahedral intermediate that undergoes hydrogen transfer facilitated by the aspartyl and histidine functional groups acting as general base catalysts. The result is addition of water across the amide bond, generating two peptide fragments. These so-called *serine* proteases are particularly well suited for laundry applications owing to their alkaline pH optima, stability to detergents, and broad substrate specificity.

Table 15.1 pH profiles of common detergent enzymes

Enzyme	*Source (species)*	*pH Optimum*	*Range (>80% maximal activity)*
Alcalase	*Bacillus licheniformis*	8.5	8.0–1 0.0
BPN′	*Bacillus amyloliquefaciens*	8.5	8.0–9.5
Savinase	*Bacillus lentus*	10.0	8.0–11.0
Esperase	*Bacillus alcalophilus*	10.0	9.0–12.0

Transmission electron microscopy-immunocytochemical (TEM-ICC) analysis of residual soils remaining on fabrics after washing provides more detailed clues on the mechanisms of detergent protease action. Table 15.2 shows the frequency and mean particle sizes (MPS) of two types of protein—chlorophyll-binding protein (CBP) and β-lipoprotein (BLP)—and four types of carbohydrate—glucose/mannose, glucosamine, fucose, and galactose/galactosamine—remaining on residual grass stains on cotton and polycotton fabrics after washing in a liquid detergent with and without protease. The general trend is for the protease to reduce the frequency but not the size of the stain components. For example, with addition of protease, the frequency of occurrence of CBP on cotton drops from 2 of every 5 fibers to 1 in 10, and on polycotton, it drops from 4 markers in every 5 fibers to 1 in 6, with little or no change in particle size. Similar effects are observed for BLP and the carbohydrates. Although the subtilisin proteases are stable to most of the surfactants used in typical laundry detergents, they are susceptible to oxygen bleaches and calcium sequesterants.

Table 15.2 TEM-ICC data for grass removal

Component	*Fabric*	*Without protease*		*With protease*	
		Frequency	*MPS (μm)*	*Frequency*	*MPS (μm)*
Chlorophyllbindingprotein	Cotton	2/5	0.5	1/10	0.3
	Polycotton	4/5	0.3	1/6	0.3
β-Lipoprotein	Cotton	1/2	0.2	1/5	0.3
	Polycotton	3/4	0.3	1/3	0.3
Glucose/mannose	Cotton	1/7	0.2	1/10	0.2
	Polycotton	1/4	0.2	1/15	0.2
Glucosamine	Cotton	1/7	0.5	1/10	0.2
	Polycotton	1/5	0.4	1/10	0.3
Fucose	Cotton	1/25	0.2	1/30	0.3
	Polycotton	1/20	0.3	0/10	—
Galactose/galactosamine	Cotton	1/10	0.5	1/8	0.3
	Polycotton	1/10	0.3	0/10	—

Oxidative attack by peroxide or peracids on the methionine residue adjacent to the catalytic serine results in a nearly 90% loss of enzymatic activity. Replacement of the methionine with an oxidatively stable amino acid such as alanine significantly improves the stability of the enzyme toward oxygen bleach. In addition, the subtilisin proteases strongly bind at least one calcium ion, which contributes to maintaining the overall three-dimensional configuration of the enzyme. The calcium sequestering agents used in many laundry products to control water hardness can remove this calcium, resulting in decreased thermal and autolytic stability. Introduction of negatively charged residues near the calcium binding site can increase the binding affinity of the enzyme for the calcium, resulting in improved stability toward calcium sequesterants. Because proteases degrade proteins, there is some risk associated with washing garments constructed from natural protein fibers such as wool and silk in protease- containing detergents.

Proteolytic attack against such fibers leads to increased wear and reduced tensile strength. The degree of damage is dictated by the type and level of protease as well as the wash process. Accordingly, many detergents advise against use on wool and silk garments. In powder detergents, proteases are added in an encapsulated or granulated form, which protects them from other detergent ingredients and eliminates the potential for them to degrade each other (autolysis) or other enzymes (proteolysis) in the formulation. However, in aqueous-based liquid detergents, autolysis and proteolysis are key problems requiring the formulation of reversible protease inhibitors that can shut down the activity of the enzyme in product, but are diluted away from the enzyme when the detergent is added to the wash. Boric acid and polyols, such as propylene glycol, are commonly used to inhibit protease activity in liquid detergents.

Detergent proteases can be generally classified into two main groups, alkaline and high alkaline. Alkaline proteases, such as those derived from *Bacillus licheniformis,* show optimum activity around pH 8, making them ideal for liquid laundry products, which typically have wash pH's between 7.0 and 8.5. Several high alkaline proteases are available and are typically characterized by a pH-activity optimum around 10.0. These enzymes find widespread use in powder laundry detergents and automatic dishwashing formulations. A number of analytical methods are available for assaying proteolytic activity in product and in the wash. Most methods use colormetric detection of the hydrolysis of a specific reagent. Typical reagents include dimethylcasein, azocasein, azocoll, or hemoglobin. By far the most common assay method is based on hydrolysis of small synthetic peptides functionalized with the *p*-nitroanilide chromophore.

For most detergent proteases, the synthetic tetrapeptide Suc-AAPF-pNA, or succinyl-alanylalanylprolyl(phenylalanine)-*p*-nitroanilide (PNA), is used. Hydrolysis of the anilide bond liberates the *p*-nitroaniline group, resulting in the appearance of a yellow color that can be followed spectrophotometrically. This assay is particularly useful for establishing pH and temperature profiles of proteases and following their stability through the wash or during in-product storage. However, in general, PNA activity does not correlate with wash performance. Activity against larger substrates, such as casein, which more closely resemble the soil proteins found in common laundry stains, shows a better correlation with protease wash performance.

Amylopsin Enzymes

Amylopsin or amylases facilitate the removal of starch-based soils, particularly food soils, by catalyzing the hydrolysis of the glycosidic linkages in starch polymers. This general class of enzymes encompasses several enzymatic activities responsible for starch hydrolysis, including α-amylase, β-amylase, pullulanase, isoamylase, and glucoamylase. The α-amylases (1,4-α-D-glucan-glucanohydrolase) are most commonly used in detergents. They catalyze endohydrolysis of 1,4-*a*-D-glucosidic bonds in high molecular weight starches of the type commonly found in food soils. As such, amylases find utility in both laundry and dish detergent formulations. Current commercial detergent α-amylases are derived from various species of *Bacillus subtilis*. The α-amylase from *B. amyloliquiefacians,* is particularly well suited for detergent applications because of its thermal stability, broad pH activity, and resistance to proteolytic attack. The detergent α-amylases in use today have highly homologous primary amino acid sequences and a tertiary structure comprised of three domains, A, B, and C.

The active site is typically located in a cleft between domains A and B and is usually comprised of acidic amino acids such as aspartic and glutamic acids. Like the subtilisin proteases, α-amylases contain an essential calcium ion that is important in maintaining the tertiary structure. Accordingly, the *a*-amylases show some degree of sensitivity to calcium sequesterants, with solution stability severely compromised in well built (low calcium) environments. In addition, most wild-type α-amylases are sensitive to oxygen bleach. Recently, protein engineering has been used to improve bleach stability of amylases. Both Novo and Genencor offer bleach-stable versions of their α-amylase.

The use of amylase and protease combinations for improved stain removal and general cleaning was fairly well established in the mid-1980s and practiced in a number of European and U.S. brands. The use of these two enzymes in compact detergents is now fairly standard, especially in Europe. As with proteases, there are a number of methods available for analyzing amylase activity. Probably the simplest, and most common, involves monitoring the color change during hydrolysis of iodine-stained starch. Iodine staining of a standard starch substrate results in a blue color that changes to reddish brown upon hydrolysis with α-amylase. Additional methods include hydrolysis of synthetic substrates like *p*-nitrophenylmaltoheptoaside (pNP-G7). Amylolytic attack on pNP-G7 produces pNP-G3 and pNP-G4. Reacting the pNP-G3 with an excess of α-glucosidase generates glucose and the characteristic yellow color of *p*-nitrophenyl. Alternative methods involve detection of the reducing sugars formed during starch hydrolysis using 3,5-dinitrosalicylate, alkaline ferricyanide, or alkaline copper.

Lipase Enzymes

Lipases are fairly recent entries into the detergent enzyme market. Their use in detergents is illustrated in several patents from the 1970s. However, the lack of a viable commercial source of alkaline-active lipases with stability toward anionic surfactants essentially prohibited practical application for almost 10 years until the late 1980s when Novo Industries announced the commercial sale of the first detergent-compatable lipase, Lipolase, obtained by cloning the lipase gene from *Humicola lanuginosa* into a high-expression fungal host. Lipolase first appeared in compact powders in Japan and has since found fairly broad application in the global detergent market. Lipases (triacylglycerol acylhydrolases) degrade triglycerides by catalyzing the hydrolysis of the ester linkage between the glycerol moeity and the constituent fatty acids, generating glycerol; free fatty acids, mono- and diglycerides.

Triglycerides are common components of some of the most-difficult-to-remove soils and stains, including greasy foods, cosmetics, and body soils. Owing to their generally hydrophobic nature, triglycerides are difficult to remove from fabric and may be present even after several wash cycles. As the residual triglycerides age they can oxidize, resulting in the generation of a rancid malodor as well as an overall yellowing or discoloration of the fabric. Lipolytic attack increases the water solubility of the original triglyceride and decreases the melting point of the fat, allowing for increased soil removal during the washing process. Lipases are most active at the water–substrate interface. This phenomenon is referred to as interfacial activation. Lipase activity does not follow classical enzyme kinetics, which generally are only valid for reactions that occur in a homogeneous phase. Rather, lipase activity appears to follow a two-step process.

In the first step, lipase physically binds to the surface. This is believed to involve activation of the enzyme via a structural change that exposes the hydrophobic binding domain to the substrate.

The second step involves formation of the enzyme–substrate complex, leading to hydrolysis of the ester bond. Because physical adsorption to the interface is key to lipase activity, anything that interferes with this process, such as surfactants, can dramatically affect substrate hydrolysis. Exclusion from the soil–water interface is believed to be one of the key mechanisms involved in surfactant inhibition of lipase activity. Given the tendency of surfactants to accumulate at an oil and water interface, lipase can be denied access to the substrate by electrostatic repulsion and steric hindrance of the aggregated surfactant.

Accordingly, when formulating lipase into detergents it is important to select a surfactant system that minimizes this effect. Generally, nonionic surfactants are less inhibiting than anionics. Most commercial detergent lipases require more than one wash cycle to show a benefit. It is generally thought this traces to deposition of the lipase onto the hydrophobic stain during the wash such that, during the drying cycle, when the water content is decreased, the enzyme is activated and can hydrolyze triglycerides in the stain, facilitating removal in the second wash cycle. For Lipolase, maximum enzymatic activity against a standard lipid soil such as olive oil is obtained when the moisture content of the fabric is 20 to 30% by weight, suggesting that significant decomposition of the oily soil occurs during the drying process; hence the need for a second wash cycle to remove the hydrolytic fragments. The most commonly used method to measure lipase activity is a titrimetric assay in which the fatty acid liberated during hydrolysis of an emulsified substrate, such as Triolein, is titrated with a standardized base solution. The milliequivalents of base required to maintain the solution pH at a given value is directly related to the amount of fatty acid generated. Alternate methods rely on colormetric detection of the hydrolytic products generated from synthetic substrates such as *p*-nitrophenylbutyrate.

Enzyme Cellulases

Cellulases are unique among the hydrolase enzymes used in detergency. Rather than attacking substrates found in soils and stains, cellulases hydrolyze the cellulose in cotton and cotton blends to deliver cleaning benefits such as whitening and dingy cleanup as well as fabric care benefits such as color maintenance and fabric restoration. Like lipases, cellulases have only recently been introduced broadly into detergents. The first use of cellulases in detergents was described by Murata in 1981. Commercial application began in 1987 with the introduction of the alkaline cellulase from *Bacillus* sp. KSM-635 in Kao's compact detergent Attack. Since 1991, a number of European and North American detergents have incorporated cellulases. In 1993, Procter & Gamble introduced Novo-Nordisk's Carezyme, an enzyme reported to help keep cotton and cotton blends looking newer longer by removing the pills and fuzz that build up on cotton garments over time, making them look worn and faded. Cellulases deliver a number of benefits including cleaning (via removal of particulate and oily soils), fabric softeness, color clarification, and depilling.

The cleaning benefits are believed to derive from the removal of surface fibrils that serve as anchor points for soil. Softness benefits are believed to result from the ability of cellulases to loosen surface fibrils, preventing them from intertwining and forming a rigid network with negative hand feel. Color clarification and depilling derive from the same mechanism. Repeated washing and wearing of cotton garments generates fuzz and pill buildup as surface fibrils loosen and become entwined. The fuzz and pills scatter light, making garments look worn and faded. Cellulases weaken the attachment points of the fuzz and pills, which subsequently break off in later wash cycles. The result is brighter, more colorful garments. Cellulases (1,4-β-D-glucan-4-glucanohydrolases) catalyze the endohydrolysis of 1,4-β-D-glucosidic linkages in cellulose. Cellulose is a linear, unbranched polymer of anhydroglucose units linked through β-1,4 bonds. Each polymer contains between 10,000 and 15,000 glucose units. Up to 60 polymers are arranged in parallel bundles, called elementary fibrils, and stabilized through extensive hydrogen bond and van der Waals interactions, which promotes crystallinity of the bundle.

The elementary fibrils can further aggregate to form microfibrils that, in turn, aggregate into macrofibrils of up to 500 nm in diameter. Cotton fibers are formed by the accumulation and aggregation of macrofibrils. Disruption of the hydrogen bonding network within these structures, either through chemical or mechanical means, leads to the formation of amorphous regions. In new cotton garments, the ratio of crystalline to amorphous regions is about 70:30. In regenerated cellulose fabrics such as rayon, the structure is much more disordered, leading to a shift in the crystalline amorphous ratio to about 30:70. The amorphous regions are the site of cellulytic attack owing to the greater accessibility and flexibility of the substrate. As a result, aged garments that have experienced a number of wash and wear cycles and, hence, have more amorphous regions, tend to show greater sensitivity to cellulase than new garments. Cellulose hydrolysis is an acid-catalyzed process. The O-glycosidic bond is cleaved by donation of a proton from the carboxyl group of a glutamic acid residue in the active site of the enzyme, with the resulting cabonium-ion intermediate stabilized by an aspartic acid or other nucleophilic residue.

The formation of a carbonium-ion intermediate opens the possibility of stereochemistry, with the configuration of the anomeric carbon atom of the product either being retained or inverted. The configuration of the active site of the particular cellulase enzyme determines the sterochemistry of the hydrolysis products, leading to a classification of cellulases as being either retaining or inverting types of enzymes. Current detergent cellulases are derived from fungal sources, such as *Humicola insolens*, or bacterial sources, such as the *Bacillus* species. The fungal enzymes typically represent a mixture of *exo-* and *endo*-glucanase activities with pH optima in the acidic to neutral range. This mix of activities facilitates the degradation of crystalline cellulose. Celluzyme from Novo-Nordisk is derived from *H. insolens* strain DSM 1800 and consists of a mixture of *endo*-glucanases, cellobiohydrolases, and β-glucosidases.

Carezyme, on the other hand, is a monocomponent alkaline-active *endo*-glucanase with low affinity to crystalline cellulose. The detergent cellulases from alkalophilic *Bacillus* species are generally monocomponent *endo*-glucanases with good stability against proteases. Relative to the fungal cellulases, bacterial cellulases tend to show somewhat reduced fabric care benefits and increased detergency effects on pigmented dirt and sebum. Generally, cellulases are multidomain structures comprised of a core catalytic domain, containing the active site, and a cellulose binding domain, or CBD, which is responsible for anchoring the enzyme to the cellulose substrate. Typically, the CBD is attached to the catalytic domain through a short peptide linker. In the absence of the CBD, the core will show catalytic activity against soluble substrates but drastically reduced activity against insoluble cellulose. Because the benefits derived from cellulase clearly rely on the enzyme binding to fabric surfaces, the CBD is a critical structural component. Like lipases, adsorption of cellulase onto fabric is the first step of the hydrolytic process, corresponding to a phase transfer of enzyme from solution to the fabric surface.

Those enzymes that adsorb tightly are generally the most efficient with respect to cellulose hydrolysis. Thus, the CBD is an important structural feature controlling the efficacy of a particular cellulase with respect to detergency benefits. Table 15.3 shows activity on soluble and insoluble substrates and the depilling benefit for cellulases with (EG V, Carezyme), and without (EG 1, Endolase) a CBD as well as the catalytic core of an EG V enzyme. The importance of the CBD both to hydrolysis of insoluble substrate and in delivering a depilling benefit is apparent in the higher activity of the CBD-containing Carezyme on crystalline cellulose (Avicel) and the more than fivefold increase in depilling benefit versus the Carezyme core.

Table 15.3 Activity and performance of various cellulases: role of the cellulase binding domain

Enzyme	*CMCU/mg (soluble substrate)*	*Red-Avicel/mg (insoluble substrate)*	*Relative depilling benefit*
EG I (Endolase)	80	0.2	4
EG V(Carezyme)	100	22	100
EG V core	180	8	19

A number of analytical methods are available for assaying cellulase activity. The two most common are the viscometric method and measurement of reducing sugars. Because of their tendency to cleave internal glycosidic linkages, detergent cellulases cause a rapid decrease in chain length of soluble cellulose polymers, resulting in a drop of solution viscosity. This change in viscosity can be measured via standard viscometeric methods and, when compared to a standard sample, can be used as a relative measure of cellulytic activity. Substituted, water-soluble cellulose substrates are commonly used in this assay, such as carboxymethylcellulose (CMC) or hydroxyethylcellulose. Measurement of the reducing sugars formed upon hydrolysis of a cellulose substrate is often used in conjunction with the viscometric method to assess cellulase activity. Reducing sugars can be detected colorimetrically by ferricyanide.

Role of Peroxidase Enzyme

This represents the newest class of enzyme developed for detergent. There is currently only one commercially available peroxidase for detergents, Guardzyme from Novo-Nordisk A/S. Guardzyme was isolated from the inkcap mushroom, *Coprinus cinereus,* and has been cloned into a production host for large-scale production. It is a heme-containing protein that, in the presence of hydrogen peroxide, can mediate the oxidation of fugitive dyes in solution, thereby providing a dye-transfer inhibition benefit. Peroxidases are a subclass of the general class of enzymes known as oxidoreductases. Oxidoreductases catalyze oxidation–reduction reactions on a broad range of substrates, including alcohols, carboxyl groups, alkenes, and amines. Peroxidases specifically catalyze the reaction:

$$\text{Donor} + H_2O_2 \rightarrow \text{Oxidized donor} + 2H_2O$$

where the donor molecule can be anything with an oxidizable group. Specifically, if the donor molecule is a fugitive dye that has bled off a colored garment, peroxidase can be used to bleach the dye in solution, effectively eliminating the transfer of the colored dye species to other fabrics.

For Guardzyme, the dye-transfer inhibition system consists of three components: the enzyme, a mediator, and hydrogen peroxide. The mediator is generally a substituted phenolic compound that shuttles electrons from the substrate to peroxidase. Through selection of the appropriate mediator, a catalytic oxidation cycle can be established as follows:

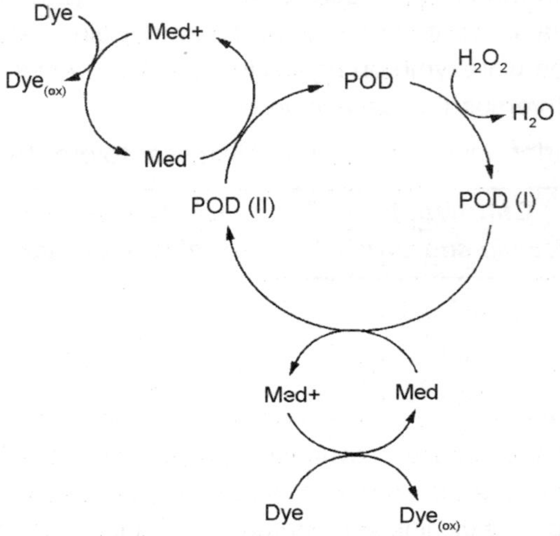

Dye
Med+
$Dye_{(ox)}$
Med
POD
H_2O_2
H_2O
POD (II)
POD (I)
Med+
Med
Dye
$Dye_{(ox)}$

where POD (I) and POD (II) are oxidized forms of the peroxidase. The peroxidase-based dye-transfer inhibition system essentially eliminates transfer of dye from a blue bleeder to a white cotton tracer.

FORMULATION INGREDIENTS

The key challenge to the use of enzymes in detergents is stability. This is particularly true in liquid detergents, where the enzyme is in intimate contact with the formulation ingredients. For powder detergents, enzymes are typically supplied as a coated granulate, thereby separating the enzyme from other detergent ingredients and generally delivering a stable composition. Liquid detergents, however, must be formulated to minimize damage to the enzyme. Denaturation by surfactants and builders, unfavorable interactions with perfume ingredients, and degradation by protease are the principal processes responsible for loss of enzyme activity upon storage.

Liquid Detergents

Most modern liquid detergents contain fairly high concentrations of anionic surfactants for cleaning. These surfactants are particularly good at denaturing proteins, including enzymes. In addition, most heavy-duty liquids contain multiple enzyme systems, including proteases, amylases, lipases, and cellulases. Proteolytic degradation of these enzymes by the added protease poses a significant long-term stability issue. In order to achieve acceptable shelf stability and ensure delivery of a high level of cleaning performance over time, detergent manufacturers and enzyme suppliers have taken three basic approaches to improve the stability of enzymes in liquid detergents:

Improving enzyme stability

A number of ingredients can be added to the liquid detergent to improve enzyme stability. Calcium is known to stabilize both proteases and amylases. The added calcium serves to tie up detergent actives that could, otherwise remove the essential calcium from the enzyme, resulting in a loss of the catalytic three-dimensional structure. The half-life of the enzyme under stressed storage conditions increases as calcium concentration is increased. In addition, short-chain carboxylates have been shown to promote protease stability through a noncompetitive-inhibition mechanism. Accordingly, calcium and sodium formates are commonly added stabilizing agents in enzyme-containing liquid detergents.

Reversible protease inhibitors can also be added to shut down the activity of the protease in product. The combination of a polyol with vicinal hydroxyl groups, such as propylene glycol, and a boron compound, such as boric acid, results in formation of a borate-diol complex that effectively inhibits proteolytic activity. Upon dilution of the product into the wash, the inhibitor is diluted away from the protease yielding a fully active enzyme. Substituted boronic acids and peptide derivatives have also been shown to be effective reversible inhibitors.

Classical screening methods

Enzymes with enhanced stability to detergent ingredients can be obtained by classical screening methods. For example, the Kao Corporation in Japan has reported isolation of protease-resistant cellulases from cultures of *Streptomyces* strains. Kawai et al. identified a number of new cellulases from *Bacillus* species with good stability to surfactants, proteases, and chelants. Similarly, new α-amylases have been found that do not appear to be substantially affected by anionic surfactants. Natural isolate screening for new wild-type lipases with improved detergent compatibility has met with limited success to date, although lipases from *Fusarium culmorium* have been isolated and reportedly show improved activity versus Lipolase in the presence of alcohol ethoxylate surfactants.

Detergent engineering

Another way to improve stability of enzymes for detergents is to engineer in specific stabilizing mutations. For example, replacing a neutral or positively charged amino acid in the calcium- binding domain of an enzyme with a negatively charged amino acid would be expected to increase electrostatic interactions, thereby minimizing stability loss by removal of the essential calcium ion. Aehle et al. used this approach to modify the calcium binding site of an α-amylase. The resulting variant showed a 2.5-fold increase in stability versus the native enzyme in low calcium environments. Similarly, amino acid substitutions in the vicinity of a hydrophobic domain in the alkaline *Bacillus lentus* protease are reported to improve storage stability of the resulting variants in both liquid detergents and soap bars.

Schulein et al. reported a number of modifications to Carezyme that improved its alkaline activity and detergent compatibility. In addition to stability, aesthetics are another major consideration for the formulation of enzymes into liquid products. Odor, color, and clarity are key aesthetic signals for consumers. Enzyme raw materials, being products of microbial fermentation, often contain a number of smelly by-products, including volatile amines and low molecular weight carboxylic acids, that must be removed to avoid consumer malodor complaints. Likewise, fermentation-derived by-products may cause turbidity, flocculation, or color changes in liquid detergent formulas. Therefore, the formulation of enzymes into liquid detergents requires extensive evaluations of odor, color, and clarity in order to select enzyme formulations that minimize aesthetic negatives.

Washing Powders

Many of the same considerations for the formulation of enzymes in liquids apply to powders. Odor, color, and stability are important considerations although to a somewhat lesser extent owing to the physical separation of the enzyme from the bulk detergent via granulation. In addition, particle size, shape, flowability, and bulk density become important when considering mixing and segregation of enzyme particles with powdered detergents. Probably the most critical factor in formulating solid enzymes is dust generation during detergent manufacture. This was clearly evident in the experiences of the detergent industry in the late 1960s and early 1970s, with the raw powdered enzyme formulations originally supplied for detergent use. These formulations led to high airborne dust levels in the manufacturing plants and an increase in worker sensitization. As a result, enzyme manufactures replaced these early powder and agglomerated enzymes with various granulate formulations, the most important of which are summarized below:

Role of dry enzymes

In this process, dry enzyme is dispersed into a molten wax, nonionic surfactant, or polymer matrix, then sprayed into a cooling tower to form solid, spherical particles. The molten phase is typically comprised of a water-soluble or water-dispersable material with a melting point above 50°C. The key advantages of prilling are high throughput and the ability to recycle particles that fall outside the desired size range. The key disadvantage is the relatively poor physical strength of the particle, leading to break up and high dust generation in subsequent processing. Overcoating with an inert, mechanically tough material such as polyethylene glycol can be used to improve physical strength and reduce dustiness. However, this typically is not sufficient to ensure no break up of the particles during detergent processing and, as a result, enzyme prills are not in widespread use today.

Extruded granulations

By combining an enzyme powder or liquid concentrate with binders, such as clay, sugar, or starch, an extrudable dough can be produced that can then be pressed through a perforated metal plate. Upon drying, the extrudates are reasonably durable. In order to produce spherical particles of the appropriate size the wet extrudate is fed into a Marumerizer, where a rotating textured plate breaks up the noodles into smaller particles and rounds the corners to yield a more spherical shape. After sieving, the particles are coated with pigments, such as TiO_2, and a protective outer layer to achieve the desired appearance and further improve granulate integrity. Key disadvantages of this method are the high capital investment in a multistep process and the sensitivity of the process to variation in feedstock moisture and composition.

High shear granulation

This method is used by Novo-Nordisk to produce their T-granulate products. In this process, enzyme is mixed with controlled amounts of water, binders, such as polyethylene glycol, ethoxylated fatty alcohols, fatty acids, bentonite, kaoline, or other clays, and other granulation ingredients to form a low-moisture agglomerate. The agglomerate is fed into a high-shear mixer that breaks it up into smaller particles. The particles are dried in a fluidized bed and coated with a final protective layer of polymers and pigments. By incorporating cellulose fibers into the original agglomerate, particles with high mechanical stability and physical strength can be produced. Enzyme T-granulates are widely used in dry laundry products and are among the lowest-dusting enzyme raw materials available. The process is reasonably flexible, especially with respect to enzyme feedstock concentration, allowing granulation of a given enzyme, such as Savinase, across a wide range of enzyme payload.

Spraying technique

In this process, liquid enzyme feedstock is sprayed onto a solid, inert support, or core, such as NaCl or a sucrose-starch nonpareil. Once the enzyme layer has dried, additional coatings can be applied, including stabilizers, chelants, antioxidants, and pigments. The outer coating typically consists of a film-forming polymer, such as polyvinyl alcohol, and a pigment, such as TiO_2. Genencor International uses this granulation process to deliver their Enzoguard line of dry enzyme formulations. The key advantage of fluid bed coating is that a single piece of equipment, the fluid bed coater, can be used to produce the final product.

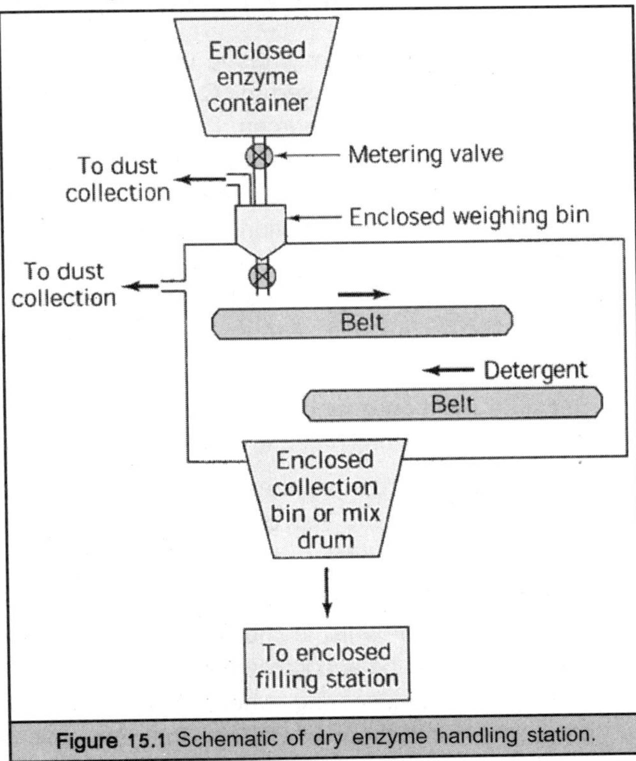

Figure 15.1 Schematic of dry enzyme handling station.

Clinical Aspects

Controlling protein dust from enzyme raw materials is a key consideration in the manufacture of enzyme-containing laundry products because enzymes are proteins, like animal dander or pollen, with the potential to act as respiratory allergens. Enzyme formulations, such as the granulate compositions described above, that minimize protein dust or reduce the tendency of enzyme protein to form aerosols are critical to protein dust control. Just as important is the need for appropriate industrial hygiene practices and in-plant exposure control technology to further minimize the potential for exposure to enzyme protein. This involves compliance to operational guidelines that require no visible dust in the manufacturing environment, no recurring enzyme spills or spills of enzyme-containing detergent, no gross skin contact with raw material enzyme, and cleanup processes that minimize dust and aerosol generation. Design of dust handling equipment should follow current best practices for enclosure and local exhaust ventilation.

Spill-containment systems should be in place for those areas where either spills of raw material enzyme or enzyme-containing product are likely. Additionally, an appropriate preventive maintenance program should be implemented to ensure the equipment performs as intended. Strict adherence to guidelines such as these has resulted in a drop in airborne enzyme dust levels in North American and Western European detergent manufacturing plants by more than three orders of magnitude versus 1970 values. Enzymes are received from the supplier in an enclosed container that can be coupled to a metering system for accurate dosing. From the weigh station, enzyme is mixed with a stream of detergent base granules in a dust-proof enclosure and sent to either a mix drum or holding bin in preparation for pack-out to cartons. The entire process, including packaging of the final product into boxes or cartons, is housed in dust-proof enclosures to minimize airborne protein dust. Systems such as these ensure low risk for enzyme handling.

COMMERCIAL PRODUCTION OF DETERGENT ENZYMES

Nearly all detergent enzymes in use today are produced through large-scale fermentation of microorganisms. As discussed in "Current Detergent Enzymes", detergent enzymes are derived from bacterial and fungal sources. Large-scale fermentation is typically carried out with optimized strains of appropriate bacterial (*Bacillus* or *Pseudomonas* strains) or fungal species (*Aspergillus*) that have been genetically engineered to express high levels of the target enzyme, excrete that enzyme into the bulk fermentation medium for ease of recovery, and minimize production of other, unwanted proteins. After fermentation, the enzyme is recovered from the broth by removal of the whole cells and other particulates through centrifugation or filtration, concentrated via ultrafiltration and diafiltration techniques, and, if necessary, further purified by selective precipitation of nonactive proteins.

Because multiton quantities of low-cost enzymes are needed to support the requirements of the global detergent business, enzyme manufacturers closely adhere to the following principles:

1. The production organism must be capable of secreting the enzyme extracellularly into the bulk fermentation medium. This greatly facilitates recovery because the cost, both in time and money, to recover intracellular enzymes is to great to support the needs of the detergent industry. Enzyme secretion can be maximized by appropriate optimization of fermentation conditions.
2. The production organisms should be able to be tuned to give the highest possible yields. Strain optimization can be accomplished either through classical mutagenesis and screening methods or via genetic engineering, where multiple copies of the gene coding for the enzyme of interest are inserted into the host DNA so that each microbe in the strain can produce several copies of the enzyme of interest.
3. The number of steps in the downstream recovery process should be minimized to avoid yield losses and keep overall processing costs down.

4. The production organisms should produce the target enzyme in relatively high purity with no contaminating side activities or proteins. This ensures that only basic recovery methods, such as filtration and centrifugation, are needed to obtain the final product and eliminates the need for high-cost process steps like solvent precipitation. By deleting the genes that code for unwanted enzymes and proteins, strains can be prepared that secrete only the enzyme of interest into the bulk media.

ENVIRONMENTAL FACTORS

The influence of temperature on the structure and function of proteins has been extensively studied. However, it is still not possible to determine in any definitive sense which factors underlie the vast differences in thermostability and thermoactivity between seemingly similar proteins. The discovery of hyperthermophilic microorganisms (those with growth temperatures of 90°C and above) and the subsequent purification and characterization of their extremely thermostable and thermoactive enzymes have underscored the complex nature of protein stability. It is clear now that differences of 100°C or more may separate the temperature optima of enzymes carrying out identical biocatalytic functions. Information available on amino acid sequences of enzymes with widely varying temperature optima, and even three-dimensional structural data, may not provide much insight into how seemingly similar enzymes can respond to thermal energy so differently. Thus, enhancing the thermostability of a given enzyme to a significant extent remains an elusive objective left to fortuitous improvements through random mutagenesis, or to extensive screening efforts from natural high temperature biotopes. In cellular environments, protein stability is a property that has important physiological implications.

Protein architecture must be consistent with in vivo requirements. For example, the free energy of stabilization of an enzyme (ΔG), a balance between large but opposing entropic ($T\Delta S$) and enthalpic forces (ΔH), is equivalent to only a few weak bonds, presumably so that it can be regulated in response to changing cellular conditions. Once an enzyme has performed a needed cellular function that is no longer necessary or becomes detrimental, it may denature, be deattenuated through a regulatory response, or be proteolytically inactivated. Although enzymes are desirable in industry because of their specificity and availability, in order to be useful for commercial purposes, some degree of stability is required to make economic sense. In fact, proteins may need to not only withstand thermal forces, but also be resistant to non-aqueous solvents, high ionic strengths, detergents, and extremes of pH as well. In addition to stability, it is important that the enzyme be functional in the specific extreme environment in which it is to be used.

Thus, a highly thermostable and thermoactive enzyme may not be useful in applications at very low temperatures, even though it may be very stable under these conditions. Clearly, the maintenance of an enzyme's structural integrity and its catalytic efficiency need both be considered for possible application of these biocatalysts in an industrial setting. Even though site-directed and random mutagenesis have yet to become a routine approach for conferring thermostability on an enzyme, nature has fortunately provided a source of thermostable biocatalysts in extremely thermophilic microorganisms (optimal growth temperatures above 75°C). An ever-increasing list of such enzymes shows that most significant activities from mesophiles (optimal growth around 37°C) have high-temperature, and presumably thermostable, counterparts. The biochemical and biophysical characteristics of many of these enzymes are currently being examined for both scientific and technological purposes. In this article, the methodology being used to discover extremely thermostable enzymes will be discussed in addition to information on their function and use. It is expected that the extraordinary stability of these enzymes will make them candidates for replacing many currently used biocatalysts in addition to opening up many new applications.

IDENTIFICATION OF ENZYMES

The methodology for identifying enzymes with technological importance has gone through considerable changes in recent years with the ever-expanding use of molecular biological tools. Concomitantly, the search for enzymes from microorganisms in extreme environments has also expanded and has made use of new discovery techniques. Gone are the days when industrial enzymes were exclusively mined from microorganisms isolated from soil samples and grown in pure culture. Now, only the gene encoding an enzyme of interest is, in principle, necessary if it can be overexpressed in an appropriate host. Another development that is sure to impact enzyme discovery is the increasing availability of genome sequences for a variety of mesophilic and thermophilic microorganisms.

Geothermal Habitats

Expeditions to geothermal habitats, as accessible as hot springs in Yellowstone National Park and as inaccessible as deep sea geothermal vents, have led to the isolation of an array of high-temperature microorganisms covering a diversity of physiological types. Typically, isolates are obtained through serial dilution techniques because high temperatures preclude cultivation on solid media in many cases. Most microorganisms classified as extreme thermophiles ($T_{opt} \geq 75°C$) are anaerobic, which is not surprising considering the low solubility of O_2 at elevated temperatures. Notable exceptions are members of the order Sulfolobales, many of which are thermoacidophiles. A subset of the extreme thermophiles, the hyperthermophiles (capable of growth at 90°C or higher) are almost exclusively anaerobic, and many reduce sulfur to sulfide during growth. Cultivation of extreme thermophiles on a scale sufficient to provide biomass for enzyme purification efforts presents many challenges. In short, lack of knowledge of microbial physiology, low biomass yields, generation of copious amounts of hydrogen sulfide, explosive substrate and product gases, salt-laden media, anaerobic conditions, and cell sensitivity to shear are all potential obstacles.

A few heterotrophic hyperthermophiles, such as *Pyrococcus furiosus, Thermococcus litoralis,* and *Thermotoga maritima,* can be cultivated without sulfur and have, thus, been the primary focus of direct purification efforts. These organisms can be grown to biomass yields of 1 kg (wet) or higher in 600-L fermentation systems. From such efforts, more than 20 enzymes have been purified from the biomass generated from a single fermentation. Once sufficient biomass has been generated, enzyme purification can proceed along conventional lines using multistep protocols involving various types of liquid chromatography. Cells can be broken in a variety of ways including freeze-thaw cycles, French pressing, or sonification. Because many extreme thermophiles are members of the domain archaea, which have unusual cell membrane features, including a lack of peptidoglycan, lysozyme treatments are typically not effective. Most types of prep-scale chromatography have been used in the purification of these enzymes. These include ion exchange, hydrophobic interaction, gel filtration, and affinity methods.

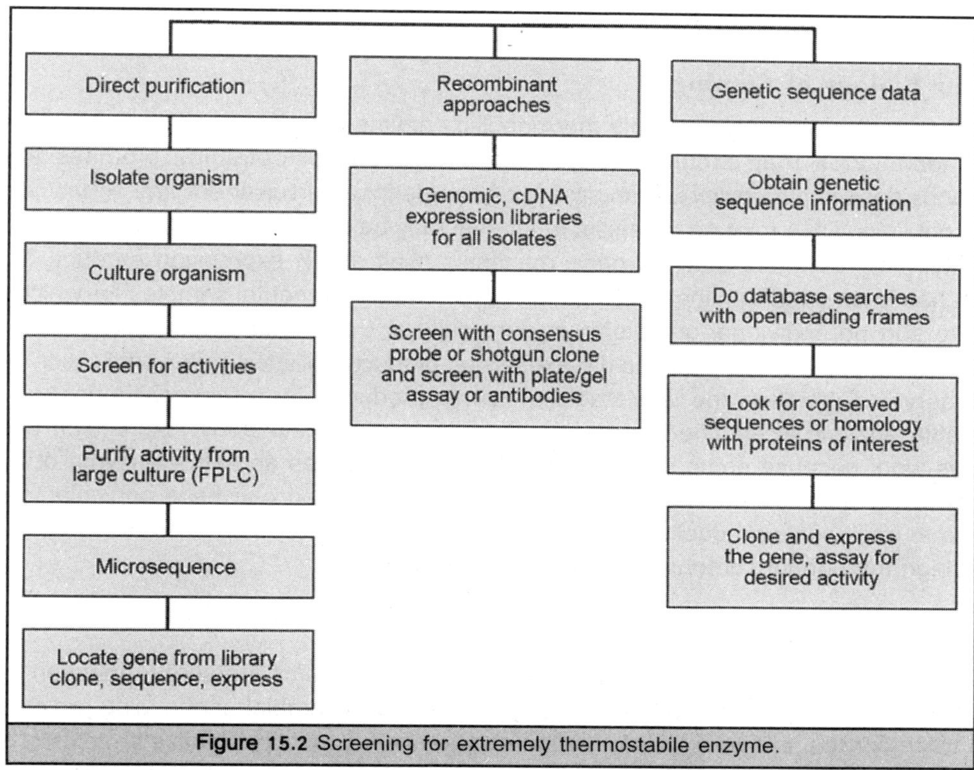

Figure 15.2 Screening for extremely thermostabile enzyme.

The extreme thermostability of these enzymes does provide a convenience: Purification can be carried out at ambient temperatures rather than in refrigerated environments. On the other hand, thermostability does present some difficulties in assessing the homogeneity of products of the purification process. Conventional sodium dodecylsulfate-polyacrylamide gel electrophoresis (SDS-PAGE) can be misleading because extremely thermostable enzymes may not be sufficiently denatured in this process. The result is that both the native and partially denatured forms of multimeric proteins may appear on SDS-PAGE, making assessment of M_r (relative molecular mass) and homogeneity difficult.

Unfortunately, extended periods of heating to ensure complete denaturation may also lead to covalent modifications of labile amino acid residues. Because it is economically impractical to produce large amounts of extremely thermostable enzymes in their native hosts, recombinant approaches must be used. The purified enzyme may be N-terminally or C-terminally (or both) sequenced, and conventional methods for identifying the corresponding gene in genomic preparations by nucleic acid probe hybridization or polymerase chain reaction (PCR) techniques can be used. Overexpression of genes encoding extremely thermophilic enzymes is still in its earliest stages, and most efforts to date have used a variety of vectors in *Escherichia coli*. Although many efforts have yielded a recombinant enzyme seemingly identical to the native at low expression levels, it is not known whether existing expression systems will be suitable to produce the large amounts of enzyme needed for certain industrial uses. This is a critical issue for the development of applications making use of these enzymes.

Molecular Biology of Cloning

Although not specific to extremely thermophilic enzymes, molecular biology has provided an array of techniques that circumvent the tedious process of obtaining biomass from large fermentations done in pure culture. If an assay is available for a particular enzyme activity, expression libraries containing DNA from an organism of interest may be screened to isolate the gene encoding this activity. This assumes that the gene manifests itself as an expression product. This same approach has been used to mine enzymes directly from environmental samples for which no effort is made to sort out individual organisms in pure culture. For extreme thermophiles, the intrinsic thermostability of the enzyme made as the expression product facilitates partial purification. A heating step will serve to denature the less-stable enzymes produced by a mesophilic host so that the thermostable enzyme is enriched in the resulting mixture. Unfortunately, expression cloning will not always work because some genes may not be expressed as an active enzyme or a suitable screening assay may not be easily used. However, this method can be a particularly attractive alternative to large-scale production of extreme thermophile biomass and can also be used as the basis for high throughput screening and directed evolution.

Method of Analyzing the Sequence of Genomic DNA

Recent efforts to sequence entire genomes for microorganisms are sure to have an impact on the enzyme discovery process. Because many of the initial microbial sequencing projects involve extreme thermophiles, this bank of information can be used to identity putative proteins of industrial importance. The *Methanococcus jannaschii* genome has been sequenced, and many other extreme thermophiles are currently being sequenced, such as *Archaeoglobus fulgidus, Aquifex aeolicus, Methanobacterium thermoautotrophicum, Pyrobaculum aerophilum, Pyrococcus furiosus, Pyrococcus horikoshi, Sulfolobus solfataricus,* and *Thermotoga maritima*. Computer databases can be searched using amino acid or nucleotide sequences corresponding to known motifs, and resulting homologous sequences can be the basis for cloning the gene of interest. At this time, however, 50% or more of the genome sequence obtained from specific microorganisms contain open reading frames (ORFs) that are not attributable to a protein whose function is known. Furthermore, there are relatively little sequence data available for many hydrolytic enzyme types of potential commercial significance so that identification of homologous versions of certain enzymes can be an imprecise exercise. Nevertheless, as entire genome sequence data expand, this approach should not be ruled out because it may be difficult to purify a particular enzyme directly from the native source.

BIOCHEMISTRY OF THERMOPHILIC ENZYME

Enzymes of all known classes have been purified from extremely thermophilic microorganisms and characterized biochemically. Many of these proteins are intrinsically thermostable. In many cases, the genes encoding these enzymes have been sequenced and in some cases expressed in active form in a foreign host, usually *E. coli*. With the availability of genome sequence data, we also now know that these microorganisms contain thermostable counterparts to many proteins previously characterized from mesophilic sources, many of which are linked to central metabolic pathways. Also, because most existing enzyme applications make use of hydrolytic enzymes, these enzyme types have been sought in extreme thermophiles.

Table 15.4 Types of enzymes purified from extreme thermophiles

Enzyme	M_r	T_{opt}	$t_{50\%}$	*Function*	*Source*
		Hydrolases			
Aminopeptidase	320 (α_4)	75°	NA	Peptide hyrolysis	Ss
Carboxypeptidase	170 (α_4)	85°	<15 min 95°	Peptide hydrolysis	Ss
Archaelysin	52 (α)	98°	1.5 h/95°	Peptide hydrolysis	DM
Protease PfpI-H	66 (α_3)	105°	33 h/98°	Peptide hydrolysis	Pf
Protease PfpI-CI	113 (α_6)	85°	<0.5 h/98°	Peptide hydrolysis	Pf
Protease pyrolysin	150 (α)	115°	9 h/95°	Peptide hydrolysis	Pf
Tetrabrachion	130 (α)	>90°	NA	Structural/proteolytic	Sm
Thiol protease	44 (α)	110°	1 h/100°	Peptide hydrolysis	P KOD1
Proteinase	68 (α)	85°	22 h/90°	Peptide hydrolysis	Ts
Thermopsin	46 (α)	75°	90 min/90°	Peptide hydrolysis	Sa
Proteinase I	108 (α_2)	90°	5.7 h/92°	Peptide hydrolysis	Ss
Proteasome	640 ($\alpha_{14}\beta_{14}$)	95°	>24 h/95°	Protein turnover	Pf
α-Glucosidase	125 (α)	110°	48 h/98°	Maltose hydrolysis	Pf
α-Glucosidase	80 (α)	105°	3.0 h/95°	Maltose hydrolysis	Ss
Amylase	132 (α_2)	100°	2 h/120°	Starch hydrolysis	Pf
Amylase	70 (α)	100°	6 h/100°	Starch hydrolysis	Pw
Amylopullulanase	140 (α)	118°	20 h/98°	Starch degradation	ES
β-Glucosidase	230 (α_4)	105°	13 h/110°	Cellobiose hydrolysis	Pf
Sucrose α-glucohydrolase	114 (α)	105°	48 h/95°	Unknown	Pf
β-Mannosidase	240 (α_4)	105°	1.1 h/98°	Mannobiose hydrolysis	Pf
β-Mannosidase	200 (α_2)	87°	17 h/85°	Mannobiose hydrolysis	Tn
β-Mannanase	55 (α)	92°	34 h/85°	Mannan hydrolysis	Tn
α-Galactosidase	66 (α)	105°	9 h/85°	Galactomannan hydrolysis	Tn
Endo-β-1,4-glucanase	27 (α)	95°	>6 h/80°	Glucan hydrolysis	Tm
Exo-β-1,4-glucanase	29 (α)	95°	>6 h/80°	Glucan hydrolysis	Tm
Xylanase A	120 (α)	92°	45 min /90°	Xylan/xylose hydrolysis	Tm
Xylanase B	40 (α)	105°	125 min/95°	Xylan/xylose hydrolysis	Tm
Exo-1,4-β-cellobiohydrolase	36 (α)	105°	70 min/108°	Cellulose hydrolysis	TF

Enzyme	M_r	T_{opt}	$t_{50\%}$	Function	Source
β-Xylosidase	174 (α_2)	80°	2 h/80°	Xylobiose hydrolysis	TF
α-Arabinofuranosidase	92 (α)	80°	4 h/98°	Arabinoxylan hydrolysis	TF
Fumarate hydratase	170 (α_4)	85°	15 min/90°	Sugar metabolism	Ss
N^5N^{10}-methenyltetrahydro-methanopterin cyclohydrolase	41.5 (α)	95°	>1 h/90°	Methanogenesis	Mk
		Oxidoreductases			
Aldehyde OR	85 (a)	>95°	6 h/80°	Glycolytic enzyme	Pf
Pyruvate OR	120 ($\alpha\beta\gamma\delta$)	>95°	0.3 h/90°	Pyruvate oxidation	Pf
Formaldehyde OR	280 (α_4)	>95°	2 h/80°	Unknown	Tl
Indolepyruvate OR	180 ($\alpha_2\beta_2$)	>95°	6 h/80°	Peptide fermentation	Pf
Gly. 3-phosphate OR	63 (α)	>95°	NA	Glycolytic enzyme	Pf
Isovalerate OR	220 ($\alpha\gamma\beta\delta$)$_2$	>95°	0.6 h/90°	Peptide fermentation	Pf
2-Ketoglutarate	220 ($\alpha\gamma\beta\delta$)$_2$	>95°	0.3 h/90°	Peptide fermentation	Pf
Pyruvate OR	200 ($\alpha\gamma\beta\delta$)$_2$	>95°	15 h/80°	Pyruvate oxidation	Tm
Quinone OR	280 ($\alpha\gamma\beta\delta\varepsilon\zeta\eta$)	65°	5 min/75°	Electron donor	Af
Glutamate dehydrogenase	270 (α_6)	95°	10 h/100°	Glutamate oxidation	ES
GAPDH	150 (α_4)	NA	0.7 h/100°	Glycolysis (?)	Pw
GAPDH (NAD)	196 (α_4)	NA	0.3 h/100°	Unknown	Tt
GAPDH (NADP)	156 (α_4)	NA	0.5 h/100°	Unknown	Tt
Sulfide dehydrogenase	92 ($\alpha\beta$)	>95°	12 h/95°	Sulfur reduction	Pf
Alcohol dehydrogenase	200 (α_4)	80°	2.0 h/85°	Unknown	Tl
Isocitrate dehydrogenase	90 (α_2)	NA	NA	Sugar metabolism	Ss
Quinol oxidase	149 ($\alpha\beta\gamma\delta\varepsilon$)	NA	NA	Proton gradient	Aa
Methyl coenzyme M reductase	270 ($\alpha\beta\delta$)$_2$	NA	NA	Methanogenesis	Mk
N^5N^{10}-methylenetretrahydrom-ethanopterin reductase	300 (α_2)	90°	>1 h/100°	Methanogenesis	Mk
N^5N^{10}-methylenetetrahydrom-ethanopterin dehydro-genase	170 (α_4)	>90°	Very labile	Methanogenesis	Mk
Adenylylsulphate reductase	160 (α_2)	85°	NA	Electron donor	Af
		Hydrogenases/redox proteins			
Hydrogenase	118 ($\alpha\beta$)	>90°	1 h/98°	H_2 production	Pf
Hydrogenase	280 (α_4)	>95°	1 h/90°	H_2 production	Tm
Ferredoxin	7 (α)	>95°	>24 h/95°	Electron transfer	Pf, Tl
Rubredoxin	5 (α)	>95°	>24 h/95°	Unknown	Pf
Thioredoxin (glutaredoxin)	24.8 (α)	NA	>3 h/90°	Reduces disulfides	Ss, Pf

Enzyme	M_r	T_{opt}	$t_{50\%}$	*Function*	*Source*
			Isomerases/invertases		
Xylose isomerase	204 (α_4)	97°	2.3 h/95°	Sugar metabolism	Tm
Triose phosphate isomerase	100 (α_4)	NA	4.7 h/104°	Unknown	Pw
Prolyl *cis/trans* isomerase	18 (α)	NA	NA	Unknown	Ss
			Nucleic-acid modifying enzymes		
DNA polymerase	93 (α)	>75°	20 h/95°	Replication	Pf
Endonuclease SuaI	NA	65°	30 min 80°	Restriction	Sa
Reverse gyrase	200 ($\alpha\beta$)	NA	NA	Supercoiling	Mk
DNA topoisomerase II	230 ($\alpha_2\beta_2$)	80°	35 m/80°	DNA relaxation	Ssh
DNA topoisomerase III	108 (α)	NA	NA	DNA relaxation	Damy
DNA toposisomerase V	110 (α)	75°	NA	DNA relaxation	Mk
RNAse p2	13 (α_2)	NA	<15 min/90°	RNA hydrolysis	Ss
			Synthetases		
ACS I	140 ($\alpha_2\beta_2$)	>95°	18 h/80°	Acetyl-CoA formation	Pf
ACS II	140 ($\alpha_2\beta_2$)	>95°	8 h/80°	Aryl-CoA formation	Pf
Carbamoyl-P synthetase	46.6 (α)	>90°	3 h/95°	Pyrimidine pathway	Paby
PEP-synthase	2000 (α_7)	NA	NA	Metabolism	Sm
			Transferases		
Aromatic aminotransferase	90 (α_2)	>95°	NA	Peptide fermentation	Pf
Aspartate transcarbamoylase	NA	70°	>6 h/90°	Nucleotide synthesis	Paby
Propylamine transferase	110 (α_3)	90°	>1 h/100°	Amino acid metabolism	Ss
Formyltransferase	35 (α)	90°	>1 h/90°	Methanogenesis	Mk
			Other		
Glucokinase	93 (α_2)	105	3.7 h/100°	Sugar metabolism	Pf
Enolase	180 (α_4)	>95°	0.7 h/100°	Glycolysis	Pf
Chaperonin	920 (α_{16})	80°	NA	Protein refolding	Ss
PEP carboxylase	240 (α_4)	88°	>2 h/80°	Metabolism	Ms
5′Methylthioadenosine phosphorylase	160 (α_6)	120°	15 min/130°	Nucleoside catabolysis	Ss

Evolutionary Placement of Enzymes

Because of the evolutionary placement of extreme thermophiles, based on 16S rRNA phylogeny, interest in the origins of metabolic pathways has led to the isolation and characterization of enzymes directly involved in intermediate metabolism and bioenergetics. These include a number of oxidoreductases, hydrogenases, dehydrogenases, and redox proteins. Many such enzymes from the most studied hyperthermophile *P. furiosus* appear to be intrinsically thermostable and function in vitro at the organism's optimal growth temperature of 98 to 100°C. The study of these enzymes in vitro presents certain challenges because assays to detect activity may involve thermally labile substrates, anaerobic conditions, and coenzymes.

For example, even though *P. furiosus* contains several dehydrogenases, these catalyze reactions involving NADH as a cofactor, which itself is not very thermostable. This organism is also known to contain a modified Emden-Myerhoff-Parnos pathway for glycolysis, which implies that glucose is processed at high temperatures through several intermediate steps. The mechanisms by which simple sugars, which may be sensitive to thermal modifications, can be processed at high temperatures are not known.

Proteolytic Enzyme

Proteolysis (hydrolysis of peptide bonds) is essential to cellular function and is implicated in nutrition, heat shock response, immune response, and certain housekeeping roles. Historically, proteases have found important industrial uses in detergents, leather bating, food processing, and much more, collectively comprising the largest volume use of commercial enzymes. But in the past 10 years, these enzymes have also been shown to be active in organic solvents for peptide synthesis and are often selective enough to produce pharmaceuticals. Recently, protease inhibitors have received much attention in medical applications for combatting the HIV virus and some cancerous tumors. There are many potential benefits to be derived from the use of proteases from extreme thermophiles. If mesophilic proteins are to be hydrolyzed, it is very likely that at elevated temperatures their denatured forms will facilitate access of thermostable proteases to specific peptide bonds. Also, at higher temperatures, many substrates are more soluble, solution viscosity is reduced, and diffusion rates are higher, which can lead to higher reaction rates.

Because these enzymes appear to also be more stable to denaturing forces other than heat, proteases from extreme thermophiles should be amenable to use in organic solvents for synthetic reactions. Many proteases have been purified and characterized from extreme thermophiles. From an evolutionary viewpoint, these proteases are of interest because they presumably predate those found in less thermophilic cells and organisms in terms of structure and function. Some of these proteases appear to be novel with no previously known mesophilic counterparts, whereas others represent more stable forms of known proteases. Proteolytic activities produced by the hyperthermophilic archaeon *P. furiosus* have been the most extensively studied among the extreme thermophiles. This organism appears to produce a number of proteases, which is consistent with its excellent growth on peptide- based media. The properties of several proteases from this organism are discussed below.

Protease I (PfpI) of P. furiosus

P. furiosus protease I (PfpI) is a major intracellular proteases in *P. furiosus*. Previously, PfpI was named S66 because, after heating at 95°C in 1% SDS for 24 hours, it migrates to a molecular mass corresponding to 66 kDa on a denaturing gel. It was subsequently found to occur in at least two active forms (hexamer and trimer composed of 18.8-kDa subunits) in vitro and appears to have at least one active larger form. The calculated pI (from amino acid sequence data) of the 18.8-kDa subunit is 6.1, whereas the experimentally determined pIs of the hexamer and trimer are 6.1 and

3.7 to 3.9, respectively. If all asparagine and glutamine residues are changed to their corresponding acidic residues, the pI of the monomer is calculated to be 4.8. Thus, at high temperatures, it is possible that the protein is deamidated, especially in its trimeric form, explaining in part the difference from the measured and predicted pI values. Upon long exposures to heat (24 to 48 hours), the predominant band on an overlay gel containing gelatin is in fact the smaller form. PfpI is specific toward basic and bulky, hydrophobic P1 amino acid residues in peptide substrates and most active toward N-succinylalanine-alanine-phenylalanine-7-amido-4-methylcoumarin (AAF-MCA).

Large proteins, such as azocasein and gelatin, are also degraded by PfpI, although specific cleavage sites in these proteins have not yet been identified. The gene for PfpI (*pfpI*) has been expressed in *E. coli* using the T7 promoter and infection with λ phage CE6. Vector constructs are often unstable, and toxicity has been a problem in producing recombinant PfpI. Expression levels were highest (1 mg per 100 mL culture) when a histidine tag was used to make a fusion protein, which could then purified using an affinity column. The amino acid sequence of PfpI was found to be highly homologous to putative proteins of unknown function contained in representatives of all three domains of life (e.g., *E. coli* (bacteria), *M. jannaschii* (archaea), *Arabidopsis thaliana* (eukarya)). However, the role of this protease in *P. furiosus* or in other cells and organisms is not clear at this time.

Enzyme pyrolysin of P. furiosus

Pyrolysin is a cell-envelope-associated protease that can be purified from *P. furiosus* membrane fractions. Two versions of this enzyme, one based on a 150-kDa subunit and one that was apparently cleaved by carboxy-terminal autoproteolysis, have been reported. The *pls* gene encoding the subunit has been cloned and sequenced. The gene corresponds to a 4194-bp open reading frame, containing a prepro sequence, whereas the mature protein is encoded by a 1249-bp region. The N-terminus shows high homology to subtilisin-like serine proteases, whereas multiple glycosylation sites are proposed at the carboxy terminus. Pyrolysin shows highest homology at the amino acid level with eucaryl tripeptidyl peptidases (TPP), although it has different cleavage specificities; TPPs are typically exopeptidases. Pyrolysin appears to be a version of the surface protein assembly (designated tetrabrachion) identified in another hyperthermophile, *Staphylothermus marinus*, which contains a protease with trypsin- and chymotrypsin-like activities.

Eukaryotic proteasome

Several thermophilic archaea contain a version of the eukaryotic proteasome. *P. furiosus* produces a version of the archaeal proteasome that appears to be composed of α (25 kDa) and β (22 kDa) subunits with an overall M_r of 640 kDa ($\alpha_{14}\beta_{14}$). The *P. furiosus* proteasome has a temperature optimum of 95°C, and preliminary biochemical analysis showed that methionine was preferred in the P1 position for hydrolysis of short peptides, although aromatic residues were also cleaved. Because the purification fold of the proteasome from *P. furiosus* cell extracts was approximately 16, it appears that it is a major cytosolic protease. Studies to assess its physiological role are under way. Unlike PfpI, no evidence for smaller active forms of the proteasome in *P. furiosus* was detected.

Proteases of Pyrococcal species

There have been reports of other proteases produced by pyrococcal species. A prolyl endopeptidase was identified in *P. furiosus,* the gene for which was cloned and expressed in *E. coli.* The recombinant enzyme was found to be optimally active between 85 and 90°C. The gene encoding the enzyme corresponds to a polypeptide of 70 kDa, which was consistent with SDS-PAGE analysis of the recombinant version. A thiol protease from *Pyrococcus* strain KOD1 was isolated and characterized. This protease, one of at least three extracellular proteases produced by this strain, had a molecular mass of 44 kDa and was optimally active at pH 7.0 and 110°C. Its threefold activation in the presence of 8 mM cysteine led to its classification as a thiol protease.

Enzyme Hydrolases from Extreme Thermophiles

The hydrolysis of natural polymers, such as cellulose, hemicellulose, and starch, by glycosyl hydrolases is the focus of a significant portion of all current commercial uses of enzymes as biocatalysts. It is desirable to hydrolyze natural polymers with these enzymes at elevated temperatures because of increased substrate solubility and lowered solution viscosity, which improves enzyme access to susceptible bonds. It is somewhat surprising that the source of these thermostable enzymes, extreme thermophiles, are thought to be primitive life forms and presumably predate the emergence of higher life forms, such as plants, which produce most naturally occurring polysaccharides.

Nonetheless, these organisms have proved to be a fertile source of an array of glycosyl hydrolases, many of which have potential as commercial enzymes. Classification schemes for glycosyl hydrolases have advanced considerably in recent years. Because these enzymes are often capable of hydrolyzing a range of glycosidic bonds, classification in terms of substrate specificity can be confusing. Amino acid sequence information has been used to fortify traditional classifications schemes so that glycosyl hydrolases have now been arranged in over 60 families. As three-dimensional structures and detailed kinetic mechanisms are resolved for enzymes within these families, a comprehensive basis for distinguishing among various glycosyl hydrolases can be developed.

Enzymes from heterotrophic thermophiles

Starch, glycogen, pullulan, and maltose have proved to be good growth substrates for a variety of heterotrophic extreme thermophiles. These microorganisms contain a variety of *exo*-acting and *endo*-acting α-specific glycosyl hydrolases that are apparently used to hydrolyze complex carbohydrates to glucose. *P. furiosus,* for example, produces an extracellular amylopullulanase that can hydrolyze either α-1,4 or α-1,6 linkages, both intracellular and extracellular α-amylases, and an α-glucosidase.

Versions of these enzymes have been found in many other heterotrophic extreme thermophiles. In fact, glycosyl hydrolases may be more widely distributed among the extreme thermophiles than previously thought. The genome sequence of *M. jannaschii* indicates that this methanogen also produces an intracellular α-amylase and glucoamylase, although this organism has not been reported to grow on starch-based carbon sources.

Hydrolysis of polysaccharides

Because of the natural abundance of cellulose and hemicellulose, enzyme systems that are capable of hydrolyzing these polysaccharides to simple sugars have long been sought to facilitate recycling of renewable resources. The structure and molecular composition of β-linked natural biopolymers are highly variable, often necessitating the use mixtures of enzymes to get substantial hydrolysis. On the other hand, unsubstituted cellulose and hemicellulose (e.g., xylan and mannan) are sparingly soluble in water, and enzymes capable of hydrolyzing these compounds often contain binding domains used to pry open the structure for enzymatic access. *Exo*-acting, β-specific glycosyl hydrolases have been found among extremely thermophilic bacteria and archaea. *P. furiosus* produces an intracellular β-glucosidase and an intracellular β-mannosidase; although the former appears to be induced by cellobiose in the growth media and has broad substrate specificity, the function of the latter is not clear.

β-glucosidases, β-galactosidases, β-xylosidases, and cellobiohydrolases have also been identified in other extreme thermophiles, which suggests that biopolymeric substrates associated with these specificities are present in geothermal environments. The wide occurrence of *exo*-acting, β-specific glycosyl hydrolases among the extreme thermophilic archaea suggests that *endo*-acting enzymes capable of hydrolyzing β-linked polysaccharides are also present. However, no such *endo*-acting, β-specific glycosyl hydrolases have been identified to this point in the extremely thermophilic archaea. Although some evidence for *endo*-acting hemicellulase activity toward xylan-based polysaccharides among the extremely thermophilic archaea has been mentioned, a xylanase corresponding to this activity has not been isolated. The hyperthermophilic bacterial genus *Thermotoga*, however, has been a rich source of *endo*-acting, β-specific enzymes. Glucanases, xylanases, and mannanases have all been purified and characterized from various *Thermotoga* species.

FEATURES OF THERMOPHILIC ENZYMES

Although there has been much progress in isolating and characterizing the biochemical features of extremely thermophilic enzymes, little is yet known about the intrinsic basis for their astounding thermostability. It was only a few years ago that the amino acid sequences of these enzymes were eagerly awaited, because it was anticipated that they could be used as templates for the design of stable forms of industrially important, but thermally labile, enzymes. However, as more and more genes encoding hyperthermophilic enzymes have become available and their amino acid translations scrutinized, it is evident that the present sophistication of protein chemistry provides little encouragement that strategies for protein stabilization in a generic sense are at hand. In fact, three-dimensional structures are now available for four hyperthermophilic proteins but these have not yielded specific mechanisms for stabilization.

At present, it appears that the most fruitful approach to obtaining extremely stable proteins is to isolate them from extreme thermophiles or to express the genes encoding these proteins in mesophilic hosts. But even in these cases, it is important to examine both intrinsic and extrinsic factors to ensure that the recombinant protein so produced resembles its natural counterpart. Much has been done to examine intrinsic factors that lead to the thermostabilization of hyperthermophilic enzymes. However, much less has been done to examine extrinsic factors, such as small molecules and other proteins present in the microenvironment in which the protein folds. By mimicking the intracellular environment encountered by these enzymes, their in vitro stability and function might be optimized. In addition, such conditions may be useful in improving the stability of less thermophilic proteins, particularly those that are used industrially where their stability is a limiting factor.

Effect of Intrinsic Factors

A number of thermostable enzymes from extremely thermophilic microorganisms have been discovered during the past decade. In most cases, it is not unusual to find the $t_{1/2}$ of these enzymes to be on the order of a day or more at temperatures near or even above 100°C. Just as is the case with mesophilic organisms, enzymes from a particular hyperthermophile vary in their catalytic efficiency (k_{cat}/K_m), T_{opt}, and thermostability because of factors presumably related to their physiological roles. However, the intrinsic bases for protein thermostabilization at high temperatures are no better understood than the bases for stabilization at moderate conditions. No one feature seems to underlie the fact that proteins exist in stable conformations as a result of a delicate balance of large, opposing enthalpic and entropic forces. Additional stabilization arises from the efficient packing of amino acid residues, optimization of charged interactions within the structure, and minimization of exposure of hydrophobic elements to the solvent. This all implies a degree of cooperativity within the protein structure, which is difficult to attribute in a quantitative way to its component parts.

Effect of Extrinsic Factors

Extrinsic factors clearly play an important role in protein stabilization. Environmental conditions such as temperature and pressure have a universal effect on the collection of proteins within a given cell, but other extrinsic factors that influence protein thermostability are more specific. Some proteins benefit from the presence of critical concentrations of certain cations, whereas others function best and are the most stable in concert with coenzymes and cofactors, such as ATP. It is difficult to be definitive about how the in vitro stability of an intracellular protein relates to its in vivo stability, because the latter can be influenced by so many intracellular factors. This assessment may be especially difficult for extremely thermophilic proteins because in vitro, they are subject to an array of potentially damaging chemical reactions, involving covalent modifications to thermally labile amino acid residues, such as deamidation and oxidation. Protection against such destabilizing forces is presumably provided in the intracellular environment and may or may not be important in vitro. There are ways that protein stability can be influenced in vitro. Of course, some conditions that optimize enzyme activity can be readily determined, such as pH, ionic strength, temperature and enzyme concentration. Other factors that may be important in vivo are less easily reproduced. These include the presence of compatible solutes or molecular chaperones. Each of these will be discussed with regard to extremely thermophilic proteins.

Thermoprotection

Although enzymes purified from extreme thermophiles are intrinsically more thermostable than those purified from their less thermophilic counterparts, the intracellular environments of extreme thermophiles thus far examined contain high levels of certain unusual solutes. These compounds are thought to play a role in the thermoprotection of at least some of their constituent biomolecules in vivo. High concentrations of a novel phosphodiester, di-*myo*-inositol-1,1′-phosphate (DIP), have been found in a number of hyperthermophiles, and the potassium salt of this compound has been shown to offer a significant degree of thermoprotection to at least one hyperthermophilic enzyme in vitro. The mechanisms for protein stabilization and the pathways used to synthesize DIP have not been determined, nor have the factors that influence its intracellular production.

These studies are necessary to understand the intrinsic basis for the function of organisms at extremely high temperatures as well as to determine factors responsible for the stabilization of proteins in a general sense. Scholz et al. reported that the S°-dependent hyperthermophile *Pyrococcus woesii* (T_{opt}~100°C) contains high concentrations (~0.6 M) of DIP, and Ciulla et al. subsequently found that the hyperthermophilic methanogen *Methanococcus igneus* (T_{opt}~80°C) also produced DIP when grown at temperatures above 80°C or when exposed to high external salt concentrations. They also showed that DIP is not used as a metabolic intermediate.

Martins and Santos determined the intracellular concentration of DIP to be ~60 mM when *Pyrococcus furiosus* (T_{opt} 100°C) is grown at 95°C, but this increases to above 1.0 M at supraoptimal temperatures. In addition, they found that at high salinity, the organism accumulated the unusual sugar derivative *β*-mannosyl glycerate (~30 mM). In contrast to DIP, the latter compound had previously been found in some moderately thermophilic bacteria, suggesting that it may play a role in both osmolysis and thermal protection in these organisms. Moreover, the extraordinary increase in DIP concentrations in *P. furiosus* at the higher growth temperatures strongly suggests that it plays some role in thermoprotection. Accordingly, Hensel and Jakob showed that a crude preparation of DIP (~115 mM) dramatically increased the thermal stability of glyceraldehyde-3-phosphate dehydrogenase from *P. woesii* (the time for a 50% activity loss at 105°C increased by 10-fold). They also showed that compounds related in structure to DIP, such as citrate and phosphate, stabilized two hyperthermophilic enzymes.

Folding process of chaperones

Although many proteins have been shown to fold spontaneously into their native "correct" structures in vitro, it is also clear that molecular chaperones often assist in the folding process in vitro and, presumably, in vivo. In one sense, the temperatures at which extreme thermophiles grow would seem to create a perpetual situation of thermal stress, yet these organisms respond to heat shock in many of the same ways as less thermophilic microorganisms. Members of the extremely thermophilic archaea have been shown to produce heat shock proteins (HSPs) at supraoptimal temperatures. Kagawa et al. showed that *Sulfolobus shibatae* produces an HSP60-like protein, similar to the one identified in bacteria, that is a double-ring complex formed from two subunits.

Similar structures have been isolated from another hyperthermophilic archaeon, *Pyrodictium occultum* and from the moderately thermophilic archaeon *Thermoplasma acidophilum*. All these proteins appear to have ATPase activity and have been shown to be involved in the folding process in vitro. The genes for the archaeal HSP60s that have been sequenced show that they are closely related to a family of eukaryotic proteins termed TCP1s. Although the response to thermal stress in extreme thermophiles thus far appears to be similar to the stress response of more conventional organisms at lower temperatures, the degree to which molecular chaperones are involved in protein folding in vivo is not clear.

One general characteristic that has been proposed that differentiates extremely thermophilic proteins from others is increased hydrophobicity. If so, this may mean that the tendency for extremely thermophilic proteins to aggregate during the folding process is enhanced. Thus, molecular chaperones that can bind to the nascent protein as it folds may take on added importance relative to mesophilic proteins. Aggregation will be especially problematic when high protein concentrations are present and no means to prevent collisions between interactive protein surfaces exist. Hence, approaches that would prevent aggregation of extremely thermophilic proteins through the addition of extrinsic factors are highly desirable, especially in systems where high levels of overexpression are needed.

Cloning of Genomic DNA

There have been several literature reports of the successful cloning and expression of the genes encoding extremely thermophilic proteins in foreign hosts. In addition, expression cloning of genomic DNA from hyperthermophiles, coupled with robotic screening for enzyme activities of interest, has led to the discovery of a number of thermostable enzymes. This approach, however, says little about what genes were not successfully cloned and expressed. In fact, only in a limited number of cases has the native protein been characterized and the recombinant protein purified. For example, only DNA polymerase, glyceraldehyde-3-phosphate dehydrogenase, glutaredoxin, and β-glucosidase from *P. furiosus* have been shown, thus far, to have comparable thermal stability and molecular and catalytic properties in their native and recombinant forms. In contrast, native glutamate dehydrogenase (GDH) from *P. furiosus* has a half-life at 100°C of about 10 h, whereas the recombinant form has a half-life of minutes at 100°C, suggesting that the protein is incompletely folded or not assembled.

The latter was supported by the fact that the native enzyme is a homohexamer (α_6), whereas the recombinant form was comprised of inactive monomers. Interestingly, some conversion of the latter into the active hexameric form was accomplished by heating (85°C for 15 min). The opposite situation appears to exist with the α-amylase of *P. furiosus*, as the recombinant form of this enzyme, although not purified, was of higher molecular weight than the native form, suggesting some type of aggregation had occurred. Last, the recombinant form of *P. furiosus* citrate synthase was reported to have a half-life of only 1 min at 100°C. Unfortunately, the native enzyme has not been purified but one could anticipate that it would be much more stable than that produced by *E. coli*. Clearly, some recombinant extremely thermophilic enzymes are as stable as their native counterparts, but it is also apparent that this is not a general phenomenon.

APPLICATION OF THERMOSTABLE ENZYMES

The increasing variety of enzymes purified from extreme thermophiles supports the prospect of their future use in industrial applications. In some situations, existing enzyme-based approaches could potentially be improved by the use of substantially more thermostable versions of the same enzyme type. This presumes that the increase in enzyme stability achieved with an extremely thermophilic enzyme is not offset by a loss of enzyme activity at the desired operating temperature; this may or may not be the case.

The fact that extremely thermophilic enzymes tend to be less active at ambient temperatures may be desirable in some instances, especially in cases where it is necessary to have strict control over the timing or extent of the reaction. In other instances, strategic opportunities that make use of improved biocatalyst stability and also advantageously use higher operating temperatures are desirable. Probably the most intriguing aspect is the use of extremely thermophilic enzymes to catalyze reactions that otherwise were not amenable to biocatalysis. Examples of these cases are presented below.

Revolutionary Approach

The revolution in molecular biology has to a significant extent relied on the ability to amplify small amounts of DNA to levels that facilitate cloning, sequencing, and expression of particular genes of interest. This can be accomplished through the use of thermostable DNA polymerases that are produced by thermophilic microorganisms. To be effective in catalyzing the DNA polymerization reaction, the polymerase must be able to withstand repeated thermal swings from low to high temperatures so that the DNA template can denature to be copied and then reanneal before the next cycle.

Although the DNA polymerase from a moderately thermophilic bacterium, *Thermus aquaticus,* was initially used in PCR applications, a host of polymerases from bacterial and archaeal sources are now available. Recent advances include the ability to replicate long sequences (20 to 40 kb) by adding to the reaction mixture an additional DNA polymerase with proofreading (3′ to 5′ *exo*-nuclease activity) capability to repair mistakes that might otherwise terminate the replication process. PCR is an example of the case in which protein thermostability is used strategically because thermally labile polymerases would not survive the repeated exposures to high temperatures, thus requiring addition of enzyme at each cycle.

Role of Isomerase Enzymes

The biocatalyzed conversion of glucose to fructose for the production of high-fructose corn syrup (HFCS) by xylose (glucose) isomerase represents the largest existing industrial application of an immobilized enzyme. Because of enzyme thermostability limitations, this process is currently operated at approximately 60°C, at which temperature the equilibrium conditions limit the HFCS concentration of fructose to a maximum of 40 to 42%. Because 55% fructose concentration is desired for sweetening properties, a chromatographic separation step is subsequently used to enrich the mixture to this level by the removal of unreacted glucose. However, because the equilibrium yield of fructose increases with temperature (55% fructose estimated at 95°C), thermostable versions of glucose isomerase (GI) that could be used at higher temperatures are desirable to minimize the need for the glucose removal step. Brown et al. reported the presence of xylose isomerase in members of the hyperthermophilic eubacterial genus *Thermotoga*. For example, *T. neapolitana* produces a version of GI with an optimal temperature of at least 95°C, which was also found to be highly active over a broad pH range. Compared to many other GIs that have been used commercially, the *T. neapolitana* GI possesses a 50 amino acid insert at the N-terminus, the function of which is unknown.

The catalytic efficiency of this enzyme was superior to those determined for other GIs from less thermophilic sources at their respective optimal temperature ranges. The GI from *T. neapolitana* again illustrates the potential strategic benefits that arise from the availability of thermostable versions of enzyme used in current industrial applications. By operating at elevated temperatures, the separation step needed for fructose enrichment can be minimized or eliminated because of the improved reaction yield. However, many other issues must be resolved before this enzyme can be considered as a replacement for the one used in existing processes. For example, consideration must be given to the levels of overexpression that can be attained in a foreign host, the impact of and approach used for immobilization, problems with potential side reactions of sugars at high temperatures, and regulatory approvals that might be required in the production of a substance for human consumption.

Hydrolysis of natural biopolymers

Interest in renewable resources has driven efforts to identify enzyme systems capable of hydrolyzing natural biopolymers, such as cellulose and hemicelluse, to more readily usable saccharides. This may be done in a nonspecific way so that the biopolymer is extensively hydrolyzed to simple sugars or in a specific way so that the properties of the biopolymer are systematically modified. As mentioned earlier, solutions containing natural polymers are less viscous at high temperatures compared to ambient conditions so that enzyme access to specific sites for hydrolysis is facilitated. Therefore, natural polymers may be hydrolyzed more readily by extremely thermophilic enzymes.

Additionally, because these enzymes are typically stable at temperatures much lower than their optimal temperatures for hydrolysis, these hydrolases may retain their function at suboptimal temperatures for extended periods of time. In the stimulation of oil and gas wells by hydraulic fracturing, aqueous solutions of natural polymers such as guar gum (a molecule consisting of a mannan backbone substituted with galactose through α-1,6 linkages) are used to transport particles to the site of the fracture. The objective is to prop open fissures in the bedrock perpendicular to well bore and, hence, promote the flow of oil or gas to the point where it is readily recovered. The viscous properties of the guar gum solution used advantageously to transport the particles become problematic once the particles are in place. Thus, various methods, including the use of hemicellulases, are used to hydrolyze the guar to reduce viscosity.

The high temperatures encountered in deep wells reduce the effectiveness of conventional enzyme systems but create an opportunity for extremely thermophilic enzymes that can withstand temperatures above 100°C. An additional advantage of using thermostable enzymes is their lower activity at ambient conditions that otherwise can lead to premature hydrolysis of the guar. Such an enzyme system, consisting of β-mannanase, α-galactosidase, and β-mannosidase from the hyperthermophile *T. neapolitana,* has been shown to work to hydrolyze guar at high temperatures but much less so at moderate temperatures.

Further Scope

To date, extremely thermophilic enzymes have been considered mostly as replacement enzymes to improve existing bioprocesses. However, their high degree of stability to thermal and other denaturing forces makes them candidates for catalyzing reactions not previously amenable to biocatalysis. This includes reactions conducted in non- aqueous environments such as organic solvents and supercritical fluids. The fact that their operational ranges are consistent with conditions used in chemical processing environments presents some intriguing possibilities for biocatalysis. It remains to be seen if these enzymes will be used for these purposes.

Certainly, any insights gained into the intrinsic or extrinsic basis for enzyme stability that can be translated into bioprocessing improvements for enzymes from any source would be a welcome development. Extremely thermophilic microorganisms have proved to be a rich source of thermostable enzymes. The diversity of such enzymes and the associated biochemical properties are only beginning to be understood. The discovery process has made good use of the tools of molecular biology, and these will also be critical for producing sufficient enzymes for various applications. The promise that these enzymes hold for the expansion of uses of biocatalysis is considerable but will require input from those outside of this field to find the most suitable applications. Nonetheless, biocatalysis has added a new dimension with the increased availability of extremely thermostable enzymes, which bodes well for the development of more efficient and effective bioprocesses.

CONCLUDING REMARKS

A number of market trends and consumer needs are currently driving research on new enzyme technologies for detergent application. Two critical changes occurring in the marketplace today that are expected to have a major impact on future enzyme technologies are the continued move to cooler wash temperatures and fundamental changes in washing machine design leading to reduced water consumption. In the United States, there has been a slow but steady decline in wash temperatures since the early 1980s. This trend has been matched in Europe where more loads are being washed at 40°C. Asia and Latin America have historically been cold-water-wash geographies. In order to deliver consumer valued benefits under these conditions, enzymes with improved activity at low temperatures will be required, particularly for those laundry stains and soils that rely on thermal energy for removal, such as mechanical or food grease and oil.

In response to anticipated legislation on energy consumption, U.S. washing machines will become more energy efficient, work at lower temperatures, and require less water per load. The net effect will be a much higher ratio of fabric and soil to wash liquor. These new machine designs will likely require enzymes with increased stability to other detergent ingredients in low water environments. In addition, changes in enzyme raw material formulation may be necessary for improved solubility.

There is a trend in the detergent industry toward development of technologies that can deliver fabric care benefits. This ranges from enzymes that can maintain the color of a garment over successive wash cycles to systems capable of reducing the wear associated with the laundry process. As a result, there is increased emphasis on identification and development of enzymes that can extend the useful wear life of fabrics. Cellulases are probably the best example of this for cotton garments. Recently, proteases have been considered as a way to deliver comparable benefits to washable wools and silks.

Finally, there is growing concern among consumers about communicable disease and the ability of laundry and cleaning products to eliminate this risk. The trend toward colder wash temperatures is expected to decrease the efficacy of traditional detergent ingredients, especially oxygen bleaches, toward microbial kill. Recent patent literature has cited the use of *endo*-glycosidases alone for removal of bacteria from fabric and, in combination with traditional antimicrobial actives, for improved microbial kill during the wash process. Antimicrobial detergent compositions in which the active is a peroxidase or laccase have also been disclosed. Current trends in the development of new detergent enzymes are focused on improved activity in cold water, increased stability and compatibility with other detergent actives, and the identification of new enzymes that can deliver new benefits such as fabric care and santization.

16

RECOMBINANT COAGULATION PRODUCTS

For the past 40 years, therapeutic preparations of plasma coagulation factors have been derived from plasma. The plasma fractionation industry developed during and after the second World War and was often a partnership between a governmental or not-for-profit organization that obtained the blood from volunteer donors and the nascent commercial plasma industry. Since the beginning of blood banking and the production of plasma products, the regulatory oversight has been done by the national organizations responsible for biologics control. The authority and the procedures of these organizations are different than those for pharmaceutical regulation. Regulation of the plasma products in the United States was based on the Public Health Service Act of 1902. This act was promulgated to regulate the production of vaccines and therapeutic serums. The latter phrase was construed to include blood and plasma products when this industry developed later on.

Compliance was monitored by the Public Health Service Laboratories, later to become the National Institutes of Health. In 1972, the regulation of U.S. biologics was moved to the Food and Drug Administration (FDA). One of the major efforts in any of the biological control agencies was the development of standards and assays for impure and poorly characterized natural products. This major component of biologics regulation is seen in the name of the former U.S. control organization, Division of Biologics Standards at NIH, and in the U.K., the National Institute of Biological Standards and Control (NIBSC). Although regulation has properly belonged to national regulatory bodies, there are other inputs to the regulatory process from supranational organizations, such as the European Community and the International Conference on Harmonisation (ICH). In the United States, the FDA requires that an advisory committee votes for approval of licensure.

These committees include representatives of advocacy groups that have voting membership in the FDA blood product advisory committee (BPAC). The recommendations of professional organizations are also quoted in regulatory documents or regulatory decisions. The Scientific and Standardization Committee of the International Society of Thrombosis and Haemostasis has provided recommendations for the hemophilias and thrombotic disease. In addition, civil liability law has an impact on all therapeutic producers, particularly in the United States. Regulatory documents of the FDA are published in the Code of Federal Regulations (CFR), and updated guidance documents can be obtained from the FDA World Wide Web site. European documents are found in the European Pharmacopeia, and guidance documents are obtained from the Committee on Proprietary Medicinal Products (CPMP).

AGGLUTINATION REACTION

Blood coagulation results from a series of proteolytic reactions that first activate a procoagulant protein, and then the resulting coagulant is inactivated either proteolytically or by interaction with an inhibitor. Although presented as the coagulation cascade, implying amplification, the complexity results in many points where there is sensitive biochemical control. Individual components are identified by roman numerals, and the activated form is indicated by a lowercase "a." Factor IX, the defective protein in hemophilia B, is the proenzyme of a serine protease that is cleaved to activated factor IX (IXa). Factor VIII, the defective or deficient protein in hemophilia A, is a protein cofactor that circulates bound to a large polymeric protein, von Willebrand's factor, which is defective in von Willebrand's disease. In the bound state the factor VIII remains in the plasma; in the presence of small amounts of thrombin, it is released from the von Willebrand factor and activated. In the presence of factors IXa and VIIIa there is activation of the downstream enzymes in the coagulation cascade. Absence or defects of either factor IX or factor VIII leads to inability to form fibrin clots. Absence of the carrier protein, von Willebrand's factor, leads to low circulating levels of factor VIII as well as decreased recruitment of platelets to damaged endothelium and leads to a hemorrhagic state. The factor VIII gene product is ~250,000 kDa derived from a gene of 186 kb in 26 exons. The gene product is glycosylated intracellularly, and the tyrosine at position 1680 must be sulfated. During passage through the Golgi apparatus there is excision of much of the central portion of the protein (B-domain), resulting in a heavy and light chain held together by Ca^{2+}. This is the form found in the circulation.

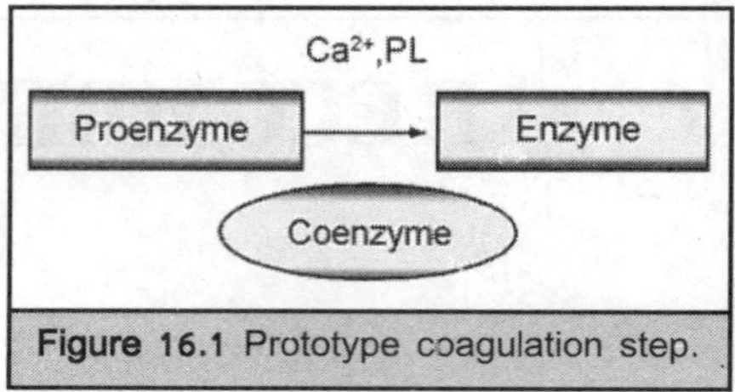

Figure 16.1 Prototype coagulation step.

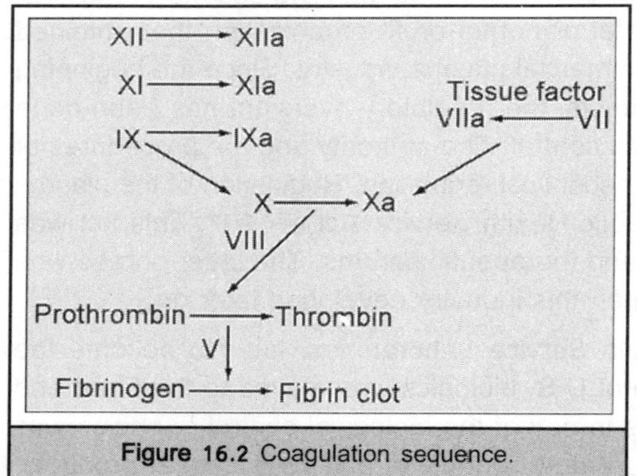

Figure 16.2 Coagulation sequence.

Hereditary Bleeder's Disease

Hemophilia A and hemophilia B are the result of mutation of genes located on the X chromosome. As such they are sex-linked, and sons of carrier mothers have a 50% likelihood of carrying the defective gene and presenting with the clinical phenotype of abnormal hemorrhage. This leads to crippling and death in those severely afflicted. Treatment of these diseases began with plasma and evolved to more purified and concentrated plasma products. Adequate treatment for those with severe hemophilia involves lifetime intravenous administration given approximately weekly. Although the treatment is effective, there are two major safety issues with plasma products: blood-borne viral infections and the development of antibodies that inhibit the therapeutic effect. Hemophilia A and B were one of the early targets of the biotechnology industry. The patient population is small, with 16,000 patients with hemophilia A (deficiency of factor VIII) and hemophilia B (deficiency of coagulation factor IX) in the United States. Of these, about 8,000 have severe disease and need frequent treatment.

Treatment with plasma products temporarily improved the well-being of this patient population until the HIV epidemic; 6,600 U.S. patients were infected. Patients with severe hemophilia A need frequent and lifelong treatment; this group uses ~50,000 to 100,000 factor VIII units annually. Viral infection from plasma products has been the major problem in the treatment of the congenital bleeding disorders. Infection with most blood-borne viruses has been minimized during the past 10 years by viral inactivation. The most common procedure is the destruction of the lipid-coated viruses (HIV, hepatitis B and C) by extraction with a solvent-detergent mixture. Another procedure is the use of heating either in solution or in the dry state.

Many regulatory authorities are demanding two viral inactivation and removal steps for recombinant and plasma-derived factor VIII and factor IX. The second problem for the treatment of hemophilia A is that 15% of patients will develop inhibiting antibodies to factor VIII and become refractory to standard treatment. Although most patients who develop antibodies may be genetically predisposed, in other cases this is because of neoantigenicity induced in the factor VIII during production and viral inactivation. A third hemorrhagic disease, von Willebrand's disease, is the result of heterogeneous mutations of the gene for a polymeric carrier protein of factor VIII, von Willebrand's factor.

REGULATORY RECOMBINANT COAGULATION FACTORS

Statutory drug law is based on the principles of safety and efficacy. Regulatory oversight of pharmaceutical production is done in large part through good manufacturing practices (GMPs) and good laboratory practices (GLPs). This is a body of regulation and policy that can change rapidly. In addition, in the regulation of biologics there is a need for product consistency whether the product is derived from animals or from cells. The particular origin of biologics leads to special needs in the control of the starting materials, including cells, genes, etc. This is best exemplified in the regulations for the establishment of intensively characterized master cell banks for cell-derived products. This cell bank will be expanded in number under standard conditions to produce a consistent product over decades. The advent of the biotechnology products for coagulation deficiencies (i.e., cell-derived rather than plasma-derived coagulation factors) raised few unique regulatory issues.

Traditional biologics regulation had already dealt with cell-derived products, including the potentially dangerous live virus vaccines. A regulatory framework for the characterization and control of cell lines had already been constructed, and in 1984, a workshop sponsored by the Bureau of Biologics dealt with many of the residual issues. This was followed by a WHO report in 1987. This was followed by the development of *Points-to Consider* for products derived from mammalian cells and the EEC equivalent. The *Points-to-Consider* are not legal regulatory documents, but they inform those who are dealing with the agency of the current views of the regulatory authority. There is an effort to have as much commonality as possible between the international regulatory agencies, and extensive efforts to harmonize the guidelines are ongoing by the ICH. Of particular concern originally was the quality and quantity of residual DNA in the final product; a fear that has diminished with time. The FDA does not produce specific documents on coagulation or fibrinolytic products. In contrast, the European Community has specific and detailed regulatory documents for the coagulation products.

The basis for regulation of recombinant coagulation factors is similar to plasma-derived coagulation factors: the assessment of safety, efficacy and consistency. More particularly, there is intensive and extensive evaluation of the cell production systems, the control and standardization of the final product, and clinical effectiveness and safety. The main point of the original regulation for cell-derived coagulation factors was that the products should be identical with plasma-derived products. More recently, products with similar but not identical structures have been received by regulatory authorities with little additional concern. The GLPs and GMPs are in principle identical around the world; however, there can be significant differences in detail. There has been rapid change in pharmaceutical science during the past decade, including filling procedures, air and water systems, and analytical methods.

Approved production facilities of today may not meet GMPs in 5 years. In addition, regulatory policy can change without any formal notification or publication. For these reasons, it is important that the manufacturer maintain constant communication with the regulatory agency from inception to final approval. Many facilities for recombinant products will be designed and maintained as single product manufacturing areas. This will include raw materials, cell lines, expansion of cell lines, fermenters, purification equipment, and the filling line. It will only be for the postfilling operations that there will be contact with other products. Some useful products may have such small demand that a few lots will meet needs for several years. Such a case may be recombinant activated factor VII (VIIa) for use in hemophilics with inhibitors not responsive to other modes of therapy. In this case, the plant must have a validated procedure for producing a different product.

Points to Consider

All cells used for the production of recombinant coagulation factors must either meet the *Points-to-Consider* documents by the FDA or have information indicating that there is no compromise to the safety and efficacy of the product if some of the points are modified. The relevant *Points-to-Consider* are those involved with production of the master cell banks and the *Points-to-Consider* for products derived from mammalian cells. The European equivalents are CPMP *Note for Guidance: Production and Quality Control for Medicinal Products Derived by Recombinant DNA Technology* and the ICH documents on the analysis of the expression construct. The cells must not only synthesize the desired protein, but efforts must be made to simplify the purification process and minimize the presence of adventitious infectious agents by simplification of the growth medium. In particular, the use of fetal calf serum is avoided if at all possible to minimize a significant source of microbial and antigenic contamination. It is likely that regulatory agencies in the future will require more effort to develop culture systems free of serum or other animal protein. During the production time, there must be no evidence of either microbial contamination or of change in the cell or the cell product. Both the U.S. and European documents require characterization of the vector to demonstrate the stability of the inserted gene over the time of a production run or longer.

The development of PCR technology has allowed the characterization of cDNA and RNA. Frequent sampling and testing of the conditioned media during production and the final purified product are a requirement. The cells must be capable of appropriate post-translational modifications. In the case of factor IX, there must be carboxylation of the 10 to 11 glutamic acids so that the molecule has full function. This has been accomplished by inclusion of a gene for a protease to remove the propeptide and allow carboxylation of the glutamic acids at the NH_2 terminal of the factor IX molecule. Other significant post-translation modifications include the formation of a β-OH aspartate residue, sulfation, phosphorylation, and glycosylation. In addition, there must not be the production of any protease that will degrade or activate factor IX. A similar constraint is true for cells producing factor VIII. This is a heavily glycosylated molecule, and each production system could yield different products, some of which might be immunogenic. Although this is a concern, no such immunogenic derivatives have been found as yet in factor VIII produced by recombinant technology. Von Willebrand's factor must be produced in cells allowing polymerization of the monomer and secretion of these virus-sized protein molecules into the media. The cell culture system must also be able to sulfate a specific tyrosine residue to allow binding to factor VIII and thrombin activation of the factor VIII.

Purification of Coagulation Products

Purification of coagulation products revolves around three issues: the removal of foreign allergens, the removal of DNA, and the removal of adventitious agents originating from the host cell and the media. The issue of allergens is of importance because the recipients need lifelong therapy. In contrast to the older plasma products, the purification process using mouse immunoglobulin affinity columns may be a new hazard, and at this time, monoclonal antibody purification is the universal purification method for factor VIII. The monoclonal antibodies do not have to meet the full requirements for a therapeutic monoclonal antibody but must have been examined for murine viruses. There are no absolute requirements or regulations for protein purity within the U.S. or European regulations. However, the manufacturer is obligated to have satisfactory test systems to detect nontherapeutic proteins derived either from the cell source or the purification procedure (e.g., leached monoclonal antibodies). The ultimate test will be in recipients of the product.

The factor IX product being developed not only is grown in media without protein but the purification steps do not involve monoclonal antibodies. This strategy simplifies the purification process and is also deemed desirable in terms of the introduction of viruses either in tissue culture or in the final product. Although this is apparently the first coagulation product from cells grown in the absence of protein in the media and purified without the use of monoclonal antibodies, it may set a new standard. The same procedures for the removal and inactivation of viruses that are required for cell-derived products are required for the plasma products. Although the use of cell-derived products in contrast to plasma-derived products will only rarely contain viruses transmissible to humans, the newer knowledge of virology and the public perception demands extreme measures. The manufacturer must show that the purification methodology has the capability to remove or inactivate viruses of different types. However, the European guidelines are much more specific than the comparable U.S. documents. One of the first issues for cell derived products was the establishment of limits for DNA. This was set at 10 pg/dose for all products in the original FDA documents, but the current WHO level of 100 pg/dose seems to be the actual figure used by the FDA. Products used for the treatment of hemophilias are infused 1,000 times in a lifetime; however, there is no change in the dosage for products used a few times as opposed to 1,000 times in a lifetime. The requirements for DNA removal revolve around the theoretical ability of cell-derived DNA to be incorporated into somatic cells in the recipient.

Obtaining the Desired Products

The structure of the gene product must be extensively studied. This will include extensive peptide mapping, molecular weight analysis by electrophoretic assessment, and perhaps mass spectrometry. In some cases such as factor IX, the complete amino acid structure is feasible, whereas for factor VIII the total amino acid sequence is more onerous than useful in the light of the established gene structure. Disulfide bonding should be characterized and shown to be identical to plasma-derived coagulants. As noted earlier, there are specific post-translational modifications for given products important for their function, such as glutamyl carboxylation of factor IX, that should be characterized. The final product will include degradation products of the gene product. These should be identified, analyzed for their sequence, and discriminated from other protein contaminants. A specific issue for the coagulation factors is that they are in the appropriate state (i.e., nonactivated for factor VIII or IX). For von Willebrand's factor, the polymeric forms must be functional. As noted earlier, the circulating form of the coagulant proteins should be the proenzyme; factor IXa can induce thrombus formation. In the case of the cofactor, factor VIII, the uncleaved, nonactivated form is necessary; the presence of the activated form yields inaccurate potency measurements and is rapidly cleared. Thus, the purification system should remove the activated forms and be done in a manner that does not produce these forms.

Testing of Recombinant Products

In general, the preclinical testing of recombinant products is similar to plasma-derived products and includes routine pyrogen, safety, and potency testing. Specific and accurate potency measurements for labeling and dosing must be available.

Assessment of Potency

The assumed efficacy of the final product is assessed by the potency assay, the results of which yield the labeled value for treatment. The standard assay is a bioassay with intrinsic variability. The unit definition for coagulation products is the functional activity found in 1 mL of normal plasma containing a 10th volume of anticoagulant. This operational definition is translated into WHO standards for coagulants by the preparation of plasma or plasma fractions that originally were compared to normal plasmas. All WHO coagulation standards have been prepared by the NIBSC in the United Kingdom. These are calibrated in multicenter studies organized by the Scientific and

Standardization Committee of the International Society on Thrombosis and Haemostasis. New standards are compared to previous WHO standards and also to plasma standards. As both factor VIII and factor IX therapeutic products become more purified, there are inconsistencies with older standards. A further complication is the development of what are apparently more accurate quantitative potency tests based on better biochemical knowledge but that can show very substantial differences with the standard reference tests

The European Pharmacopia designates a reference methodology for the measurement of coagulation factors. In general, this seems to be the assay used by the control laboratory rather than that with which there is the most experience. At this writing, the reference methodology for factor VIII in the European Pharmacopiea is the chromogenic assay. In contrast, the FDA has no designated reference method for coagulation factors but does distribute national references. There is a consistent difference in the pharmacokinetics when assessed by different methods. The development of variant proteins such as the B-domain deleted factor VIII aggravates the difference seen between the different methods of measuring potency.

The specific activity of the B-domainless variant is reported as being substantially higher than plasma-derived factor VIII. The in vitro potency assays are approximately equivalent on a molar basis. Whether they are the same therapeutic quantity must be answered. A recombinant von Willebrand factor is in development. Von Willebrand's factor has several different functional attributes of importance to its efficacy. These include the support of platelet aggregation, binding to collagen, and the binding to factor VIII. At this time there is no consensus as to the appropriate potency test for a therapeutic product to treat von Willebrand's disease. The WHO standard for von Willebrand's factor is plasma and not a purified preparation.

Testing of Factor VIII in Animals

Animal models add useful information before clinical testing of the products. A dog model of hemophilia A has been used extensively in the preclinical testing of factor VIII. One of the most immediate questions was the comparison of the pharmacokinetics of recombinant factor VIII with the plasma product. If the proposed product was more rapidly removed from the circulation than the plasma product, this would raise serious doubts about the efficacy of the product. Experiments in the dog hemophilia model showed normal pharmacokinetics and rapid binding of factor VIII to canine von Willebrand's factor. Various approaches have been and are being assessed to determine abnormal immunogenicity with animal models: induction of immune tolerance in newborn animals in addition to laboratory comparison of the epitopes of the plasma and recombinant product. Thrombogenicity of the factor IX preparations has been assessed by both laboratory testing and assessment of the ability to form clots in animals. Although this concern arises from thrombosis associated with the use of older, less-purified products, there is still concern that even the purified products and recombinant factor IX can produce thrombosis.

Animal pharmacokinetics comparing plasma-derived and recombinant factor IX are a necessary prelude to clinical studies. Dog models of hemophilia B are available, but the animal pharmacokinetic studies are generally performed in smaller laboratory animals. Because there is endogenous factor IX coagulant activity in the samples, the plasmas are assayed for the quantity of human factor IX antigen present over time. Animal models for coagulation factors have been developed over the years for specific known adverse effects. In addition, animal testing is of value for identifying unsuspected effects.

Efficacy and Safety

Clinical trials are the final step in the assessment of the efficacy and safety of a new therapeutic product. The FDA regulations outline the information needed to begin human studies and the type of studies needed to establish safety and efficacy. European regulations are similar. Efficacy of treatment of the hemophilias has never been established in a placebo-controlled study because observational studies coupled with historical data clearly established the effectiveness originally of both factor VIII and factor IX replacement therapy for preventing or treating hemorrhagic episodes. This type of data was acceptable to the regulatory authorities around the world in the mid-60s and was consistent with the biologics definition of efficacy based on the pharmacologic and pharmacokinetic studies. Current policy would, however, require comparison with an approved efficacious product for the factor VIII and factor IX products.

The FDA does not prescribe an exact protocol but describes the general types of clinical trials that could give efficacy data. In contrast, the EEC guidelines describe in detail the clinical trials to be undertaken. Many of these details are derived from articles promulgated by a nonregulatory professional organization, the International Society on Thrombosis and Haemostasis (ISTH). In addition to establishing efficacy, clinical trials are designed to detect adverse events. The most common concern is for known or unknown infectious agents. Recommendations of the ISTH serve as guidelines for some of the European community policies. The original protocols for recombinant factor VIII started with pharmacokinetic trials in older patients with frequent need for treatment. The strategy was to keep these patients solely on recombinant factor VIII for several years. This would give ample opportunity to test any abnormal immunogenicity of the product. In addition, there would be interim pharmacokinetic studies as the most accurate way of assessing low-level inhibitors. A trial was also done in previously untreated patients (PUPs) independent of any request from a regulatory agency. This study yielded valuable results on the treatment of hemophilia in very young patients without the confounding of immunologic or viral infection by other products. This set a precedent for the evaluation of all future recombinant products and has been incorporated into the EEC requirements for all recombinant factor VIII and factor IX. This will probably also be required by the FDA in the future.

Although the immunogenicity of recombinant factor VIII and plasma factor VIII is a major concern, it is not clear what information comes from the routine PUP studies, because any excess immunogenicity should be observed in older patients. There are no specific European clinical guidelines for recombinant coagulation factors. There are the general clinical guidelines for the coagulation products, which specifically exclude recombinant products. Based on the historical knowledge and precedent, there is no requirement for a blinded efficacy trial. Efficacy assessment is based on the comparison of pharmacokinetics with an approved product. For factor VIII, the requirements are to study 12 pretreated (PTP) subjects over 1.5 days with eight sampling times. A minimum of three lots are to studied. These subjects should be maintained solely on the experimental product for a further 3 to 6 months and then have a repeat pharmacokinetic study. Before approval, a study of pharmacokinetics in 20 PUPs must be started using the same protocol as in the other patients.

There is a requirement for at least 10 surgical procedures in a minimum of 5 patients and clinical assessment of 30 immunocompetent patients after licensure. In addition, information should be submitted on 30 patients who have been treated on at least 10 days with at least 6 months of follow-up. The factor VIII inhibitor titer must be repeated every 3 months. After approval, there is a periodic update on the laboratory parameters of at least 50 patients treated for at least 2 years. With plasma-derived factor IX, the major adverse event other than viral disease transmission was thrombosis associated mostly with the earlier less-purified plasma fractions containing large amounts of the other vitamin K coagulants.

The clinical trials for a recombinant factor IX must test for evidence of in vivo coagulation with currently available tests for thrombin–antithrombin complex and prothrombin fragment 1.2. In addition, there will probably be required testing for the enzymatically active form of factor IX (IXa). Although the FDA does not specifically deal with this issue, this still must be addressed for U.S. licensure. The European guidance documents call for tests for in vivo activation of coagulation. PUP studies will be required by the FDA and by the European Community for factor VIII and factor IX products. The European community requirements for plasma-derived factor IX require pharmacokinetic studies in 10 subjects with 10 sampling times. Treatment with the same product should be maintained for 6 months, and repeat pharmacokinetic studies should be performed in 3 to 6 months.

Current published clinical experience with recombinant factor IX are limited to abstracts indicating pharmacokinetics and clinical effectiveness similar to plasma- derived factor IX. These studies are not inconsistent with the European guidelines and were accepted by the BPAC. Von Willebrand's factor is a new molecule to deal with, and there are no national regulatory comments addressing the therapeutic use of this particular protein. Although the pharmacokinetic analysis of the clearance of the coagulation activity of factors VIII and IX is accepted throughout the world, there is no accepted or validated functional assay for von Willebrand's factor. Should the functional assay be based on the ristocetin cofactor activity or the ability to bind to platelets or the content of multimers or the content of intact monomers or the ability to bind factor VIII? One can anticipate that the regulatory agencies will use the SSC/ISTH guidelines for appropriate clinical studies. The current FDA policy is to request comparative clinical data with another product, but there is no product licensed in the United States for the treatment of von Willebrand's disease.

Although von Willebrand's disease is a common genetic deficiency, the clinical phenotype is extremely variable, and for the most part, the hemorrhages are less severe and less frequent than in hemophilia A or B. This leads to difficulty in recruiting patients into a structured protocol. Heretofore, the clinical trials of von Willebrand's factor have been uncontrolled, and current treatment strategies are based on overall experience. The surrogate markers that have been used traditionally are the level of factor VIII, the shortening of the bleeding time, and the functional level of the von Willebrand factor as assessed by the ability to support platelet aggregation in the presence of ristocetin. Inhibitor development is an issue in von Willebrand's disease, and this will have to be assessed. With recombinant von Willebrand's factor, the factor VIII level will be endogenously regulated and will increase to peak levels 10 to 20 hours after infusion. This is in contrast to the current practice of infusing patients with plasma products containing von Willebrand's factor and factor VIII with instantaneous increase in the factor VIII level. Two recombinant factor VIII products have been licensed in the United States and in Europe. Recombinant factor IX is in clinical trials and the license was submitted to the FDA in 1996. Recombinant factor VIIa, a product proposed for use in factor VIII and IX patients with inhibitors, is being reviewed by the FDA (1997) and has been licensed in Europe and Canada for the treatment of patients who acquire inhibitors to factor VIII. A recombinant von Willebrand factor is undergoing preclinical development.

First Fibrinolytic Agent

The in vivo formation of blood clots is usually a process limited in time and space. The formation of thrombi, pathologic blood clots, are the result of subacute or chronic disease processes resulting in blockage of blood vessels. The development of fibrinolytic agents, enzyme that digest fibrin clots, was the first major advance in the treatment of thrombotic diseases in 50 years. The biologics used for this purpose are activators of the endogenous proenzyme, plasminogen. Tissue plasminogen activator (tPA) (alteplase) was the first fibrinolytic agent produced by recombinant technology.

Use of Roller Bottle

One technical issue was emphasized: the effect on product of altering culture conditions. Early production of tPA in roller bottles was used for the initial clinical investigation, including the initial pharmacokinetic studies. When there was change to fluid-phase culture, there was a significant decrease in the pharmacokinetics, leading to the need for an increase in the clinical dosage. It was originally proposed that this was due to the chain structure of the tPA but that has been ruled out. Despite investigation of the carbohydrates and other possible structural changes, the post-translational change resulting in the pharmacokinetic change is still a mystery.

GENE THERAPY

Gene therapy is the procedure in which a new gene is transplanted into the body of a host to improve its health. Recombinant proteins are produced by placing the desired gene in a cell system and producing the desired gene product in vitro. Ex vivo gene therapy is accomplished by transplanting cells containing the desired gene into the host; these cells will then propagate and produce the desired biologic effect. In vivo gene therapy injects a gene construct that will enter the cell of the host and produce the desired protein. Hemophilias A and B are prime targets for gene therapy. The ideal of a one-time or infrequent treatment as opposed to weekly intravenous injections is attractive, but despite substantial advances there are many hurdles to overcome. Gene therapy procedures are changing rapidly, and this is noted in the regulatory documents with the suggestion that there be constant contact between the regulators and developers of any product proposed for gene therapy. Many of the molecular biology issues are covered by the regulatory documents on recombinant products.

Vectors

Vectors and gene product are characterized in the same manner as for cells producing rDNA proteins. For smaller viral vectors this would include sequencing the entire viral sequence; for larger viral vectors this may not be feasible. The gene product should be completely characterized, however, because the vectors used for in vivo gene transfer will not be expressed in the same cell, and many of the post-translational changes may differ. As with other cell-derived products, the development and characterization of master cell banks will be required. In addition, there must be establishment of a vector bank, that is, the final product for most of the proposed vectors is the result of the interaction of a cell and the vector. The vector bank should be tested for the presence of adventitious agents. Stability of the host cell–vector system must be established. The most promising vectors under study for treatment of the hemophilias include replication of defective retroviral and adenoviral vectors, adeno-associated virus, and nonviral vectors. Retrovirus vectors have promise for in vitro exchange with a variety of cells, including fibroblasts, endothelium, and myoblasts. The nonviral vectors would seem to raise fewer issues albeit much lower in vivo expression. The strategy issues to be resolved include choice of vector and in vitro or in vivo gene transfer.

Regulatory Issues

There are different regulatory issues and pathways for the licensure of vectors used in vitro and in vivo. Products for in vivo transduction will be produced purified and packaged in the final form for administration to the patients in a facility meeting all the standards for biologics production. For vectors where the transduction will be done in vitro, the producing institution will be responsible for the production of the vector and will have to meet all GMPs; however, the actual product administered will be done after expansion of the patient's transduced cells and will be a service for an individual patient similar to treatment with blood. Current standards for the production of pharmaceutical products encourage the development of single product production facilities. However, the target group for gene therapy in the hemophilias is small, and only a few production runs may be needed if the hope of long-lasting or permanent effectiveness is achieved. This would lead to a multiuse production area with stringent validation of the facility after changeover from one product to another to prevent cross-contamination. There are no specific guidelines as yet for the degree of purification for a gene therapy product; rather, it is left to the producer to show adequate purification and justify the adequateness of the procedure.

The consistency of the purification method must be demonstrated. Published purification methods for viral vectors involves density gradient ultracentrifugation; it is unlikely that this will be implemented as a production method. Published purification procedures give little information on the degree of purity achieved. Because the desired outcome with gene therapy would be one-time or infrequent exposure to the agent, the presence of small amounts of cell-derived or process-derived protein is less of an issue than for recombinant proteins administered frequently. The fidelity of reproduction of the vector is an issue because an error in replication could yield a toxic gene product. Where possible the correct nucleotide sequence should be verified. Lot release testing suggested by the FDA includes:

1. Test for RNA or DNA content
2. Test for homogeneity of size and structure of RNA/DNA
3. Test for contamination with RNA or host DNA
4. Test for noninfectious viruses
5. Tests for toxic materials
6. Identity tests by restriction enzyme mapping
7. Sterility including testing for adventitious agents and replication competent viruses
8. Potency

The only test that would be unique for the coagulation products would be the potency test. It will be the responsibility of the manufacturer to design and validate a potency test during the development phase. It is likely that the standard coagulation tests of conditioned media produced by cells treated with a given amount of vector would be adequate for the validation of a lot. Whole animal methods could be developed but could be anticipated to be less quantitative and too time consuming to be useful for lot release.

Mutagenesis and Tumorigenicity

Viral mutagenesis and tumorigenicity are of concern even in replication of incompetent vectors because the genes could be inserted into chromosomes and disrupt growth regulation genes. Identification of the insertion mode and testing for replication competent viruses is recommended for the master cell bank, the working cell bank, and lot release testing on the adenoviral vector products. The FDA and EEC guidelines specifically are concerned with the ability of replication-incompetent viral genomes to form recombinants with cellular viral components.

Strategies suggested include separation of the structural and enzymatic genes, removal of complementary viral sequences from the packaging cell lines, and prevention of infection of the packaging cell line with wild-type viruses that could lead to the formation of replication-competent recombinant viruses. Specific issues with viral vectors are the pathology induced by the viral components. As many pathologic viral sequences as possible should be removed. The induction of virus antibodies preventing retreatment is an issue that can be modulated by minimization of the viral structure and perhaps regulation of the administered dose.

Preclinical Tests in Animals

Preclinical animal testing will be of importance not just for determining efficacy but for supporting safety and studying tissue distribution. All viral vectors will have to be examined by appropriate routes of administration and dose effects in the most sensitive animal models. Species will depend on sensitivity to particular viral toxicity. There is no requirement or indication that regulatory authorities will require primate studies. Studies are mandated of the vector distribution in animals. A specific issue will be whether there is uptake into germ-line cells. Although this can be studied in animals by study of ovarian and testicular cells, it is not clear whether this will be a reliable indicator of human cell or organ location. Although there is public and political concern about germ-line modification, the FDA does not have any specific prohibition of a product with germ-line modification. The organ of synthesis for both factor VIII and IX is the liver, and this is the site of uptake of the majority of adenoviral vectors administered intravenously. The cell or organ location is of importance to the final gene product; post-translation modification will be affected by the cell type transduced.

There is at this time, however, no set experimental protocol for such studies, and the manufacturer will be expected to use all appropriate tools and develop specific tools as necessary for a given product. Retroviral transduction and implantation of the ex vivo transduced cells should give an identifiable localized expression of the gene product. There may, however, be specific cell-cell interactions. Presence of functional clotting should ensure adequate in vivo function. Human factor IX and factor VIII have been maintained in the therapeutic range for up to a half a year in various animal models with various vectors. The most promising are the ex vivo retrovirus-transformed fibroblasts or muscle cells and the adenovirus vectors with in vivo transfer. As noted above, measurable shifts in clearance occur with recombinant products. Although the steady-state levels are measurable in both animals and humans, it will be difficult to measure clearance in patients, although this might be a regulatory issue.

A particular problem with factor VIII may be the need to have the gene product secrete directly into the plasma compartment. In the absence of the stabilizing effect of von Willebrand's factor, it is possible that factor VIII will be proteolyzed to an inactive form before reaching the plasma compartment where it functions. Persistence of the transgene depends on absence of antibody to the vector-infected cell or the gene product. The antibody response to the vector and the toxicity of the vector can be limited by introducing lower quantities of the viral DNA. More recent experience indicates that immunity to the adenovirus may be circumvented. Adverse effects of the vector and the implanted gene may be detected before patient exposure only in animal models, and at this time there have been no unexpected events. As expected, there are reports of liver damage with the use of adenoviral vectors, which can be minimized by dose adjustment. Adverse effects of the vector and the implanted gene may not be detected in an animal model.

Approved Issues

The initial decision to attempt gene therapy in hemophilia will be difficult and controversial and will be reviewed by regulatory groups, patient advocates, lawyers, and politicians. In the United States, any clinical use directly or indirectly involved with government funding formerly was approved by the NIH's Recombinant DNA Advisory Committee (RAC) and subsequently by the Office of Recombinant DNA Activities after approval of the originating institution's internal review board. The results of these reviews were placed in the public domain, and it was understood that all adverse effects would be open information. In addition, the RAC and the FDA often had parallel review processes for other projects. In 1996, the role of the RAC was made more limited, and the FDA assumed all regulatory review and approval responsibility. Certain principles in regard to informed consent have already been established by the RAC and the FDA. These include many of the usual points such as the trial goal and known risks.

In addition, there must be clear discussion of the methodology and the duration of the therapy and of follow-up. One specific point mentioned is the provision for autopsy. The patient must have a commitment and understanding of the trial and the importance of each patient's value to the study as well as the advantages he or she may achieve. The hemophilia patient who would benefit most would be the patient without access to adequate means of alternative treatment. Because 80% of the hemophilic population in the world does not have access to treatment, one can argue that this is a potential target group. Outside the United States, there has been an effort to treat human factor IX deficiency by gene therapy in China. A second group to be considered for early treatment with gene therapy might be patients with inhibitors (antibodies) to factor VIII. These patients are already undergoing frequent exposure to induce tolerance.

Treatment with gene therapy would simply be a modification of tolerance therapy, which could have great benefit for patients with inhibitors. Assessment of therapy would be more difficult if there is no measurable plasma factor VIII. Alternatives could be evidence of a decrease in the inhibitor titer, similar to that seen in classic tolerance regimens, or evidence that there is long-term production of gene product or mRNA. A third group would be those without adequate venous access for administration of current products. The group that might benefit the most would be the very young or newborn hemophiliacs who, in principle, would not have any of the damage caused by lack of factor VIII or factor IX. It can be anticipated that gene therapy treatment of this group will not be approved in the absence of substantial long-term safety data from older patients. There will be a regulatory requirement for long-term follow-up: how long and for what are not discussed in the *Points-to-Consider*.

The obvious points are evidence of functional expression, unusual immunologic outcomes, viral shedding if a viral vector is used, and cell trafficking. Semen samples will be collected to assess germ-line effects. The patients currently undergoing gene therapy have been informed of the importance of autopsy information. It is easy to state that the clinical trial must be designed to assess clinical safety and efficacy; it is far more difficult at this time to anticipate with precision the scientific or regulatory requirements, such as the number of patients to be included, the duration before licensure by a regulatory agency, and the statistical design and the number of patients in a monogenic disease where the assessment of bleeding reduction can be done with retrospective data. The first major effectiveness issue will be the expression level. Clinical efficacy initially will be a plasma level deemed sufficient to minimize hemorrhage, 5 to 10% of the normal plasma level. The need for other replacement therapy will be the ultimate test of efficacy.

The initial trials will begin with a dosage based on the animal data or some fraction thereof, and there may well need to be adjustment. Expression should be seen in days to weeks; duration of expression must be monitored frequently. There are no historical data to support toxicity of supranormal levels of factor VIII in the patient with hemophilia. However, nonhemophiliac subjects with supranormal levels of factor VIII have increased incidence of venous thrombosis. The immunologic issues, inhibitor formation, and immune complex disease will have to be studied. The worst possible outcome would be the induction of chronic immune complex disease, which has never been reported in the 30-year history of the treatment of hemophilia despite studies indicating increased levels of immune complexes. The other side of the issue is that for years large daily doses of factor VIII have been used to develop immune tolerance in patients with factor VIII inhibitors, and no chronic immune complex disease has been reported. If recipients of gene therapy do develop immune complex disease, it would be observed within days to months and perhaps should be part of the normal treatment protocol as well as play a role in the investigative use of gene therapy in hemophilia. Although hemophilia patients with inhibitors are routinely treated with factor VIII, these are patients with no intrinsic synthesis of the factor VIII.

A concern raised by the FDA is that the cells producing the gene product could be destroyed by an immune cytolysis. This potential outcome would argue against the use of known inhibitor patients for the initial trials. Patients without hemophilia A who develop antibodies to factor VIII have not been reported to have any organ damage despite the presence of cells producing normal amounts of factor VIII. The development of inhibitors usually occurs with the first 20 treatments, but the current European guidelines request prolonged inhibitor follow-up, and it is likely that long-term follow-up will be requested in gene therapy. Follow-up for the development of antibodies to the vector may also be suggested. If therapy is by a modified viral vector, there may be limited antibody formation. Methods should be developed to detect antibodies to this vector in the presence of wild-type virus infection.

Surveillance of the impact of the vector will be mandatory. The older hemophilia patient population is infected with a variety of chronic viral diseases, and many already have liver disease. Adenoviral vectors targeted to the liver will at least temporarily induce some inflammatory reaction. Other unknown adverse effects resulting from the vector may occur with this novel and dramatic therapeutic approach. Gene therapy science is moving rapidly, and as old problems are solved, new areas of concern arise. The developers of any such product must be aware of all the issues. Preclinical and clinical studies must be rigorous; at this time, the regulatory guidelines are appropriately rigorous. The best updated source of information on the regulatory issues may be found at the FDA's Center for Biologics Evaluation and Review (CBER) internet site.

GENETICALLY ENGINEERED PRODUCTS

The term *transgenic animal* is used for animals with either somatic cell DNA alterations or germ-line alterations. The potential value of transgenic animals for the production of therapeutic proteins would involve the production of animal strains with germ-line DNA alterations, a regulatory base exists both in the United States and Europe. The use of transgenic animals to produce proteins secreted into milk is an attractive alternative to the large investment in a sophisticated plant to produce therapeutic proteins. At this time, there is little evidence of commercial interest in this approach. Many coagulation proteins have been produced in the milk of transgenic animals. The regulation in the United States of transgenic products would involve not only the FDA drug or biologics regulatory groups but in addition the veterinary medicine and food regulators of the FDA in addition to the Department of Agriculture.

Many of the issues of blood-derived products would be involved, including screening for viruses, assessing the health of the donor, etc. The possibility of adapting unknown animal viruses to a human host will not be a trivial issue and will have to be addressed by the producer. The actual gene and the regulatory element characterization are covered by the *Points-to-Consider* for cell-derived products. The regulatory elements are of particular interest, and extensive characterization of the enhancers, promoters, suppressors, and presence of dominant control regions should be reported. There must be a full record of all the procedures in obtaining the ova, methods for in vitro fertilization, and introduction of the altered DNA. After homologous recombination, it is important to show loss of function of the wild-type gene. Both the history of the donors of the gametes and the recipient (foster parent) animals need to be documented in detail. Intensive veterinary evaluation of health with particular emphasis on diseases related to the species and the breed will be required.

The founder animal must be studied for the presence of the introduced gene and for the expression of the gene product in addition to the absence of functional product from the inactivated wild-type gene. The characterization of the gene expression should be evaluated as a function of age, season, etc. The acceptable range for the gene product should be established to allow for all the systematic variations. Anatomical localization and function of the transgene should be evaluated, and that the transgene is at the desired location must be verified. There is particular concern about the post-translational modifications in a species and at a tissue site different from the normal human product, which could lead to functional and immulogic problems. The maintenance of a given transgenic line is presumed desirable, and methods similar to the master working cell bank for cell-derived products is suggested.

In addition to study of the transgene in the founder, there must be extensive study of the transgene stability in the succeeding generations until stability of the gene and the gene expression are demonstrated. In the founder, there may well be multiple gene copies inserted which over succeeding generations will be rearranged or deleted. It is desirable to demonstrate at the founder stage that there is integration at a single chromosomal site. If this is not possible, studies of the chromosomal integration must be done in the later generations. In addition, the RNA transcript should be examined for quality and quantity and identification of the tissues or cells producing the transcript. The animals must be kept isolated from sources of known and unknown infections during their productive life. Animals would only be acceptable in the absence of disease and with a consistent level of production of the gene product.

17

COFACTORS AND COENZYMES

There are about 3500 known enzymes, of which about 70% are cofactor dependent and about 15% are commercially available. Cofactors such as adenosine triphosphate (ATP), nicotinamide adenine dinucleotide (NADPH), acyl coenzyme A, *S*-adenosylmethionine (SAM), sugar nucleotides, and 3´-phosphoadenosine-5´-phosphosulfate (PAPS) are involved in stoichiometric amounts in cofactor-dependent enzymatic reactions. The very high cost of these cofactors prohibits their use as stoichiometric reagents. For economic reasons, regeneration of the cofactors from their reaction products is an essential requirement in making use of cofactor-dependent enzymes in preparative synthesis of organic compounds. Using coenzyme recycling, the expensive cofactor is needed only in catalytic amount, so there is a drastic reduction in cost of the respective reactions. Additionally, coenzyme regeneration can accomplish three major objectives. First, it can influence the position of equilibrium of a reaction.

A thermodynamically unfavorable reaction can be driven by coupling with a favorable cofactor-regeneration reaction. Second, regeneration affords only catalytic amounts of coenzyme, preventing the accumulation of cofactor by-product, which may inhibit the reaction of interest. Third, coenzyme regeneration can simplify the downstream processing of the reaction mixture. The following chapter deals with methods for the regeneration of nicotinamide coenzymes, because this type of cofactor is the most significant. Hitherto, more than 650 oxidoreductases are known, of which about 90 are commercially available. For the majority of these redox enzymes, about 80% and 10% require nicotinamide adenine dinucleotide (NADH) or its respective phosphate (NADPH) respectively.

COENZYME NICOTINAMIDE

Nicotinamide adenine dinucleotide (NAD) and the analogous 2′-phosphate (NADP) are involved in redox reactions catalyzed by alcohol dehydrogenases. Two electrons and a proton (hydride) are transferred from the reduced coenzyme NADPH to the carbonyl compound in a stoichiometric reaction. Using coenzyme recycling, the expensive cofactor is needed only in catalytic amounts, leading to a drastic reduction in the cost of dehydrogenase-catalyzed reactions. This chapter discusses methods for regeneration of nicotinamide coenzymes.

Commercial Value

Nicotinamide adenine dinucleotides (NADH, NADPH) are expensive compounds. If these coenzymes were used in stoichiometric amounts, enzymatic reductions with dehydrogenases would be prohibitively expensive. Therefore, the application of dehydrogenases on the industrial scale requires effective methods for coenzyme regeneration. The economic barrier to large-scale reactions posed by cofactor costs has been recognized for many years. The turnover number (TN) is defined as the number of moles of product formed per mole of cofactor or enzyme per unit time. This number is indicative of the productivity of the process.

The total turnover number (TTN) is defined as the total number of moles of product formed per mole of coenzyme or enzyme during the course of the entire reaction. Hence, the TTN reflects the loss of coenzyme and enzyme throughout the process, and thus it is a useful estimate of the operational cost of the coenzyme or the enzyme. In order to overcome economic limitations, total turnover numbers of greater than 100 are desirable. For a batch synthesis, the cost of the coenzyme or enzyme is calculated on the basis of its initial cost. In contrast, for a continuous process, the cost may be calculated on the basis of the activity loss over the entire process time. When estimating the cost of a given reaction per unit time, the rate of the reaction must also be considered. The regeneration cost is defined as the cost of the components (enzymes, reagents, and coenzymes) required to regenerate one mole of coenzyme per unit time. The highest TTN can be achieved as follows:

1. The substrate concentration is as high as possible, determined by the substrate solubility, product inhibition, and enzyme stability.
2. The concentration of coenzyme and enzyme to achieve acceptable reaction rates is as low as possible.
3. The coenzymes and enzymes are highly stable.

Thus, both the initial costs of the cofactor and the enzyme, and the efficiency of their regeneration and utilization, determine their contribution to the cost of the product.

Historical Background

Nicotinamide adenine dinucleotide (NAD) is presently isolated from yeast, and its 2′- phosphate NADP is normally synthesized by enzymatic phosphorylation of NAD using NAD-kinase and ATP. NADPH can be prepared from NADP by three different methods: chemical, enzymatical, or microbial reduction. Additionally, a combined chemical and enzymatic route for the synthesis of NADP has been developed that may be useful for the preparation of NADP analogs.

Economical Values

Chenault and Whitesides compiled the following criteria for an ideal coenzyme regeneration system:

1. Compatibility with the synthetic reaction
2. Commercial availability of inexpensive and stable enzymes with high specific activity
3. Simple and cheap auxiliary substrates; the corresponding products should not interfere with the isolation of the product of interest. They should not influence the stability of enzymes and coenzymes.
4. High turnover numbers (TN) and total turnover numbers (TTN).
5. The equilibrium constants of the coupled coenzyme regeneration system should be as high as possible.
6. The analysis of the yield should not be effected by the substrates or products of the coenzyme regeneration.

As a rule of thumb, 10^3-10^4 cycles are sufficient for laboratory-scale redox reactions, whereas at least 10^5-10^6 cycles are desirable for industrial-scale synthesis. The exact turnover number required depends on the initial cost of the coenzyme and the value of the product of interest. A high turnover number requires a high selectivity for the formation of enzymatically active coenzyme. 1,4-dihydro-NAD^+, or $NADP^+$, is the only enzymatically active form; 1,2- and 1,6-dihydro species are not active. For instance, if 50% of the original coenzyme activity is to remain after 1,000 cycles, the regeneration reaction must be 99.3% regioselective. In the case of industrial applications of dehydrogenases, 10^6 turnovers are desired, resulting in a requirement for the regioselectivity of 99.99993%. Chemical, electrochemical, photochemical, or enzymatic methods show different selectivities in the formation of 1,4-dihydro NADP. Electro- and photochemical methods especially suffer from poor regioselectivities. Thus, the requirement for very high regioselectivity necessitates enzymatic catalysis, particularly in the recycling of NADH and NADPH.

Life of nicotinamide coenzymes

Nicotinamide coenzymes contain three labile bonds that may be the target of general or specific acid-or base-catalyzed hydrolysis:

1. The N-glycosidic bond between nicotinamide and ribose
2. The phosphodiester bond between the two riboses
3. The bond between ribose and phosphate (NADP)

Oppenheimer and Chenault and Whitesides compiled data on the influence of different buffers on the stability, and the possible products from the hydrolysis of NADPH. Under basic conditions, the reduced nicotinamide coenzymes are stable; under acidic conditions, they are unstable. In contrast, the oxidized coenzymes are stable under acidic conditions, and labile under basic conditions. As a compromise, NAD and NADH are used at pH 7-7.5, whereas NADP and NADPH are used at pH 8-8.5. The mechanisms of the decomposition of NADPH were studied in great detail. The N-glycosidic bond between nicotinamide and ribose is the most labile bond of the molecule. Base-catalyzed hydrolysis of this bond results in the removal of the nicotinamide ring and, thus, the coenzymatic activity is destroyed. On the other hand, the general acid-catalyzed protonation of the nicotinamide ring at position C_5 is followed by a rapid rearrangement to a cyclic ether product. Additionally, the 2′-phosphate group of NADPH catalyzes this decomposition intramolecularly. The half life for NADH and NADPH in 0.1 M phosphate (pH 7, 25°C) are 27 h and 13 h, respectively.

Organic buffers such as imidazole, Tris, Hepes, and triethanolamine, partially stabilize reduced coenzymes, resulting in a half life that is three times as long. The major pathway for the decomposition of $NADP^+$ is nucleophilic addition at C_4 of the nicotinamide ring, yielding a 1,4-dihydropyridine structure. In Tris-buffered solution at pH 7–8, $NADP^+$ is stable over at least 14 days. There is a strong dependence of the stability of the oxidized coenzymes on the concentration of the buffer. At low molarity (50 mM), the half-life in Tris buffer is lower compared with the half-life in buffer of high molarity (500 mM). Poly(ethylene)-glycolcoupled $NADP^+$ is very stable at pH 8 in Tris buffer of any molarity. Half-lifes of more than 240 days have been reported.

Coenzymes Regeneration

Nonenzymatic method

Nonenzymatic methods for coenzyme regeneration can be divided in chemical, electrochemical, and photochemical methods. Chemical reduction of $NADP^+$ using a reducing agent like sodium dithionite ($Na_2S_2O_4$) suffers from very low total turnover number of less than 100. Moreover, this agent can deactivate enzymes, presumably by modification of thiol groups in the protein. The major disadvantages of most electrochemical and photochemical regeneration methods are low regioselectivity, leading to coenzyme inactivation occurrence of side reactions, and low total turnover numbers of less than 1,000. However, Simon and coworkers stated that electro- and photochemical methods should not be underestimated. Schummer and coworkers reported the synthesis of poly-functional (*R*)-2-hydroxycarbonic acids on the preparative scale with resting cells of *Proteus vulgaris* by using electrochemical regeneration.

Several groups reported on the electrochemical regeneration of NADPH by using a bipyridine rhodium-(I) complex as an electron-transfer agent and formate as an electron donor. The bipyridine-rhodium complex can be coupled to poly(ethylene) glycol and thereby retained in an enzyme-membrane reactor. Dicosimo and coworkers described the electrochemical regeneration of NADPH using the dye methyl viologen and flavoenzymes. However, enzyme denaturation on the surface of the electrode or enzyme deactivation by redox agents may occur. In summary, enzymatic methods meet most of the requirements for an ideal coenzyme regeneration system as described earlier. The difficulties encountered with enzymatic coenzyme regeneration are mainly competitive or mixed inhibition by the substrate or product. In the following sections, the two general concepts for coenzyme regeneration using enzymes are presented. Focus will then turn to the most attractive methods for the recycling of reduced and oxidized nicotinamide coenzymes.

Substrate-coupled approach

In the substrate-coupled approach, both the main reaction and the regeneration of the coenzyme are catalyzed by a single enzyme. The auxiliary substrate (S'_{red}) of the coenzyme regeneration reaction serves as the hydride donor. In order to shift the equilibrium of the reaction to the desired position, the hydride donor, usually a low-cost alcohol, is applied in large excess. However, yield of the system is limited by equilibrium of the coupled system. For the reduction of acetophenone using 2-propanol as auxiliary substrate, the equilibrium was reached at 60% yield. In the case of the reduction of 2-acetyl-naphthaline using 2-propanol as hydride donor, the equilibrium could be shifted by the use of cyclodextrins, and thereby the total yield was increased to 93% of 1-(2-naphthyl)-ethanol. In the continuous synthesis, yields of more than 80% were obtained, and total turnover numbers for the coenzyme were in the range of 10^4. In this application the native coenzyme was not recovered from the product solution but was directly used in substoichiometric amounts.

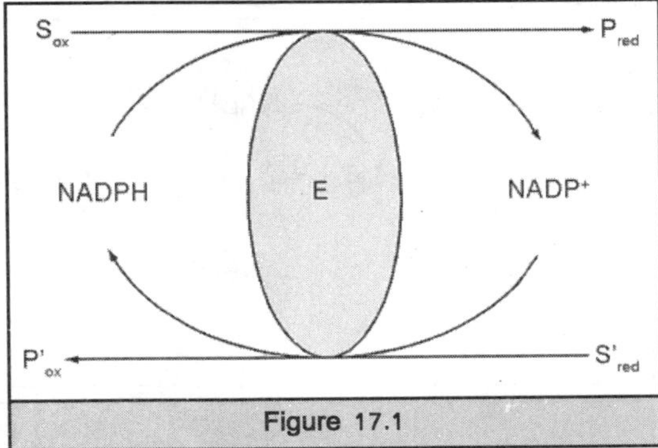

Figure 17.1

The total turnover number for the enzyme was greater than 10^6, and the productivity was 2.5 times as high as the comparable system with enzyme-coupled cofactor regeneration (formate dehydrogenase system). This demonstrates the potential of substrate-coupled systems for coenzyme regeneration. Due to the lack of a well-established method for NADPH regeneration, the substrate-coupling method is mainly used for this purpose. However, there are some drawbacks of this method:

1. The auxiliary substrate has to be present in a large excess, which may lead to complications in product recovery, or which may cause enzyme deactivation or inhibition.
2. Enzyme deactivation may also occur as the result of the accumulation of highly reactive carbonyl species like acetaldehyde or cyclohexenone.

A special enzyme reactor employing a membrane that is only permeable to gases may avoid most of these drawbacks. This type of reactor was successfully applied to improve cofactor recycling via the yeast alcohol dehydrogenase-ethanol system in the stereoselective reduction catalyzed by lactate dehydrogenase.

Enzyme-coupled approach

In the case of the enzyme-coupled approach to NAD regeneration, the reduction of the main substrate and the coenzyme regeneration reaction are catalyzed by two different enzymes. The enzymes employed should have sufficiently different substrate specificities to avoid interferences.

The enzyme-coupled cofactor recycling has the following major advantages:

1. The position of the equilibrium of a redox reaction can be influenced. Thermodynamically unfavorable reactions can be driven by coupling with a favorable cofactor regeneration reaction.
2. High turnover numbers can be achieved: >600,000.
3. The activity ratio has to be optimized for each pair of enzymes to ensure neither lack nor excess of reduction equivalents.

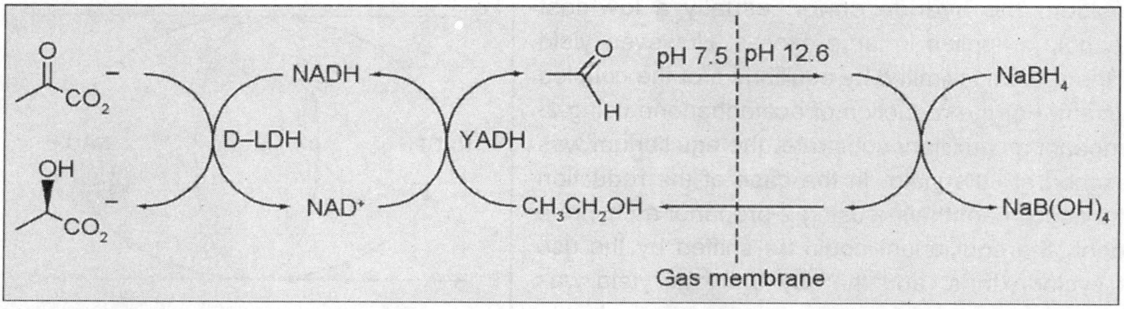

Figure 17.2

However, there are also some disadvantages to this type of cofactor-recycling method:

1. Cross inhibition between the main substrate and the coenzyme regeneration system may occur.
2. A second enzyme must be used, generating additional costs.
3. The characteristics of the second enzyme, such as pH and temperature optimum, must be compatible with the main reaction.

Observing Regeneration and Assessment

Different methods for the regeneration of NADPH have been discussed in great detail by several authors. In this chapter, only the most useful methods are discussed and assessed with respect to their applicability in batch and continuous synthesis. Furthermore, recently developed new methods for nicotinamide coenzyme recycling are highlighted.

Recovery of NADPH

For the recovery of native and poly(ethylene) glycol-bound NADH (PEG-NADH), the formate dehydrogenase (FDH) from *Candida boidinii* has been widely and successfully applied. This enzyme catalyzes the oxidation of inexpensive formate (HCOOH) to carbon dioxide (CO_2). The reverse reaction is hardly detectable. Therefore, the FDH reaction can shift the equilibrium of unfavorable reactions to the product side. Both the auxiliary substrate and the coproduct are innocuous to enzymes and are easily removed from the reaction. Therefore, product isolation is easy. FDH has a broad pH optimum for activity, so it can be easily implemented in coupled enzymatic synthesis.

The major disadvantages are the low specific activity of 4 U/mg and product inhibition by NADH. However, these drawbacks can be circumvented or minimized: the former by using immobilized or membrane-retained systems; the latter by adjusting the optimal ratio of enzymatic activities for the main reaction and the coenzyme-recycling reaction. Overall, the formate–FDH system is the most economical method for regeneration of NADH for batch and continuous processes. The total turnover numbers (TTN) range typically from 10^3 to 10^5. Unfortunately, the *C. boidinii* formate dehydrogenase does not accept $NADP^+$. Although several NADP -dependent formate dehydrogenases have been discovered, only the recently described mutant formate dehydrogenase from *Pseudomonas* sp. 101, using NADP as coenzyme, was applied in batch and continuous synthesis. Acetophenone was reduced by the NADPH-dependent alcohol dehydrogenase from *Lactobacillus* sp with this enzyme for the regeneration of NADPH, at a space–time yield of 8 g/(L day) and a conversion rate of 90-95%.

The new FDH is quite stable, showing no loss in activity over one year at 4°C. Unfortunately, this enzyme is not yet commercially available. A further disadvantage is the relatively high K_m value of the mutant FDH for NADP (320 μM). Additional work is required to engineer FDH mutants with improved kinetic properties. In conclusion, the new formate–NADP dehydrogenase has a large-potential for use as a regeneration system for NADPH. The secondary alcohol dehydrogenases from *Thermoanaerobium brockii* (TBADH) accept a broad range of secondary alcohols and $NADP^+$ as coenzyme. The enzyme is commercially available and contains a very low NADP-2′-phosphatase activity (NADP(H) is hydrolyzed at the 2′-ribose phosphate bond at a rate of 0.1% per day in the continuous synthesis of NADPH.) This enzyme has been applied in the reduction of numerous carbonyl compounds using the substrate-coupled approach for coenzyme recycling. Since this enzyme tolerates up to 30% isopropanol, the substrate-coupled regeneration of NADPH is highly favorable.

The TBADH-enzyme family is also an excellent system for the regeneration of native and poly(ethylene) glycol-bound NADPH. In a model system, glutamate dehydrogenase (GluDH) was employed for the synthesis of glutamate from α-ketoglutarate. A complete set of kinetic data was recorded and the effectivity of the regeneration of PEG-NADPH by the TBADH was studied in a batch reactor. The optimal activity ratio of GluDH/(GluDH + TBADH) was 0.6. At this ratio, 95% of the total coenzyme is reduced. The conversion of α-ketoglutarate is affected negatively only at a TBADH portion of less than 20%. For comparison, in the leucine dehydrogenase–FDH system studied by Wichmann and coworkers, only 10% PEG-NADH is present at the optimal activity ratio of 1:1. The reason for this different behavior is the product inhibition of FDH by PEG-NADH. In contrast, TBADH shows only slight product inhibitions by 2-butanone and PEG-NADPH. The loss of coenzyme in the enzyme-membrane reactor is a function of the absolute coenzyme concentration and, therefore should be as low as possible.

The K_m value of the TBADH for native and PEG-bound NADPH is 10 μM. Hence, only 0.2–0.3 mM NADPH is usually sufficient to ensure a maximum usage of the catalytic activities in the reactor. Overall, the TBADH is recommended for the recycling of NADPH for the following reasons. First, the enzyme is commercially available and is stable in a polypropylene enzyme-membrane reactor. Second, the specific activity is almost 10 times as high as that of formate dehydrogenase. Favorably, the K_m values for native and PEG-bound NADPH are low (10 μM), and the product inhibitions by 2-butanone and PEG-NADPH are also low. Third, the TBADH is compatible with high amounts of organic solvent (e.g., 30% isopropanol) and is stable at elevated temperatures.

Recovery of $NADP^+$

Oxidized nicotinamide coenzymes are used for the synthesis of carbonyl compounds from the corresponding racemic hydroxy compounds. For their regeneration, both enzymatic and nonenzymatic methods have been reported. Enzymatic methods seem to be preferred because of their compatibility with biological systems and their simplicity. The regeneration of oxidized nicotinamide coenzymes is somewhat problematic due to thermodynamics that are often unfavorable.

Additionally, product inhibition by the reduced coenzyme may cause problems. The alcohol dehydrogenase from *Pseudomonas* sp. which is not yet commercially available, has potential for the recycling of NAD^+. A range of simple ketones can be used as hydrogen acceptors, which are reduced at high velocity (up to 270 U/mg).

However, the kinetic limitations of this enzyme must be explored in more detail. For recycling of $NADP^+$, the secondary alcohol dehydrogenase from *Thermoanaerobium brockii* (TBADH) is problematic due to the low product-inhibition constant K_{ip} for NADP of 50 μM. The K_m value of NADPH is 10 μM. Therefore, the oxidation of alcohols producing the reduced coenzyme is the favorable reaction. Acetaldehyde and the commercially available alcohol dehydrogenase from baker's yeast (YADH) have also been used to regenerate NAD^+ from NADH. The total turnover numbers were 10^3 to 10^4. However, deactivation of the enzyme by the highly reactive acetaldehyde and the self-condensation of acetaldehyde outweigh the advantages of the inexpensive enzyme and the volatility of the reagents involved.

The most widely applied method for regeneration of oxidized nicotinamide coenzymes involves glutamate dehydrogenase (GluDH), which catalyzes the thermodynamically favorable reductive amination of inexpensive α-ketoglutarate to give L-glutamate. Both the substrate and the product are innocuous to enzymes. The commercially available and inexpensive GluDH accepts NADH, NADPH, and their respective poly(ethylene) glycol-bound coenzymes and has a reasonably high specific activity of 40 U/mg protein.

The products of the reductive amination of α-ketoglutarate, PEG-$NADP^+$ and L-glutamate, do not strongly inhibit the reaction. The ratio of the product inhibition constant to the Michaelis–Menten constant (K_{ip}/K_m) will determine the efficacy of the reaction. At a K_{ip}/K_m ratio greater than 1, the reaction may proceed to acceptable conversions. If this ratio is less than 1, the reaction can never proceed efficiently. The K_{ip}/K_m ratio for the substrate and product of the reductive amination of α-ketoglutarate to L-glutamate is 10.5.

For the PEG-NADPH and PEG-$NADP^+$, the K_{ip}/K_m ratio is about 30. Hence, product inhibition by L-glutamate and PEG-$NADP^+$ does not play a significant role in the efficacy of the regeneration reaction. The disadvantage of this regeneration method is that L-glutamate may complicate product isolation from the reaction mixture. This must be addressed case by case. Pyruvate and commercially available and inexpensive lactate dehydrogenase (LDH) have been used for recycling of NAD^+. The LDH is stable and has a high specific activity of about 1,000 U/mg protein. However, the redox potential is less favorable and, in contrast to GluDH, LDH does not accept NADPH. A further disadvantage is that pyruvate tends to polymerize in solution and reacts with NAD^+ in a process catalyzed by LDH. Despite these draw-backs, the pyruvate–LDH system has been applied in enzymatic oxidation reactions of 10–100 mmol of material.

Enzyme-Catalyzed Reaction

Enzyme-catalyzed reactions are often accompanied by the following disadvantages:

1. Enzymes may not be sufficiently stable under the reaction conditions employed. Some lose their catalytic activity due to autooxidation, others due to self-digestion and/or denaturation by the solvent, and others due to mechanical shear forces.

2. Isolated enzymes are water soluble and must be used repeatedly due to the economics of the processes for which they are used. They must be separated from the reaction mixture without destroying their catalytic activity.
3. The productivity, expressed as the space-time yield, is often low due to the limited tolerance of enzymes to high concentrations of substrate(s) and product(s).

Some of these problems may be overcome by "immobilization" of the enzyme. However, a special problem is encountered with NADPH-dependent dehydrogenases because the coenzyme has to associate and dissociate freely to and from the active site(s) of the enzyme(s). Therefore, coimmobilization of the dehydrogenase and the cofactor is necessary to make the system practical. Either dehydrogenase, coenzyme, or both may decompose, leading to the replacement of the immobilized system. If the K_m value for the coenzyme is in the micromolar range and the substrate-coupled approach for coenzyme regeneration is employed, a column reactor can be used with a small amount of free coenzyme dissolved in the mobile phase. This situation is particularly efficient if the reaction mechanism is an ordered bi-bi- or Theorell-Chance mechanism where the coenzyme is bound first to the active site, followed by the substrate. These mechanisms were found for a broad range of alcohol dehydrogenases. Immobilization techniques for coenzymes may be divided into coupling and entrapment methods.

Coupling of the coenzymes

For dehydrogenases, coupling methods can be further divided into methods for coupling the coenzyme on cross-linked enzymes or methods for coupling the enzyme and coenzyme on a carrier. Enzymes can be attached to each other by covalent bonds, termed *cross-linking*. Bifunctional reagents such as glutardialdehyde, dimethyl adipimidate and dimethyl suberimidate can be used. The coenzyme can be bound to cross- linked enzymes using a spacer long enough so that the coenzyme has access to the active sites of both enzymes, thereby transferring the hydride. These requirements are very difficult to meet in practice. Lortie and coworkers reported on the coimmobilization of horse liver alcohol dehydrogenase (HLADH), albumin, and NAD^+ using glutaraldehyde as the cross-linking agent.

Long-chain aldehydes were prepared using this immobilized HLADH/NAD^+ system in a packed-bed reactor (glass column). The conversions were performed in organic solvent (hexane) saturated with buffer. Because proteins and coenzymes are insoluble in the water immiscible hexane, they remained in the aqueous phase. The long-chain alcohols (substrate) diffused into the aqueous phase, underwent enzymatic conversion, and the product (aldehyde) finally diffused back into the organic phase. This system was limited by the mass transfer between the organic phase and the water-containing porous particles. Lortie and coworkers generalized this result to all types of fixed-bed enzyme reactors in biphasic systems. Importantly, the regeneration of the coenzyme was not a rate-limiting step.

Role on a carrier

Another option for immobilizing coenzymes is to bind both the dehydrogenase(s) and the coenzyme to an insoluble carrier such as cellulose, dextrans, starch, chitin, agarose, or synthetic copolymers. The activation of the hydroxyl groups of polysaccharides is usually done by reaction with cyanogen bromide, leading to the formation of reactive imidocarbonates. Synthetic polymers can be activated using partial hydrolysis of the acetate groups followed by activation with epichlorhydrin. A range of preactivated polysaccharides and synthetic polymers may be obtained from commercial sources. The enzyme is coupled via its amino groups. Also, in this case, a spacer long enough to allow the coenzyme to have access to the active sites of the enzyme(s) must be used. Horse liver alcohol dehydrogenase (HLADH) and an NADH analogue, N^6-[(6-aminohexyl)carbamoyl-methyl]-NADH, have been successfully coimmobilized to Sepharose 4B. It was shown that the coenzyme could be regenerated at a cycling rate of 3,400 cycles/h using the substrate-coupled approach. Both thermal and storage stability of the HLADH were increased substantially. It was not necessary to add soluble coenzyme to the substrate solution to maintain activity.

Coenzyme-dependent Processes

Polymer-linked coenzyme derivatives are often introduced in enzyme reactors with coenzyme-dependent processes, because they can be retained upon ultrafiltration or dialysis and are recycled. Small substrate and product molecules can freely pass through the membrane (e.g., polyamide, polyethersulfone), but large enzymes and the polymer-linked coenzyme cannot. Synthetic membranes of defined pore size covering the range between 500 and 300,000 Da are commercially available at a reasonable cost. Additionally, a variety of synthetic membrane shapes, such as membrane discs or hollow fibers, are on the market. In this simple technique, termed *membrane-enclosed enzymatic catalysis* (MEEC), the enzymes and polymer-bound coenzyme are entrapped in a dialysis bag, which is mounted on a gently rotating magnetic stirring bar. One modification of the enzyme membrane-reactor concept makes use of charged ultrafiltration membranes in order to retain the native coenzyme in the reactor.

The application of charged ultrafiltration membranes in membrane reactors is limited to the production of non-charged products. In contrast, nanofiltration membranes can be used to retain the coenzyme in the reactor, whereas the reaction product(s) may freely pass the membrane. The separation of coenzyme and reaction product is due to the differences in the molecular weight, without a negative net charge of the membrane. Many reports on the preparation of polymer-linked coenzyme derivatives are available. The dextran-bound coenzymes have been demonstrated to the serious problem of protein leakage, due to the instability of the isourea linkage between coenzyme derivatives and dextran. However, more stable dextran derivatives can be synthesized by using bromohydroxypropyl derivatives of dextrans. These derivatives are used mostly in enzyme electrodes or as stationary phases in affinity chromatography.

The polyethyleneimine and polylysine derivatives of coenzymes have the disadvantage of having electrostatic interactions between the multicharged polymers and enzymes, thus interfering with their effectiveness as active coenzymes. The relative reaction rates compared to native NAD^+ were 2-25% (polyethyleneimine) and 2-60% (polylysine), depending on the enzyme tested. The synthesis of polyacryl copolymers has the advantage that parameters like solubility, charge and the functional group can be varied just by exchange of the monomer. However, polyacryl-bound NAD showed only 30% of the V_{max} value of the native coenzyme. Another approach was established by Buckmann and coworkers using poly(ethylene) glycol derivatives with molecular weights of 3,000 to 20,000 kDa and modified coenzymes. The yield of the coupling of *N*(1)-(2-aminoethyl)-NAD with the carboxy-PEG derivative was 90%. The alkylation of NADPH with 1,4-bis-(2,3 -epoxypropoxy)-butane was investigated by Stein.

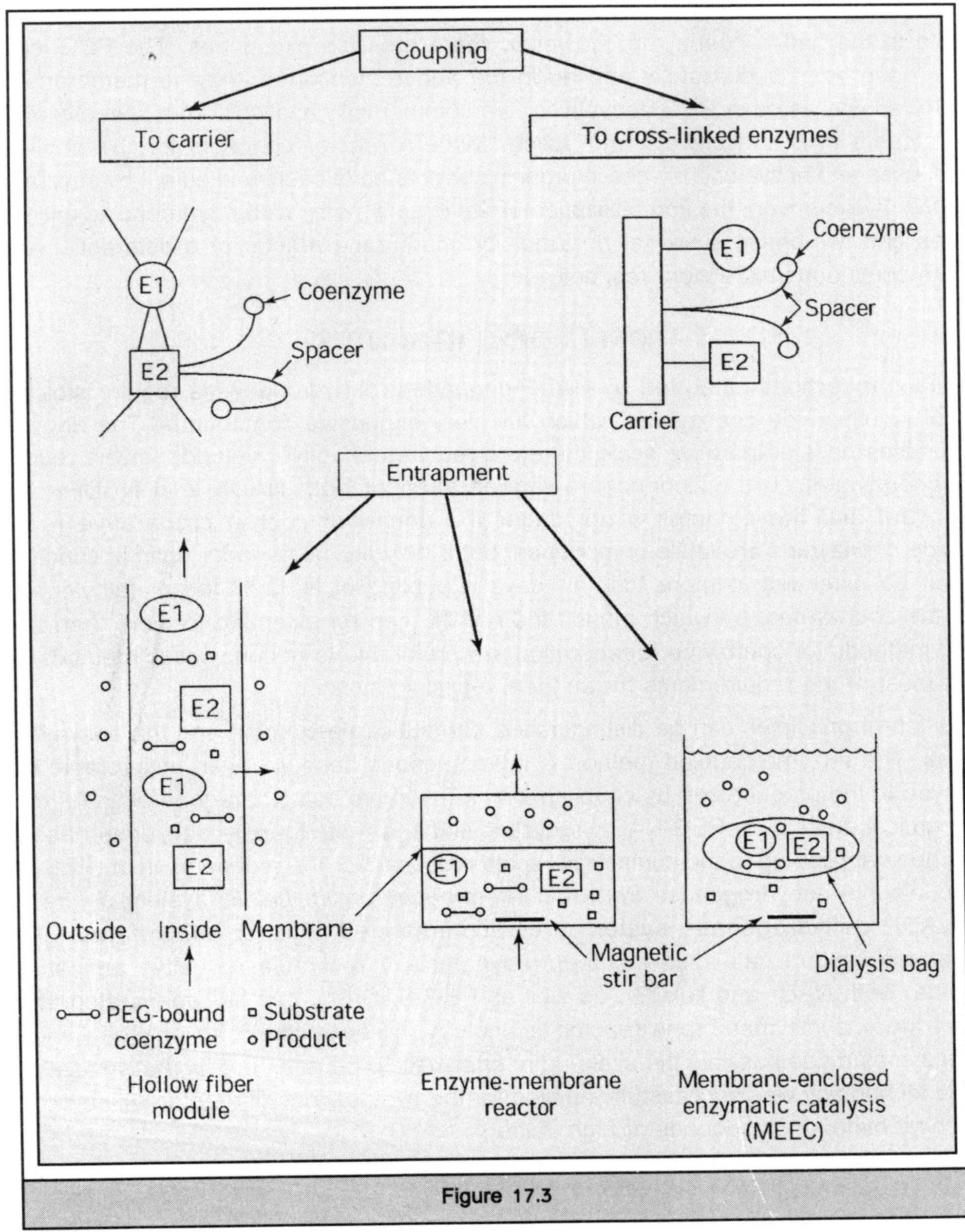

Figure 17.3

This method allows the direct alkylation of NADPH in a one-step reaction, whereas all other described alkylation methods are multistep reactions starting from the oxidized coenzymes. Moreover, the coenzymatic activities of the NADPH derivatives were almost identical to the native coenzymes. The NADPH derivatives carry a hydrophilic spacer and a terminal epoxy group, which can be coupled easily to PEG-NH_2, resulting in a stable, covalent bond between PEG and the coenzyme derivative. For most of the enzymes tested, the coenzymatic activity of PEG-NADH is comparable to the native coenzyme. However, there are also exceptions. The PEG–coenzyme derivatives have been successfully applied in the above-mentioned enzyme membrane reactor. PEG-NADH, as well as other PEG-derivatives, are commercially available from spin-offs of German research centers in Braunschweig and Julich. Modern reactor concepts for the application of dehydrogenases and native coenzymes in organic solvents have been worked out by several groups. These concepts circumvent the immobilization of the coenzyme by using hydrophobic microfiltration membranes and two-phase systems, an emulsion membrane reactor or a differential circulation reactor with continuous extraction, respectively.

CONCLUDING REMARKS

Enzymatic reductions catalyzed by NADP-dependent dehydrogenases require stoichiometric amounts of nicotinamide coenzymes, which are very expensive compounds. The application of dehydrogenases on a preparative scale therefore requires effective methods for the regeneration of these coenzymes. Today, coenzyme regeneration of both NADH and NADPH are very straightforward and the cofactor is no longer the dominant cost in preparative reductions. Nicotinamide coenzymes are labile compounds, but if they are used under optimal conditions, the half-life can be extended to more than 14 days. Coupling of NADPH to poly(ethylene) glycols stabilizes the coenzymes, by which means the half-life can be extended to more than 240 days. Enzymatic methods for coenzyme regeneration are preferable to nonenzymatic methods because they meet most of the requirements for an ideal recycling system.

Two different principles can be distinguished: the substrate-coupled and the enzyme-coupled approaches. The enzyme-coupled method is advantageous because even unfavorable reactions can be driven to the product side by coupling with a favorable cofactor-regeneration reaction. The most economical method for PEG-NADH recycling is the formate-formate dehydrogenase system, which has been applied up to the commercial-scale process. For the regeneration of (PEG-NADPH, the alcohol-alcohol dehydrogenase from the *Thermoanaerobium brockii* system or the formate-formate:NADP dehydrogenase system are recommended. The α-ketoglutarate-glutamate dehydrogenase system catalyzes the thermodynamically favorable reductive amination of α-ketoglutarate. Both NAD^+ and $NADP^+$, as well as PEG-$NAD(P)^+$, can be regenerated effectively. The use of the enzyme membrane reactor technology in conjunction with poly(ethylene) glycol-linked coenzymes made possible the scale-up of enzymatic processes to preparative scale. Among others, this technology was successfully applied for the synthesis of different L-amino acids and a range of chiral hydroxy compounds of high value.

18

PHYLOGENY OF ENZYMES

Staunch Darwinists attribute all the complexity of living things to an algorithm of mutation and natural selection. The exquisite products of this evolution algorithm are apparent at all levels, from the amazing diversity of life all the way down to individual protein molecules. Scientists and engineers who wish to redesign these same molecules are now implementing their own versions of the algorithm. Directed evolution allows us to explore enzyme functions never required in the natural environment and for which the molecular basis is poorly understood. This bottom-up design approach contrasts with the more conventional, top-down one in which proteins are tamed "rationally" using computers and site-directed mutagenesis. We will describe how molecular evolution can be directed in the test tube in order to produce useful biocatalysts. It is not possible to provide a complete account of all the methods proposed for in vitro evolution; this article therefore introduces methods and strategies used successfully in our laboratory for directing the evolution of enzymes. Some alternative methods for biocatalyst evolution are also be discussed. The reader should also be aware that there is a rather substantial and largely separate literature on combinatorial approaches to engineering binding molecules. Recent advances in the ability to create genetic diversity and to screen or select for improved functions in large libraries of enzyme variants are being combined in a robust approach to solving difficult molecular design problems. With directed evolution we now have the ability to tailor individual proteins as well as whole biosynthetic and biodegradation pathways for biotechnology applications.

HISTOLOGY

When natural enzymes are recruited for industrial applications—from serving as catalysts in chemicals synthesis to additives for laundry detergents—we discover that they are often not well suited to these tasks. Due to poor substrate solubility, breakdown of unstable products, or competing chemical reactions, the conditions for an enzyme reaction may be unsuitable for large-scale applications. Reflecting their participation in complex biochemical networks inside living cells, enzymes are often inhibited by their own substrates or products, either of which may severely limit the productivity of a biocatalytic process. Evolution is usually the culprit: enzymes are optimized and often highly specialized for specific biological functions within the context of a living organism.

Biotechnology, in contrast, needs enzymes that are stable and active over long periods of time (a feature that might clash with the need for rapid protein turnover inside a cell), are active in nonaqueous solvents (a feature probably not required in most biological milieu), and can accept different substrates (substrates not present in nature). It is possible to produce new enzymes in recombinant organisms, altering the amino acid sequence and therefore the properties through appropriate modifications at the DNA level. We are hobbled, however, by near complete ignorance of how the amino acid sequence affects every aspect of enzyme performance, from its ability to be expressed in a heterologous host to its catalytic activity in nonnatural environments. Numerous protein engineering experiments have demonstrated that changes in protein properties are brought about by the cumulative effects of many small adjustments, many of which are distributed or propagated over significant distances.

Furthermore, proteins are usually teetering on the brink of instability, with folded structures that are more stable than unfolded—and therefore inactive—ones by the equivalent of a few hydrogen bonds out of the hundreds that form. Superimposing on the need to retain this relatively fragile folded state, the additional requirements of having to fold in the first place, and the need of maintaining or even reengineering a catalytic site that is affected at some level by virtually any modification yields a design problem of such complexity that any rational design effort will require enormous inputs of structural, mechanistic, and dynamic information. Information that is available for but a tiny fraction of interesting catalysts. Even if one trait is successfully designed (e.g., enhanced stability), it is virtually impossible to predict the cost to another (e.g., catalytic activity or expression level). The relatively few examples where rational design has yielded *useful* enzymes do not negate the view that rational enzyme design is often a fruitless exercise.

Evolution at Molecular Level

All these hurdles to the rational design of enzymes are bypassed by evolution. The power of evolution as an algorithm for molecular design is perhaps best appreciated by studying its products. By constructing the evolutionary histories of today's proteins we have learned that they are highly adaptable molecules, at least on evolutionary time scales. Well illustrated by the panoply of α/β-barrel enzymes, enzymes catalyzing very different reactions have evolved divergently from a common ancestral protein of the same general structure, acquiring diverse capabilities by processes of random mutation, recombination, and natural selection. We also know that enzymes sharing a common function (for example, all catalyzing a particular step in a metabolic pathway) in addition to three- dimensional structure can exhibit widely different properties (stability, solubility, tolerance to pH, etc.), depending on where they are found. Enzymes evolve, and adapt, at the molecular level.

The structures of protein modules are conserved (although the modules themselves are often shuffled to create new, multifunctional proteins). Function, however, can vary. Specific features such as substrate specificity or thermostability vary significantly. Amino acid sequences can vary to such an extent that evolutionary relationships may no longer be apparent from sequences alone. Evolution is a powerful algorithm with proven ability to alter enzyme function and especially to "tune" enzyme properties. It is also an algorithm that can be implemented in the laboratory for redesign. The challenge is to collapse the time scale to months, or even weeks.

PHYLOGENETIC MECHANISM

Evolutionary mechanisms at work in nature assure adaptability to ever-changing environments. Evolution does not work toward any particular direction, nor is there a goal; the underlying processes occur spontaneously during reproduction and survival of the whole organism. In contrast, a directed molecular evolution experiment has a defined goal, and the key processes (mutation, recombination, and screening or selection) are controlled by the experimenter. Although there may be multiple ways to reach a defined goal (i.e., a desired enzyme function), the approach that minimizes the effort is preferred. The genetic diversity for evolution is created by mutagenesis and/or recombination of one or more parent sequences. These altered genes are inserted into a plasmid for expression in a suitable host organism (bacteria or yeast).

Clones expressing improved enzymes are identified in a high-throughput screen or by selection, and the gene(s) encoding those improved enzymes are isolated and recycled to the next round of directed evolution. Approaches for carrying out these key steps are discussed in some detail in the "Methods of Directed Evolution" section. Here we focus on more fundamental considerations that help to define workable strategies for directed evolution. To appreciate the challenge of designing and carrying out a successful directed protein evolution experiment, it is important to underscore the powerful combinatorial features of this system. A typical enzyme is a linear chain of *N* amino acids (*N* is usually several hundred), and there are 20 possible amino acids at each position in the chain. Thus the "sequence space" of possible proteins is huge beyond the imagination (N^{20}). Even in 4 billion years, nature has had a chance to explore but a tiny fraction of these possibilities. A laboratory exploration of this vast space of sequences and their corresponding functions must obviously be severely limited and carefully guided.

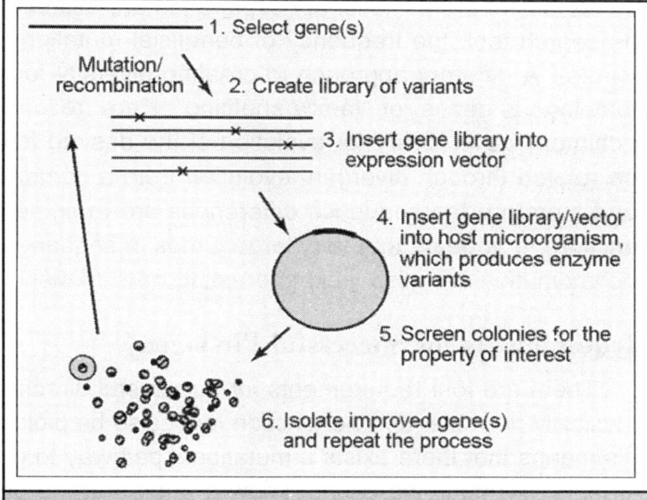

Figure 18.1 Key steps of a typical direct enzyme evolution experiment.

Because much of sequence space will be devoid of the desired function, and probably even of folded proteins, it is best to direct the evolution of an existing enzyme rather than look for function in random peptide libraries. Evolution is often referred to as a hill-climbing exercise in the fitness landscape of sequence space. The fitnesses (performance, for laboratory evolution) of the proteins in sequence space make up this landscape, whose most basic features are still quite unknown. The landscape for laboratory evolution will be different for each property or collection of properties undergoing evolution. An uphill climb in a protein landscape is more likely to be successful if it can take place in small steps (one or two amino acid substitutions). The high dimensionality of the surface (there are $19N$ one-mutant neighbors of any given sequence) offers many opportunities to find improved mutants. Although we may never reach the "global optimum," the improvements achieved by taking even a simple random up-hill walk via single amino acid mutations often yield useful results.

A widely effective evolutionary strategy, is one in which the steps are small (preferably one or two amino acid substitutions in each generation), and multiple such mutations are accumulated either sequentially or by recombination to acquire the desired function. Such an approach is compatible with a low level of random mutagenesis over the entire gene. An alternative approach is to direct a much higher level of random mutation to a relatively small region of the gene. Both approaches have their advantages. Mutagenesis over the entire gene allows discovery of unanticipated solutions (a common experience). More intense, directed mutation, however, may yield novel combinations of amino acid substitutions, combinations that would be inaccessible by single-step walks because the intermediates are unfavorable.

Details of the evolutionary exploration are ultimately dictated by a combination of the power of the search tool, the frequency of beneficial mutations (usually small!), and the choice of starting point(s). A different approach to creating diversity for directed evolution is the in vitro shuffling of homologous genes, or "family shuffling". Here, recombination of two or more parent genes yields a chimeric gene library for evolution of the desired features. Because the recombined sequences are related through divergent evolution from a common ancestor of similar structure and function (and therefore the sequence differences are to some extent neutral with respect to structure and function), it appears that very large jumps in sequence space can yield functional proteins. In vivo recombination can also yield interesting new chimeric enzymes.

Requirements for Successful Phylogeny

There are four requirements for successful directed evolution. (i) The desired function must be physically possible. (ii) The function must also be biologically, or evolutionarily, feasible. In practice, this means that there exists a mutational pathway to get from here to there through ever-improving variants. Although we cannot know a priori that the path exists, a good experiment will maximize the likelihood. (iii) One must be able to make libraries of mutants complex enough to contain rare beneficial mutations. This usually means functional expression in a suitable microorganism such as *Escherichia coli* or *Saccharomyces cerevisiae*. (iv) One must have a rapid screen or selection that reflects the desired function. Just how rapid the screen must be depends on how rare mutations leading to the desired property are and how many must be accumulated to achieve the desired result. Whether directed evolution will solve a particular problem depends to some extent on how hard natural evolution has already worked at it. If a particular trait is already under selective pressure (e.g., catalytic activity), it is unlikely that further improvements can be obtained in the laboratory by small mutational steps. However, if biological function has imposed additional constraints, for example the trait is coupled to another trait that is also under selective pressure (e.g., high thermostability), then this balance can be altered during laboratory evolution.

While selected traits are often difficult to improve, they should be relatively easy to remove (e.g., product inhibition). Many traits are not under selective pressure; they may be changing as a result of random genetic drift, they may be vestigial—reflecting the enzyme's history—or they may be coupled to selected traits. In general, non-selected traits are easier to improve, but may be more difficult to remove. It is especially easy to improve traits never required for biological function, such as stability or activity in a nonnatural environment or activity toward a new substrate; there is often much room for improvement, and small changes in sequence and function can be accumulated. As expected for any hill-climbing exercise, the number of pathways leading uphill diminishes as a peak is approached. Thus the ease with which improved mutants are identified (the frequency of improved clones) should eventually decrease as the sequences move closer to an optimum. The preceding discussion focuses only on altering existing enzyme traits.

Evolving a completely new function is a risky venture because we rarely know how far in sequence space we have to go in order to create the new function, or how frequent the solutions will be. If there is good reason to believe that a new function, for example, activity toward a substrate not accepted at all by the wild-type enzyme, can be obtained (at a level measurable during a high-throughput screen) by making one or two amino acid substitutions, then evolution makes sense. However, if the new function requires the simultaneous placement of multiple new amino acid residues, it is unlikely to appear in a random library of mutants. Such a problem is probably a better candidate for a combination of rational design and combinatorial tuning. In nearly all cases, the desired trait(s) was at a measurable, albeit low, level in the starting enzyme(s). The problems can be roughly divided into a few major categories: improving function in nonnatural or extreme environments (where activities or stabilities are low), improving activity toward a new substrate, tuning specificity (enantioselectivity), and increasing functional expression in a heterologous host.

GENETIC DIVERSITY

The first step in a directed evolution experiment is the creation of a mutant library containing an appropriate degree of molecular diversity. For this, the mutation rate must be tuned to the power of the sorting method. Useful, reasonably sized libraries can be created by introducing multiple mutations in a particular region or across a limited number of positions (e.g., by combinatorial mutagenesis using oligonucleotide cassettes). However, such an approach excludes the many useful solutions found in unexpected places. We therefore usually try to evolve the entire gene, rather than target particular positions. Whole-gene evolution requires a correspondingly lower mutation rate. Diversity is created by two operations: point mutagenesis and recombination. Most random point mutagenesis methods create mutations in single bases.

Due to the degeneracy of the genetic code, this provides access to only about 6 amino acid substitutions instead of 19, significantly reducing the potential diversity. In addition, many mutagenesis methods are not really random, further limiting the number of amino acid substitutions actually accessible in a given experiment. For example, a commonly used method, error-prone PCR, shows a strong bias for transitions over transversions. Recombination methods combine sequences from multiple parent genes. In vitro recombination methods include DNA shuffling, random-priming recombination, and the staggered extension process (StEP). In yeast, in vivo recombination is particularly simple. All these methods have a (controllable) level of associated point mutagenesis. Directed evolution can begin from multiple, closely related starting points rather than a single sequence. Recombination of existing functional sequences (i.e., homologous enzymes) will create another level of diversity that point mutagenesis cannot generate.

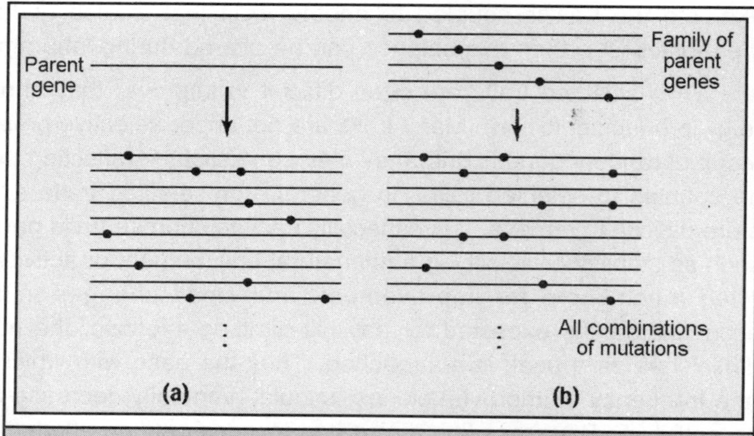

Figure 18.2 Two general approaches for generating molecular diversity: random point mutagenesis (a) and in vitro recombination (b).

Mutation Frequency

The most important factors in random mutagenesis are the mutation frequency and mutation bias. Mutation frequency is the average number of mutations per gene and is usually reported as a percentage. The optimal target mutation frequency can be calculated from the length of the DNA coding sequence and an estimate of the desired number of mutations per sequence. In deciding on the optimal number of mutations, one must balance the cost of combinatorially searching large libraries against the likelihood of finding mutants with improved properties. The number of possible variants increases drastically with the number of mutations introduced. The number of possible variants that can be produced by simultaneously substituting M amino acids in a protein of length N is $19M[N!/(N-M)!M!]$. For example there are only 3,800 possible single mutants, but 7,183,900 double mutants, and 9,008,610,600 triple mutants of a protein of only 200 amino acids. The single-mutant library of an enzyme is generally within our ability to screen exhaustively. Double-mutant libraries are already on the outer limit of current screening technology, and triple-mutant libraries are well beyond our capability for the foreseeable future.

To these numerical considerations we add the biochemical fact that beneficial mutations are rare, and combinations of beneficial mutations are extremely rare. In light of these considerations it appears that, at present, single-mutant libraries usually represent the best compromise between search effort and diversity: searching single-mutant libraries maximizes the chance of finding beneficial mutations in a reasonable amount of time. If high-throughput screening technology is available, or if significant fractions of possible mutations are neutral, then double-mutant libraries may be the optimal choice. An average of one amino acid substitution per sequence corresponds to approximately three base substitutions per gene. Thus for a 1-kb sequence, the mutation frequency should be ~0.3–0.5%.

Although numerous methods for making random DNA mutations exist, error-prone PCR is often preferable because the procedure is simple, rapid, robust, and most importantly, the mutation frequency can be precisely controlled. Error-prone PCR does have some intrinsic bias as to the location of mutations (mutations at AT base pairs occur much more frequently than mutations at GC base pairs) as well as their type (A is more frequently substituted with G). PCR modifications reduce bias, but do not eliminate it. The error-prone PCR method routinely used was originally outlined by Leung et al. and further examined by Cadwell and Joyce and Shafikhani et al. Mutation frequencies ranging from 0.11 to 2% have been obtained under different reaction conditions. In particular, the mutation frequency can be controlled simply by adjusting the concentration of manganese ions in the reaction mixture. Because polymerase fidelity also depends on the nature of the target sequences, different mutation rates will be observed on different sequences, even when the exact same reaction conditions are used.

A straightforward way to assess the overall mutation frequency and the nature of mutations is to sequence a few random clones from the amplified population. However, sequencing is time-consuming and expensive. A simple and efficient alternative is to estimate the mutation frequency from the fraction of active clones from the amplified population. The clones are sorted and plotted in descending order to create local profiles, or landscapes, of the enzyme's fitness. The higher the Mn^{2+} concentration, the higher the mutation rate, which manifests itself in a higher proportion of inactivated clones. Even though different enzymes show different activity profiles for different mutation rates, the fraction of active clones is a convenient index of mutation frequency and can be used as a diagnostic check for the successful creation of the desired randomly mutated library.

Sequential Generation

When an enzyme is evolved by sequential generations of random mutagenesis and screening, only the best variant identified in each generation is used to parent the next generation. Other improved variants are set aside and must be rediscovered in subsequent generations in order to become incorporated. This is wasteful because screening is time-consuming. If the mutation rate is high, this approach can also accumulate deleterious mutations, possibly limiting the fitness that can be reached. "DNA shuffling" and related in vitro recombination methods can overcome these limitations of the sequential approach. Recombining parental genes to produce libraries of different mutation combinations, and screening for improved variants quickly accumulates the beneficial mutations, while removing deleterious ones. The goal is to create gene libraries containing novel combinations of mutations present in the parent genes. This library can be screened in order to find the combinations of mutations giving rise to the best enzymes. Screening eliminates unfavorable combinations of mutations. For example, if two point mutations that are beneficial by themselves become harmful when combined, the variant carrying this combination will be removed during screening. We now discuss some general considerations for recombination experiments before describing different recombination methods.

Mathematical solution

When recombining genes from multiple improved variants it is important to consider the recombined library size. The various in vitro recombination methods can recombine any number of parent genes. However, the resulting libraries may be so impossibly large that the probability of finding large improvements in function is effectively zero. For the recombination of N sequences and M total mutations, the probability of generating progeny sequences containing μ mutations is equal to the number of ways a μ mutation sequence can be generated (C_{μ}^{M}) multiplied by the probability of generating any single μ-mutation sequence

$$P_{\mu} = C_{\mu}^{M}\left(\frac{1}{N}\right)^{\mu}\left(\frac{N-1}{N}\right)^{M-\mu} = \frac{M!}{(M-\mu)!\mu!}\left(\frac{1}{N}\right)^{\mu}\left(\frac{N-1}{N}\right)^{M-\mu}$$

For recombination of N single mutants the probability of generating single-mutant parents or wild-type grandparents is approximately 75%, and only 25% of the library consists of new sequences. The probability of generating individual sequences decreases precipitously with increasing numbers of parents. The least-frequent sequences (often the most desirable) are those containing the majority of mutations from the parent population. The rarest sequence will be the one containing all mutations ($\mu = M$). The probability P_M of generating this sequence is $1/N^m$.

Some degree of oversampling is required in practice to maximize the chance of discovering a given variant. The sampling S required in order to achieve a given level of confidence of having sampled the rarest variant in a library is given by

$$(1 - P_M)^S < 1 - (\text{confidence limit})$$

Generally, for 95% certainty that a specific clone has been sampled, the oversampling is between 2.6 and 3.0.

In vitro recombination

Stemmer described the first method for in vitro recombination of DNA sequences, called DNA shuffling. This method involves enzymatic digestion of the parental DNA into short fragments, followed by reassembly of the fragments into full-length genes. Because the fragments are free to associate with complementary fragments from other, similar genes, mutations from one parent can be combined with mutations from other parent(s) to generate novel combinations. The fidelity of this process is not perfect, and new point mutations not present in any of the parents will occur at a finite rate. The original method has been modified to simplify it and to yield better control over the associated mutagenic rate. The mutagenic frequency can be controlled over a wide range, from about 0.05 to 0.7%, by the inclusion of Mn^{2+} or Mg^{2+}, by the choice of DNA polymerase, and/or by using restriction enzyme digestion to prepare the starting DNA.

The finite error frequency associated with DNA shuffling has been used by Stemmer and coworkers to supply point mutations for recombination and evolution. When starting with a single sequence, we prefer to use error-prone PCR under controlled conditions to generate libraries of variants. Improved sequences identified during screening can then be recombined. An alternative method for recombinantion was developed by Shao et. al. In this method the template DNA sequences are primed with random-sequence primers, which are then extended by DNA polymerase to generate a pool of short DNA fragments. After removal of the template, these fragments can be reassembled, as with the Stemmer shuffling method, to give a library of full-length sequences containing information from multiple parents. Yet another method for in vitro recombination, the staggered extension process (StEP), does not require fragmentation and reassembly of the parent sequences. In StEP recombination the template genes are primed and very briefly extended before denaturation and reannealing. The growing fragments can reanneal to different templates and therefore pick up sequence information from different parents as they grow into full-length sequences over hundreds of very brief cycles. StEP (in principle) involves only a single relatively short protocol as opposed to the multiple steps of fragmentation, isolation, and annealing required in DNA shuffling.

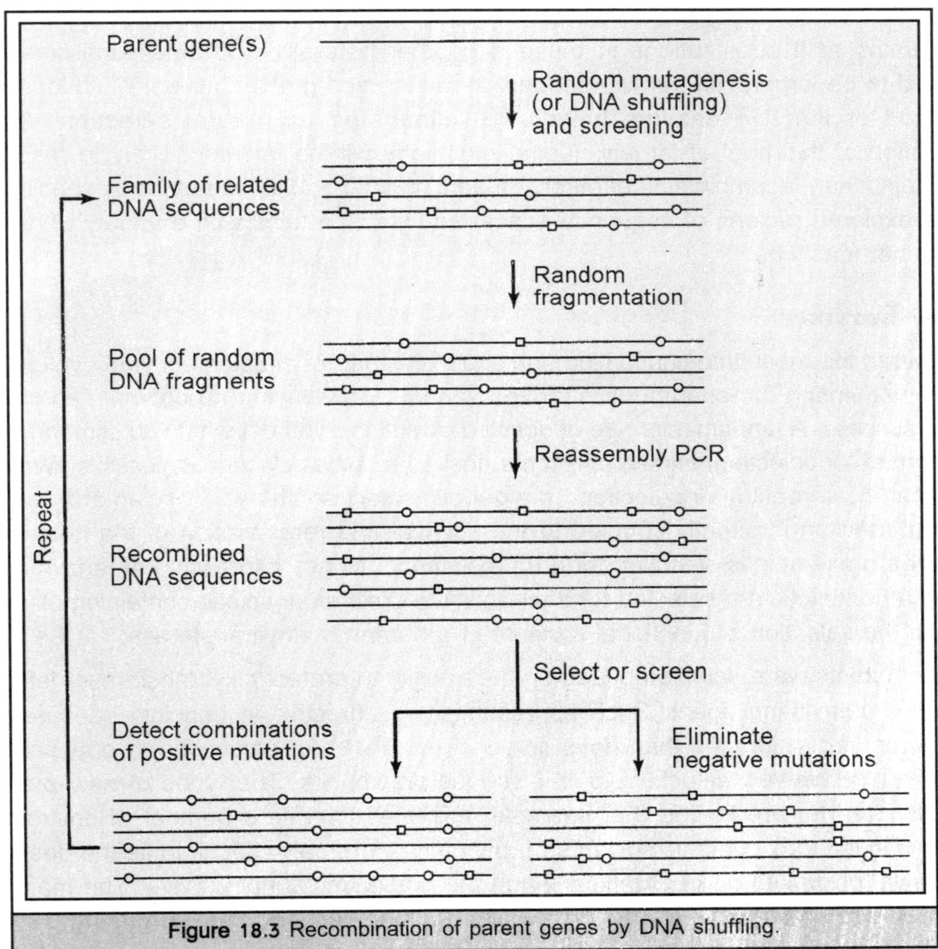

Figure 18.3 Recombination of parent genes by DNA shuffling.

Improved variants

The approaches already described (point mutation and recombination of improved variants) allow a local exploration of sequence space. Only sequences that are quite close to the parent sequence are sampled. One would like to sample distant regions of sequence space to search for new functions, but this is not readily accomplished using these methods. The simultaneous introduction of more than a few random mutations into a wild-type sequence will almost always result in variants that are inactive or unfolded. The probability that an improved enzyme will be found is usually much smaller than the screening capacity. An alternative approach to making large jumps in sequence space is by "family shuffling". In this approach, homologous genes from different organisms are recombined to create a library of chimeric molecules. Family shuffling exploits the fact that the sequence differences between two homologous proteins are *not* random. Consider the two proteases subtilisin E and subtilisin thermitase. These two enzymes differ at 157 of their amino acid positions, and yet they fold to the same overall three-dimensional structure and catalyze the same proteolytic reaction.

The amino acid substitutions in these and other naturally occurring subtilisins have been preselected to be largely neutral with respect to folding and gross function. Mutations resulting in a misfolded or inactive enzyme have been eliminated by natural selection. Accordingly, recombination of this pool of largely neutral mutations will be far more likely to result in folded, active proteins than recombination of purely random mutations. The use of family shuffling to access remote unexplored regions of sequence space may yield proteins with a variety of desirable, but not yet found, functions.

Improved Enzymes

In general, the most time-consuming and expensive part of directed enzyme evolution is setting up and implementing the search for improved variants. Developing an optimal search strategy is crucial for success. A fundamental rule of directed evolution, "you get what you screen for," specifies that the screen (or selection) should reflect the desired result as closely as possible. Mutant libraries are searched by screening or selection. In a genetic selection, the ability of an enzyme to perform the desired function is directly coupled to the survival and reproduction of the host organism. A straightforward example is selection for drug resistance: clones expressing an enzyme capable of degrading an antibiotic are selected for their ability to grow on antibiotic-containing plates. Another example is the selection of functional mutants of a human methyltransferase.

Active mutants were selected based on their ability to protect alkyltransferase-deficient *E. coli* cells from a methylating agent. Such approaches are attractive in principle because they allow one to search larger libraries than does screening (10^6–10^7 for selection, as opposed to 10^4–10^5 for screening). However, selections suffer serious drawbacks. The types of new properties that can be searched for are limited. For example, enzymes tolerate a number of environments that cannot sustain life (e.g., organic solvents). It becomes extremely difficult to tie the desired function to the survival or growth of an organism when the organism cannot survive. The main drawback, however, is the complexity of the organisms themselves and the surprising solutions that spontaneously arise during adaptation. Given our limited understanding of even simple, well-studied organisms such as *E. coli,* it is difficult to ensure that improving a particular enzyme is the *only* way for an organism to adapt to the selective pressure applied. One may go through several rounds of selection only to find that the organism has found some alternate way to solve the problem.

Various in vitro selections have been proposed based on column binding or "panning". In these methods, the proteins are displayed on the surface of a phage or a cell. Very large libraries ($>10^9$) can be searched based on the ability of the displayed protein to bind to a substrate immobilized on a column. The greatest limitation to this approach is that it is not generally applicable to screening for properties other than binding. For many problems, and especially those of practical interest, libraries of variants must be screened rather than selected. A typical screening strategy involves the construction of an arrayed enzyme library and application of a rapid assay of the desired property. The screen can be more or less sensitive, depending on the willingness of the researcher to accept false positives (and to apply additional tests). Prerequisites for an effective screening strategy include the following:

1. The screen should be sufficiently rapid that large numbers of clones can be searched in a reasonable amount of time. It should be feasible to screen at least a few hundred, and preferably a few thousand, in a day. For obvious reasons, the screen should not require purification of the enzyme. A second-level assay can eliminate false positives.
2. Only rarely do very large improvements result from a single amino acid substitution. Directed evolution most often succeeds by accumulating a number of modest improvements over several generations. Therefore a screen must be accurate enough to allow modest improvements to be identified against the background variation.
3. The phenotype (the enzyme property that is being screened) must be physically coupled to the genotype (the DNA or RNA sequence coding for the enzyme). This is most often accomplished trivially by the fact that proteins are expressed in host organisms: the same cell expressing the enzyme contains the gene coding for its sequence. Coupling can also be accomplished by display on the surface of a virus containing the genetic information in the form of DNA or RNA. Recently developed in vitro translation systems such as ribosome display or RNA–peptide fusions link the enzyme directly to the genetic information in the form of RNA without using cells. Tawfik and Griffiths have encapsulated the in vitro translation apparatus and genes within reverse micelles.

Simple visual screens are widely used when the function of interest can generate a visible signal. Various oxidases, for example, produce hydrogen peroxide as a by-product of their oxidation reaction. As illustrated, coupling this to a second enzyme that is easily assayed colorimetrically (horseradish peroxidase) provides a visual screen for the oxidase activity. Clones secreting active proteases produce a zone of clearing or "halo," the size of which is proportional to the hydrolytic activity when the organism is grown on agar plates containing casein or skim milk proteins. This simple method has been used extensively in screening protease libraries. Although visual screening based on color or halo formation is rapid and efficient, it is also nonquantitative and often relatively insensitive.

Digital imaging spectroscopy has been developed to increase the sensitivity and throughput of filter- and agar-plate-based screens. However, the 96-well microtiter plate remains the standard format for automated, high-throughput screening. Screening automation and quantification of the results are highly desirable; 96- and 384-well plates appear to be the format most compatible with currently available robotic arms, liquid handling systems, and plate readers. Data collection and analysis are greatly facilitated by computerized data acquisition. Computerized data acquisition will certainly facilitate exploitation of the enormous amounts of information available in protein libraries. Many conventional enzyme assays can be readily converted into automated formats. If activity against the desired substrate does not result in an easily detectable signal (such as a change in absorbance or fluorescence), it is sometimes possible to screen against a surrogate substrate that does. Naturally, this approach requires caution. It must be verified that activity against the replacement substrate reflects activity against the real substrate under the conditions of interest.

A successful application of this approach is described in the example provided later of the evolution of a *p*-nitrobenzyl esterase. Enzyme properties other than activity are often targeted for improvement. Duplicate 96-well plates are made from a single master plate. One is incubated at high temperature for a fixed period. Clones in both are then assayed for activity, and the ratio of activity after heating to activity without heating provides a measure of enzyme stability at the temperature of incubation. An attractive approach to screening large libraries is to couple functional complementation with screening. By requiring that at least some of the biological function of the protein is retained, functional complementation can greatly reduce the subsequent screening requirements. The initial selection asks an essentially binary question (is there any activity? yes or no), and only in the later screening stage are differences in the level of activity detected. Loeb and coworkers have applied this strategy to several enzymes of interest in gene therapy and other applications. However, the major limitation is finding or constructing an appropriate complementation system. Furthermore, retention of biological function may preclude acquisition of other functions.

Phylogenetic Output

Two major types of information can be extracted from the results of directed evolution experiments. The first and most obvious is the information contained in the sequences of the evolved proteins. The second consists of the overall activity profiles, or local fitness landscapes, that can be generated by analysis of the entire screened library. Clearly, we would like to identify the specific amino acid substitutions that are responsible for conferring new properties. In contrast to comparisons of evolutionarily related proteins found in nature, this identification can be accomplished relatively easily in directed evolution experiments. In nature, related enzymes that are separated by millions of years of evolution will generally have accumulated a large number of neutral mutations in addition to those that are responsible for changes in specific properties. In contrast, in laboratory evolution we strive to allow only functional mutations to survive and carry on to the next generation. A striking example of the difference between natural and laboratory evolution can be seen in a recent study in which subtilisin E was converted into a functional equivalent of the thermophilic enzyme thermitase.

Subtilisin E differs from thermitase at 157 positions, but only eight substitutions were required to turn it into a thermitase-like enzyme, with an 18°C increase in its temperature optimum and a 250-fold increase in its half-life at 65°C. However, it is not always possible to identify functional mutations simply by examining the sequences of the evolved enzymes. In the experimental procedures for introducing random point mutations, only the *average* error rate can be controlled. Thus, some variants will accumulate multiple mutations in a single generation. If a given variant has multiple mutations, only one of which is functional, the remaining neutral mutations can "hitchhike" their way into subsequent generations and complicate analysis of the final evolved sequences. It is hoped that detailed structural analysis of evolved enzymes will provide insight into the underlying physical–chemical mechanisms responsible for protein adaptation for new functions or environments. However, the results of single point mutations can be very subtle, and the analysis is not straightforward even when the three-dimensional structure is known. For example, both subtilisin E and pNB esterase have been successfully evolved for activity in aqueous organic solvents, but functional mutations in the evolved enzymes follow no discernible pattern: large residues are replaced by small, small by large, charged by uncharged, uncharged by charged.

Further, mutations are found both close to and far from the enzyme active sites. Numerous studies now attest to the subtlety of the effects of amino acid substitutions on enzyme structure and properties. The distribution of fitnesses (defined in terms of activity, stability, etc.) in a mutant library can provide useful information for designing evolutionary strategies. As mentioned previously, the mutation rate can be estimated from the fraction of inactive mutants. More generally, such a plot represents (at least for single- or double-mutant libraries) the distribution of fitness in the region of sequence space immediately surrounding the wild-type sequence. Most mutants will be equally or less fit than the wild type; a small number may be better. The fraction of improved mutants can be used to estimate the evolvability of the property of interest. When two or more properties are being evolved simultaneously, plots of one versus the other can be informative. Typically, single mutations that simultaneously improve two independent properties are extremely rare. However, individuals may be recombined to yield an enzyme with multiple improved properties.

PHYLOGENETIC APPLICATIONS

New lipase variants that give greatly improved residual lipid removal after one wash cycle have been developed by directed evolution. Lipases are used as detergent enzymes to remove lipid or fatty stains from clothes and other textiles. A drawback of the known detergent lipases is that they exert the best lipid-removing effect after more than one wash cycle, presumably because they are more active during a certain period of the drying process than during the wash itself. Therefore at least two wash cycles, separated by a sufficiently long drying period, are required to remove fatty stains. An intense protein engineering effort involving both site-directed and random mutagenesis yielded a large number of variant of *Thermomyces lanuginosa* lipase that were strongly improved in multicycle wash performance, but not significantly improved in one-cycle wash tests. Random mutant libraries were screened using filter assays containing detergent and low calcium concentrations, to mimic wash conditions.

Consecutive rounds of mutagenesis and screening with increasing amounts of detergent were performed. New variants with a strong one-cycle wash effect were obtained by recombining 20 variants with the best performance in multicycle wash tests and screening. The recombination was performed using a simple in vivo method in *S. cerevisiae* in which PCR fragments from the 20 variants mixed with the opened vector are used to transform competent *S. cerevisiae* cells. Upon transformation, the vector and fragments recombine and shuffle the variants. Screening yielded seven variants with a strong one-cycle wash effect; the best of these is now commercially available.

Role in Thermostability

By screening microbial cultures, scientists at Eli Lilly identified an enzyme in *Bacillus subtilis* that could remove the *p*-nitrobenzyl ester protecting group used during the large-scale synthesis of certain β-lactam antibiotics. However, the enzyme's relatively poor activity, particularly in the solvents required to solubilize the ester substrate, made it a poor competitor to the chemical catalyst for deprotection of a loracarbef synthetic intermediate. The natural function of *B. subtilis* pNB esterase is unknown. Moore and Arnold were able to evolve highly active pNB esterases that also function well in mixed aqueous–organic solvents.

A surrogate, chromogenic substrate (the antibiotic *p*-nitrophenyl ester) was used in a rapid screen to identify potential positives. Variants identified during the rapid screen could be verified during a second-level screen on the *p*-nitrobenzyl ester using HPLC. Four generations of PCR mutagenesis and two rounds of recombination by DNA shuffling yielded a clone with more than 100 times the total activity of wild type in 15–20% dim ethylformamide (DMF) toward loracarbef-*p*-nitrobenzyl ester. The enzyme's catalytic efficiency increased more than 50-fold. The total activity toward the screening substrate (loracarbef-*p*-nitrophenyl ester) increased more than 150-fold. Although the contributions of individual effective amino acid substitutions to enhanced activity were small (usually approximately two-fold increases in activity), the accumulation of multiple mutations over a number of generations significantly improved the biocatalyst for this nonnatural reaction. Moore and Arnold concluded that none of the mutations accumulated in the evolved enzyme are in direct contact with the substrate. Some are as far away as 20 .

Limiting the search for possible solutions to residues that line the substrate binding site would have overlooked important beneficial mutations (and may not have succeeded). Directed evolution has also significantly increased the therm ostability of the *B. subtilis* pNB esterase. A simple screen was developed based on retention of catalytic activity after incubation at high temperature. Positives were then subjected to differential scanning calorimetry (DSC) to verify an increase in melting temperature (T_m) relative to the parent enzyme. Accumulating mutations over eight generations of random mutagenesis and DNA shuffling yielded an increase in T_m of 17°C. This large increase in thermostability is equivalent to the difference between proteins from mesophiles and many thermophilic organisms. Furthermore, because only those variants that retained their catalytic activity as well were chosen, the increased thermostability was accompanied by a very significant increase in enzyme activity at elevated temperatures. Reflecting the fact that the thermostability screen involved activity toward *p*-nitrophenyl acetate, the resulting enzymes are highly active toward this substrate and less active toward the antibiotic substrate and *p*-nitrobenzyl esters. The most thermostable pNB esterase variant has 13 amino acid substitutions.

Enzyme Thymidine Kinase for Gene Therapy

The Loeb group has pioneered the use of directed evolution for developing improved enzymes for cancer gene therapy. Here, greater activity toward nonnatural substrates is desired, with the goal of selectively protecting bone marrow cells or sensitizing tumor cells to toxic agents. This group consistently directs intense mutagenesis of a limited region of the gene and genetic selection to identify the sequences that are still biologically functional. Screening this much smaller subset of sequences identifies those with the desired properties (altered substrate specificity, ability to protect or sensitize cells). In one study, directed evolution was used to create herpes simplex virus thymidine kinase (HSV TK) variants that enhance cell killing by the guanine analogs gancyclovir and acyclovir, which terminate DNA synthesis and prevent viral and host cell replication. Mutant HSV TKs that selectively phosphorylate gancyclovir and/or acyclovir were obtained after complete randomization of a limited set of amino acid positions (six) adjacent to the putative nucleoside binding site and previously identified to tolerate some substitution. Of the approximately 10^6 *E. coli* transformants, 426 expressed active TK, as determined by genetic complementation. These were then assayed for their ability to enhance the sensitivity of the *E. coli* to the guanine analogs. Hamster cells transfected with one of the improved variants were more than 43 times more sensitive to gancyclovir and 20 times more sensitive to acyclovir than cells expressing the wild-type HSV TK. This mutant contained four amino acid substitutions, all within the TK active site pocket.

Phylogeny of Biphenyl Dioxygenases

Biphenyl dioxygenases (BPDO) are responsible for the initial oxidation of polychlorinated biphenyls (PCBs) during their biodegradation by various organisms. In an effort to create BPDOs with enhanced ability to oxidize a wider range of PCB congeners, Furukawa and coworkers used DNA shuffling to create a chimeric library from two genes for the terminal dioxygenases of BPDO (*bphA1*) from *Pseudomonas pseudoalcaligenes* KF707 and *Burkholderia cepacia* LB400. A high-fidelity DNA shuffling protocol was found to increase the frequency of clones expressing active PBDO. The two BphA1 protein sequences are extremely similar, differing only at 20 amino acids. The genes are 96.4% identical. *E. coli* containing the chimeric genes were screened for their ability to produce yellow *meta*-cleavage products from a variety of biphenyl compounds in order to identify functional chimeras with enhanced activities and altered activity profiles.

Enzymes with greater activity toward various biphenyl compounds than either parent were identified, as were enzymes with relaxed specificity and enzymes active toward smaller aromatic hydrocarbons such as benzene and toluene. These latter compounds are poor substrates for the two parent enzymes. Sequence analysis of six shuffled functional *bphA1* genes demonstrated how the parent sequences recombined to yield these new enzymes. A very low level of point mutagenesis was observed. The specific roles of the amino acids in determining substrate specificity are difficult to pinpoint in the absence of a three-dimensional structure of the enzyme. However, the results are generally in agreement with previous studies of BPDO chimeras and site-directed mutagenesis.

INDEX